T0184943

CRYSTAL BASES

Representations and Combinatorics

CRYSTAL BASES

Representations and Combinatorics

Daniel Bump

Stanford University, USA

Anne Schilling

University of California, Davis, USA

World Scientific

NEW JERSEY · LONDON · SINGAPORE · BEIJING · SHANGHAI · HONG KONG · TAIPEI · CHENNAI · TOKYO

Published by

World Scientific Publishing Co. Pte. Ltd.

5 Toh Tuck Link, Singapore 596224

USA office: 27 Warren Street, Suite 401-402, Hackensack, NJ 07601

UK office: 57 Shelton Street, Covent Garden, London WC2H 9HE

Library of Congress Cataloging-in-Publication Data

Names: Bump, Daniel, 1952– | Schilling, Anne (Mathematician)

Title: Crystal bases : representations and combinatorics / by Daniel Bump (Stanford),
   Anne Schilling (UC Davis).

Description: New Jersey : World Scientific, 2016. | Includes bibliographical references and index.

Identifiers: LCCN 2016047620| ISBN 9789814733434 (hardcover : alk. paper) |
   ISBN 9789814733441 (pbk. : alk. paper)

Subjects: LCSH: Lie algebras. | Quantum groups. | Combinatorial analysis.

Classification: LCC QA252.3 .B86 2016 | DDC 512/.482--dc23

LC record available at https://lccn.loc.gov/2016047620

British Library Cataloguing-in-Publication Data

A catalogue record for this book is available from the British Library.

Printed in Singapore

Preface

Crystal bases are purely combinatorial objects that are analogous to representations of Lie groups or Lie algebras. They appeared in the works of Kashiwara, Lusztig and Littelmann on quantum groups and the geometry of flag varieties. In retrospect, topics from the combinatorical theory of tableaux such as the famous Robinson–Schensted–Knuth algorithm and the plactic monoid of Lascoux and Schützenberger fit into the crystal base theory. Crystal bases come up in many unexpected places, from mathematical physics to number theory.

This book originated from a plan to approach crystal base theory from a purely combinatorial point of view. It is aimed at graduate students and researchers who wish to delve into this subject.

It seems that every exposition of crystal base theory needs some powerful method behind the proofs. In existing expositions on crystal bases such as [Hong and Kang (2002)], [Kashiwara (2002)] and Littelmann [Littelmann (1997)] this has come from either quantum groups or Littelmann paths. We have taken a different path, relying on ideas of Stembridge and Kashiwara for our foundations. Thus we were able to prove everything combinatorially. In our approach, the link between crystals and representation theory is made through Demazure crystals.

It is assumed that the reader is familiar with root systems, their classifications, Coxeter groups, and Cartan types, part of which are reviewed in Chapter 2. A bit of algebraic geometry knowledge will also be helpful in Chapter 15. In order to help the reader appreciate certain analogies between crystal bases and representation theory, we have included two appendices on standard topics in the representation theory of Lie groups.

Preliminary versions of this book were used for a short lecture series at NCSU in October 2015, for a quarter special topics class at Stanford and a reading class at UC Davis in the Winter Quarter of 2016. We are grateful for lots of help from various people, as explained in the Acknowledgments.

Daniel Bump
Anne Schilling

California, August 2016

Acknowledgments

We would like to thank Graham Hawkes, Patricia Hersh, Henry Kvinge, Peter Littelmann, Neal Livesay, Molly Lynch, Shotaro Makisumi, Eric Marberg, Kirill Paramonov, Arun Ram, Ben Salisbury, Travis Scrimshaw, Peter Tingley, Kurt Trampel, and Andrew Waldron for discussion and/or helpful comments on an earlier version of this book. Special thanks to Peter Tingley for help with Chapter 15, in particular with the exercises. And we would like to thank Ms. Kwong Lai Fun and World Scientific for their help with the manuscript.

The second author would like to thank Kailash C. Misra, Daniel Nakano, Brian Parshall, and Weiqiang Wang for the invitation as a principal speaker to the 8th Southeastern Lie Theory Workshop on "Algebraic and Combinatorial Representation Theory" (October 9-11, 2015), where she gave a series of lectures on an early draft of this book.

This work was in part supported by NSF grants OCI-1147463, DMS–1001079 and DMS–1601026 (Daniel Bump), and OCI–1147247 and DMS–1500050 (Anne Schilling).

Contents

Chapter 1

Introduction

Crystal bases or *Kashiwara crystals* are combinatorial structures that mirror representations of Lie groups. Historically, crystal bases were developed independently around 1990 from two independent sources.

On the one hand, [Kashiwara (1990, 1991, 1994)] showed that modules of quantum groups have "crystal bases" with remarkable combinatorial properties. Independently, [Lusztig (1990a,b)] introduced canonical bases from a more geometric perspective. Quantum groups are Hopf algebras that are "noncommutative" analogs of Lie groups. A particular class of quantum groups, *quantized enveloping algebras*, are deformations (in the category of Hopf algebras) of the universal enveloping algebras of Lie groups. They were described independently by [Drinfel'd (1985)] and [Jimbo (1985)] to explain developments in mathematical physics. Every representation of the Lie group gives rise to a representation of the corresponding quantized enveloping algebra, and Kashiwara showed that these modules have *crystal bases* whose properties he axiomatized and proved, using deep methods from quantum groups.

On the other hand, crystals also came about through the analysis of [Littelmann (1994, 1995b)] of *standard monomial theory* ([Lakshmibai, Musili and Seshadri (1979); Lakshmibai and Seshadri (1991)]). Borel and Weil and later [Bott (1957)] showed that representations of Lie groups can be realized as sections of line bundles on flag varieties. [Demazure (1974, 1976)] had found additional structure in these modules. Inspired by work of [Hodge (1943)] on the cohomology of Grassmannians, Seshadri and Lakshmibai found convenient bases of these modules of sections that are indexed by tableaux. Peter Littelmann, in the early 1990's, reinterpreted these bases as paths through a vector space containing the weight lattice and showed that they may be organized into crystals like those found by Kashiwara in the theory of quantum groups. [Kashiwara (1996); Joseph (1995)] then proved that the crystals arising from quantum groups are the same as the crystals arising from the *Littelmann paths*.

In retrospect, some older work in the combinatorics of tableaux can be understood in terms of crystals. [Littlewood (1940)] showed that a *Schur polynomial*, which is the character of an irreducible representation of $GL(n)$, had a combinatorial definition as a sum over tableaux. In fact, both the irreducible representations of the symmetric group and the general linear group were known to have bases indexed by tableaux, and the *Robinson–Schensted–Knuth* (RSK) algorithm [Knuth (1970, 1998)] gave bijections that are combinatorial analogs of certain isomorphisms between such modules of Lie groups and the symmetric groups. Later, [Lascoux and Schützenberger (1981)] gave a multiplicative structure on the set of tableaux, called the *plactic monoid*, that is closely related to RSK. All of these topics fit into the theory of crystal bases and the connections will be discussed in Chapters 7 and 8.

Crystals appear in many other contexts from mathematical physics and combinatorics to number theory. We will not attempt to survey all of these here.

In this book, we will limit ourselves to crystals associated to finite-dimensional Lie algebras, omitting the important topic of crystals of representations of infinite-dimensional Lie algebras. Within this limited scope, we have tried to prove the essential facts using combinatorial methods. The facts one wants to prove are as follows.

Given a reductive complex Lie group G, there is an associated *weight lattice* Λ with a cone of *dominant weights*. Given a dominant weight λ, there is a unique irreducible representation of *highest weight* λ. There are two operations on these that we are particularly concerned with: *tensor product* of representations and *branching*, or restriction, to Levi subgroups.

In the theory of crystal bases, one starts with the same weight lattice and cone of dominant weights. Instead of a representation, one would like to associate a special crystal to each dominant weight. If the representation is irreducible, the crystal should be connected. There may be many connected crystals with a given highest weight, but it turns out that there is one particular one that we call *normal*. We think of this as the "crystal of the representation." More generally, a crystal that is the disjoint union of such crystals, is to be considered normal.

The operations of tensor product and Levi branching from representation theory also make sense for crystals. The usefulness of the class of normal crystals is that the decomposition of a crystal into irreducibles with respect to these operations is again normal. Moreover, the decomposition of a representation obtained by tensoring representations or branching a represention to a Levi subgroup gives the same multiplicities as the decomposition of the tensor product or Levi branching of the corresponding normal crystals into irreducibles.

There are several ways of defining normal crystals. [Kashiwara (1990, 1991, 1994)] and [Littelmann (1994, 1995b)] gave two different definitions, which then were shown to be equivalent. We give yet another definition of normal crystals, based on two key ideas: *Stembridge crystals* ([Stembridge (2003)]) and *virtual crystals* ([Kashiwara (1996); Baker (2000)]). For the simply-laced Cartan types, [Stembridge (2003)] showed how to characterize the normal crystals axiomatically. This is subject of Chapter 4. This approach does not work as well for the non-simply-laced types, but for these, there is a way of embedding certain crystals into crystals of corresponding simply-laced types. For example, to construct a normal $\mathrm{Sp}(2r)$ crystal (for the non-simply-laced Cartan type C_r) first one constructs a $\mathrm{GL}(2r)$ crystal (for the simply-laced Cartan type A_{2r-1}). Then one finds the symplectic crystal as a "virtual crystal" inside the $\mathrm{GL}(2r)$ one. This way, one may reduce many problems about crystals to the simply-laced case, including the construction of the normal crystals (see Chapter 5).

Once the normal crystals are constructed, we have a bijection between normal crystals and finite-dimensional representations of the Lie group G. In this bijection connected crystals correspond to irreducible representations. A representation has a character, and so does a crystal. If we can show that the character of the crystal equals the character of the representation, then it will follow that the decomposition of a tensor product of crystals into irreducibles has the same multiplicities as the corresponding tensor product of representations; and similarly the Levi branching rules for crystals and representations will be the same.

How does one prove that the character of an irreducible representation is the same as that of the corresponding normal crystal? The approach we follow depends on the *Demazure character formula*, which constructs the character of a representation in stages. In the Demazure character formula, given a dominant weight λ and a Weyl group element w, there is a *Demazure character* $\partial_w(t^\lambda)$. This generalizes the character χ_λ of the irreducible representation with highest weight λ, because if w equals the long Weyl group element w_0, then $\partial_{w_0}(t^\lambda) = \chi_\lambda$. For general w, the Demazure character $\partial_w(t^\lambda)$ is a *part* of χ_λ.

The corresponding construction in crystals, due to [Littelmann (1995a)] and [Kashiwara (1993)], constructs certain subsets $\mathcal{B}_\lambda(w)$ of the connected normal crystal $\mathcal{B}_\lambda$ with highest weight λ. The main fact to be proved inductively is that the Demazure characters $\partial_w(t^\lambda)$ associated with any Weyl group element w agrees with the character of the *Demazure crystal* $\mathcal{B}(w)$. This fact is called the *refined Demazure character formula*.

A direct approach to the refined Demazure character formula seems difficult, even in the simply-laced case armed with the Stembridge axioms. Instead, what works is to construct the Demazure crystals inside the crystal $\mathcal{B}_\infty$. This is an infinite crystal that contains a copy of $\mathcal{B}_\lambda$ for every dominant weight λ. We construct the Demazure crystals in $\mathcal{B}_\infty$ in Chapter 12 and then deduce the properties of their counterparts in $\mathcal{B}_\lambda$. After this, we are able to finish the proof of the Demazure character formula for crystals, and thereby establish the relationship with representations as shown in Chapter 13.

The infinite crystal $\mathcal{B}_\infty$ is itself a remarkable combinatorial object. It is, in a sense, the crystal of a representation, albeit an infinite-dimensional one, the Verma module with weight 0. As we discover in the proof of the refined Demazure character formula, a clear understanding of $\mathcal{B}_\infty$ may be the key to the finite normal crystals. Therefore after we prove the Demazure character formula we investigate $\mathcal{B}_\infty$ in more depth. Chapter 14 considers the *⋆-involution* of $\mathcal{B}_\infty$, a self-map of order two with remarkable properties that was studied by [Lusztig (1990a,b)] and [Kashiwara (1993)].

Then in Chapter 15, we turn to another combinatorial realization of $\mathcal{B}_\infty$ that arose from Lusztig's canonical bases in the theory of quantum groups. In considering the Lusztig realization of $\mathcal{B}_\infty$, we begin to see an important theme in this subject: how combinatorial maps that arise in crystal base theory are *tropicalizations* of algebraic maps. The algebraic maps that admit tropicalizations are those constructed using addition, multiplication and division, but never subtraction. (These are the operations that preserve the positive real numbers, explaining why this topic touches on the theory of total positivity.) Tropicalization replaces the algebraic operations of addition, multiplication and division by piecewise linear ones, and an algebraic map corresponds to a combinatorial one given by piecewise-linear maps. Conversely, given a piecewise-linear map, we may seek an *algebraic lifting*, an algebraic map having the given piecewise-linear one as its tropicalization. If this is done carefully, the algebraic liftings of various maps may fit together in a way that mirrors the crystal structure.

To give an example, a couple of different parametrizations of the crystal, namely the string parametrization and the Lusztig parametrization, depend on the choice of a reduced word $\mathbf{i}$ for the long Weyl group element w_0. Thus if we change to a different $\mathbf{i}'$, there is a change of basis map. In the crystal theory, this map is a piecewise linear map. However these piecewise-linear maps are tropicalizations of algebraic ones, and precisely these algebraic maps may turn out to have significance elsewhere in mathematics. For the change of basis maps in the Lusztig basis, the algebraic lifting turns out to be related to changes of basis in the maximal unipotent subgroup N of a Lie group G. Over the real numbers, these algebraic maps preserve the subset of totally positive matrices and are thus important in the theory of total positivity. Moreover, [Berenstein and Kazhdan (2000)] showed that it is possible to lift the entire crystal structure. Such *geometric crystals* have been turning up in unexpected places such as probability theory and the theory of Whittaker functions ([Chhaibi (2015); Corwin, O'Connell, Seppäläinen and Zygouras (2014)]).

Other topics that we discuss include the crystals of tableaux of [Kashiwara and Nakashima (1994)], which give explicit models for the normal crystals in the classical Cartan types A, B, C, and D (see Chapters 3 and 6). We discuss the action of the Weyl group on the crystal and the Schützenberger–Lusztig involution.

In Chapters 7 through 10, we specialize to type A to explore various aspects of tableaux theory using crystals. In particular, we discuss Lascoux and Schützenberger's theory of the plactic monoid and prove the Littlewood–Richardson rule. In the context of this discussion we emphasize analogies between the theory of crystals and representations. Two appendices on topics in representation theory of $\mathrm{GL}(n, \mathbb{C})$ have been provided to help the reader see these analogies. Roughly, Appendix A on Schur–Weyl duality is analogous to Chapter 8 on the plactic monoid, and the material in Appendix B on the $\mathrm{GL}(n) \times \mathrm{GL}(m)$ duality is analogous to

Chapter 9 on bicrystals and the Littlewood–Richardson rule. On the crystal side, the Robinson–Schensted–Knuth (RSK) insertion algorithm in its various forms plays the role of the Schur-Weyl and $GL(n) \times GL(m)$ dualities.

We have already mentioned that an important role is played in several chapters by reduced words representing the long Weyl group element. In Chapter 15 we will see that for each such reduced word $\mathbf{i}$ there is a map $v \mapsto v_{\mathbf{i}}$ of the crystal into $\mathbb{N}^N$. We will exhibit polytopes called *MV polytopes* that encode the components of the vector $v_{\mathbf{i}}$ in the lengths of various paths around the boundary of the polytope.

Thus reduced words for w_0 are important in studying crystals. In the other direction, at least for type A, [Morse and Schilling (2016)] showed that crystals can be used to study the reduced words for w_0 (or more generally any Weyl group element). As we have already mentioned the Schur function associated to the dominant weight or partition λ can be viewed as the character of the highest weight crystal of highest weight λ in type A. Another important class of symmetric functions are the Stanley symmetric functions ([Stanley (1984)]), which were introduced to study reduced expressions of symmetric group elements. Stanley symmetric functions have a positive integer expansion in terms of Schur functions. We demonstrate that these can be understood in terms of crystals by imposing a crystal structure on the combinatorial objects underlying the Stanley symmetric functions. In this case, the insertion algorithm of [Edelman and Greene (1987)], which is a variant of RSK, plays a crucial role. This is done in Chapter 10.

Given the combinatorial nature of crystal bases, they lend themselves very well to computational exploration. The category of crystals and many explicit models for specific crystals (such as the tableaux model for highest weight crystals of classical type or models for $\mathcal{B}_\infty$) have been implemented in SAGE [Sage-Combinat community (2008); Sage]. In-built examples can easily be accessed:

```
sage: C = Crystals(); C
Category of crystals
sage: C.example()
Highest weight crystal of type A_3 of highest weight omega_1
```

The highest weight crystal of type A_2 and highest weight $(2,1)$ of dimension 8 can be created as:

```
sage: T = crystals.Tableaux(['A',2],shape=[2,1]); T
The crystal of tableaux of type ['A', 2] and shape(s) [[2, 1]]
sage: T.cardinality()
8
```

An extensive introduction to crystals in SAGE can be found in [Bump, Schilling and Salisbury (2015)]. It is very beneficial to the reader to follow this tutorial along with the book, especially for some of the exercises.

Remark 1.1. Our notations differ in minor ways from some of the literature. We note a couple of differences between our notations and those of Kashiwara. The crystals $\mathcal{B}_\lambda$ and $\mathcal{B}_\infty$ in our notation are denoted $\mathcal{B}(\lambda)$ and $\mathcal{B}(\infty)$ by Kashiwara. Correspondingly, our notations for Demazure crystals are $\mathcal{B}_\lambda(w)$ and $\mathcal{B}_\infty(w)$ instead of Kashiwara's $\mathcal{B}_w(\lambda)$ and $\mathcal{B}_w(\infty)$. Another change is that the tensor product rule for us is

$$f_i(x \otimes y) = \begin{cases} f_i(x) \otimes y & \text{if } \varphi_i(y) \leqslant \varepsilon_i(x), \\ x \otimes f_i(y) & \text{if } \varphi_i(y) > \varepsilon_i(x). \end{cases}$$

For Kashiwara this would be

$$f_i(x \otimes y) = \begin{cases} f_i(x) \otimes y & \text{if } \varphi_i(x) > \varepsilon_i(x), \\ x \otimes f_i(y) & \text{if } \varphi_i(x) \leqslant \varepsilon_i(x). \end{cases}$$

We believe that for combinatorics, the convention that we employ for tensor products is more convenient. (Our convention is the default convention for SAGE.) The reader seeking to translate between our notation and Kashiwara's can simply interpret $x \otimes y$ as $y \otimes x$.

Chapter 2

Kashiwara Crystals

In this chapter, we introduce Kashiwara crystals. We start in Section 2.1 with the definition of root systems since Kashiwara crystals rely on them. This is meant as a brief review and reference as we assume that the reader is familiar with basics of root systems, their classifications, Coxeter groups, and Cartan types. For more background on these topics, see for example [Bourbaki (2002); Bump (2013); Fulton and Harris (1991); Humphreys (1990)]. We proceed in Section 2.2 with the formal definition of Kashiwara crystals together with simple examples. Further examples of crystals may be found in Chapters 3, 5, 6 and 10. One appealing property of crystals is their simple behavior under taking tensor products as explained in Section 2.3. The combinatorial rule that governs tensor products, called the signature rule, is discussed in Section 2.4. In Section 2.6, we will define the character of the crystal, which is analogous to the character of a representation, and we prove that it is invariant under the action of the Weyl group using results from Section 2.5 on root strings in the crystal. In Sections 2.7 and 2.8, we discuss twisting and Levi branchings of crystals, which will become relevant later.

2.1   Root systems

Let V be a Euclidean space, that is, a real vector space with an inner product $\langle\,,\,\rangle$ that is a positive definite, symmetric bilinear form. If $0 \neq \alpha \in V$, then the reflection r_α in the hyperplane orthogonal to α is the map

$$r_\alpha(x) = x - \langle x, \alpha^\vee \rangle \alpha, \qquad \text{where} \quad \alpha^\vee = \frac{2\alpha}{\langle \alpha, \alpha \rangle}. \tag{2.1}$$

A *root system* Φ in V is a nonempty finite set of nonzero vectors in V such that

(1) $r_\alpha(\Phi) = \Phi$ for all $\alpha \in \Phi$;
(2) $\langle \alpha, \beta^\vee \rangle \in \mathbb{Z}$ for all $\alpha, \beta \in \Phi$;
(3) if $\beta \in \Phi$ is a multiple of $\alpha \in \Phi$, then $\beta = \pm\alpha$.

Since $r_\alpha(\alpha) = -\alpha$, the first property implies that $\Phi = -\Phi$. Some authors omit condition (3) and call a root system satisfying (3) *reduced*. Since we will only encounter reduced root systems, we include it in our definition.

Elements of Φ are called *roots*, and elements of

$$\Phi^\vee = \{\alpha^\vee \mid \alpha \in \Phi\}$$

are called *coroots*. (The coroots also form a root system.)

The root system Φ it is called *reducible* if it is the union of two proper, orthogonal subsets, which are themselves root systems. If Φ is not reducible, it is called *irreducible* or *simple*. The root system is *simply-laced* if all roots have the same length.

Taking V, as we have done, to be a Euclidean space has some intuitive value. For example, it is important to know when two roots α and β are orthogonal, that is, $\langle \alpha, \beta \rangle = 0$. But orthogonality of vectors is a notion that requires an inner product. Taking the *ambient space* V to be a Euclidean space has the effect of identifying it with its dual. However, we seldom really want to think of the roots and coroots as living in the same space, and we do not make much use of this identification.

Given a root system Φ in V, a *weight lattice* is a lattice Λ spanning V such that $\Phi \subset \Lambda$, and furthermore if $\lambda \in \Lambda$, $\alpha \in \Phi$, then $\langle \lambda, \alpha^\vee \rangle \in \mathbb{Z}$. We will always fix a weight lattice as well as a root system. Its elements are called *weights*.

The weight lattice is called *semisimple* if Φ spans V. This is equivalent to assuming that the *root lattice* Λ_{root} spanned by Φ has finite codimension in Λ. See Example 2.6 for an example of a root lattice that is semisimple and Example 2.5 for one that is not.

Let us fix a hyperplane through the origin that does not intersect Φ. We call the roots on one side of this hyperplane *positive* and those on the other side *negative*. Let Φ^+ be the set of positive roots and Φ^- the set of negative roots. A positive root $\alpha \in \Phi^+$ is called *simple* if it cannot be expressed as a sum of other positive roots.

Proposition 2.1. *Let Σ be the set of simple positive roots. The elements of Σ are linearly independent. If α, β are distinct elements of Σ, then $\langle \alpha, \beta \rangle \leqslant 0$. Every positive root may be expressed as a linear combination of the elements of Σ with nonnegative coefficients.*

Proof. See [Bump (2013), Proposition 20.1] or [Bourbaki (2002), VI]. □

We will usually choose an *index set* I and index elements of Σ by I. Thus we write $\Sigma = \{\alpha_i \mid i \in I\}$. Often (but not always) we use the index set $I = \{1, 2, \ldots, r\}$. If $i \in I$, we use the notation s_i to denote the reflection r_{α_i}. These elements are called the *simple reflections*.

Proposition 2.2. *If $i \in I$ and $\alpha \in \Phi^+$, then either $\alpha = \alpha_i$, in which case $s_i(\alpha) = -\alpha$, or $s_i(\alpha)$ is another positive root. Thus s_i permutes $\Phi^+ \setminus \{\alpha_i\}$.*

Proof. See [Bump (2013), Proposition 20.1 (ii)] or [Bourbaki (2002), VI.1.6]. □

Let W be the group generated by r_α ($\alpha \in \Phi$). It is the *Weyl group* of Φ.

By a *Coxeter group*, we mean a group G generated by elements $\{s_1, \ldots, s_r\}$ of order 2 such that the relations

$$s_i^2 = 1, \qquad (s_i s_j)^{n(i,j)} = 1$$

form a presentation of G, where $n(i,j)$ is the order of $s_i s_j$. Hence any element $w \in G$ can be written as $w = s_{i_1} s_{i_2} \cdots s_{i_\ell}$. An expression $w = s_{i_1} s_{i_2} \cdots s_{i_\ell}$ is called *reduced* if ℓ is minimal, meaning that w cannot be written with fewer generators. In this case, ℓ is called the *length* of w. For finite Coxeter groups, there exists a *long element*, which we usually denote by $w_0 \in G$.

Proposition 2.3. *The Weyl group W is a Coxeter group generated by the simple reflections s_i with $i \in I$.*

Proof. See [Bump (2013), Theorem 21.5] or [Bourbaki (2002), VI.1 and V.3.2]. $\square$

Definition 2.4. We define a partial order on Λ as follows. If $\lambda, \mu \in \Lambda$, we write $\lambda \succcurlyeq \mu$ if $\lambda - \mu = \sum_{i \in I} c_i \alpha_i$, where the coefficients c_i are nonnegative.

Also, let

$$\Lambda^+ = \{\lambda \in \Lambda \mid \langle \lambda, \alpha_i^\vee \rangle \geqslant 0 \text{ for all } i \in I\}.$$

An element of Λ^+ is called a *dominant weight*. If $\langle \lambda, \alpha_i^\vee \rangle > 0$ for all $i \in I$, then we say that λ is *strictly dominant*. We introduce special vectors ϖ_i for $i \in I$, called *fundamental weights*, defined by the relation

$$\langle \varpi_i, \alpha_j^\vee \rangle = \begin{cases} 1 & \text{if } i = j, \\ 0 & \text{otherwise.} \end{cases} \tag{2.2}$$

If Φ is semisimple, this condition determines the ϖ_i, and otherwise we can choose them at our convenience subject to this condition. Although they are called "weights", it is not necessarily true that the fundamental weights are elements of the weight lattice Λ. For an example where they are not, see Example 2.8.

Assuming that the weight lattice is semisimple, we make the following definitions. In this case, the fundamental weights generate a lattice Λ_{sc} that contains Λ_{root} as a sublattice of finite index. Then

$$\Lambda_{\text{sc}} \supseteq \Lambda \supseteq \Lambda_{\text{root}}.$$

If $\Lambda = \Lambda_{\text{root}}$, we say that Λ is of *adjoint type*. On the other hand, if $\Lambda = \Lambda_{\text{sc}}$, then we say Λ is of *simply-connected type*. Thus Λ is of simply-connected type if and only if all the fundamental weights are in the weight lattice. See Example 2.8 for an example of a weight lattices that is not simply-connected and Example 2.7 for one that is. For many purposes there is no harm in enlarging the weight lattice to the simply-connected one.

The vector

$$\rho = \frac{1}{2} \sum_{\alpha \in \Phi^+} \alpha \qquad (2.3)$$

is called the *Weyl vector* and appears frequently in Lie theory. For example, ρ appears in the Weyl character formula. It is a dominant weight (see Exercise 2.13).

Now let us give some examples of root systems and weight lattices.

Roughly, the irreducible examples fall into the four infinite families of the Cartan classification: A_r $(r \geqslant 1)$, B_r $(r \geqslant 2)$, C_r $(r \geqslant 2)$, D_r $(r \geqslant 4)$ with five exceptional types G_2, F_4, E_6, E_7 and E_8. These families and exceptional classes are called *Cartan types*. See, for example, [Bourbaki (2002); Bump (2013); Fulton and Harris (1991)]. Within each Cartan type, however, there is some flexibility. For example, we will describe $GL(r+1)$ and $SL(r+1)$ weight lattices, which both correspond to the same Cartan type A_r, yet are distinct.

In the following examples, we will denote by $\mathbf{e}_i = (0, \ldots, 1, \ldots, 0)$ the usual basis vectors of $\mathbb{R}^n$, with the 1 in the i-th position.

Example 2.5. (*Cartan Type A_r, $GL(r+1)$ version*) *Let $V = \mathbb{R}^{r+1}$ with the usual inner product and $\mathbf{e}_i \in V$. Let*

$$\Phi = \{\mathbf{e}_i - \mathbf{e}_j \mid i \neq j\} = \Phi^\vee,$$
$$\Phi^+ = \{\mathbf{e}_i - \mathbf{e}_j \mid i < j\}.$$

We take $\Lambda = \mathbb{Z}^{r+1}$. A weight $\lambda = (\lambda_1, \ldots, \lambda_{r+1})$ is dominant if and only if $\lambda_1 \geqslant \lambda_2 \geqslant \cdots \geqslant \lambda_{r+1}$. The simple roots are $\alpha_i = \mathbf{e}_i - \mathbf{e}_{i+1}$ for $i = 1, \ldots, r$. For the fundamental weights we take

$$\varpi_i = \mathbf{e}_1 + \cdots + \mathbf{e}_i = (\underbrace{1, \ldots, 1}_{i \text{ times}}, \underbrace{0, \ldots, 0}_{r+1-i \text{ times}}). \qquad (2.4)$$

We will call this Λ and Φ the $GL(r+1)$ weight lattice and root system.

Example 2.6. (*Cartan Type A_r, $SL(r+1)$ version*) *Let V be quotient space of $\mathbb{R}^{r+1}$ by the subspace spanned by the diagonal vector $(1, 1, \ldots, 1)$. We take $\Phi = \Phi^\vee$ to be the image of the $GL(r+1)$ root system in the quotient space V. The weight lattice Λ is the image of $\mathbb{Z}^{r+1}$ in this quotient space.*

We call this the $SL(r+1)$ weight lattice and root system. This weight lattice is semisimple, whereas the $GL(r+1)$ root system is not. Nevertheless, the $GL(r+1)$ root system is more convenient to work with, and we will generally use it instead. The $GL(r+1)$ and $SL(r+1)$ root systems are in bijection through the quotient map.

Example 2.7. (*Cartan Type B_r, $spin(2r+1)$*) *Let $V = \mathbb{R}^r$. Let*

$$\Phi = \{\pm\mathbf{e}_i \pm \mathbf{e}_j \mid i < j\} \cup \{\pm\mathbf{e}_i\},$$
$$\Phi^+ = \{\mathbf{e}_i \pm \mathbf{e}_j \mid i < j\} \cup \{\mathbf{e}_i\}.$$

The weight lattice Λ is the set of $(x_1, \ldots, x_r) \in \mathbb{Q}^r$ such that $2x_i \in \mathbb{Z}$ and $2x_i$ are either all even or all odd. A weight $\lambda = (\lambda_1, \ldots, \lambda_r)$ is dominant if and only if $\lambda_1 \geqslant \lambda_2 \geqslant \cdots \geqslant \lambda_r \geqslant 0$. This example is not simply-laced. We call this the $\operatorname{spin}(2r+1)$ weight lattice and root system. See Figure 2.1 for the simple roots and fundamental weights.

Example 2.8. (Cartan Type B_r, $\mathrm{SO}(2r+1)$) *As a variant of Example 2.7, we can take the same roots, but take Λ to be just the lattice $\mathbb{Z}^r$, which is the root lattice. In this case, we would speak of the (orthogonal) $\mathrm{SO}(2r+1)$ weight lattice and root system. The $\operatorname{spin}(2r+1)$ weight lattice is simply-connected. The $\mathrm{SO}(2r+1)$ root system is not. We will call the elements of the spin weight lattice that are not in the orthogonal weight lattice* spin weights *and the elements of the orthogonal weight lattice* orthogonal weights.

Example 2.9. (Cartan Type C_r) *Let $V = \mathbb{R}^r$. Let*

$$\Phi = \{\pm\mathbf{e}_i \pm \mathbf{e}_j \mid i < j\} \cup \{\pm 2\mathbf{e}_i\},$$
$$\Phi^+ = \{\mathbf{e}_i \pm \mathbf{e}_j \mid i < j\} \cup \{2\mathbf{e}_i\}.$$

The weight lattice $\Lambda = \mathbb{Z}^r$. A weight $\lambda = (\lambda_1, \ldots, \lambda_r)$ is dominant if and only if $\lambda_1 \geqslant \lambda_2 \geqslant \cdots \geqslant \lambda_r \geqslant 0$. This example is not simply-laced. See Figure 2.1 for the simple roots and fundamental weights.

We will call this the (symplectic) $\mathrm{Sp}(2r)$ weight lattice and root system.

Example 2.10. (Cartan Type D_r, $\operatorname{spin}(2r)$) *Let $V = \mathbb{R}^r$. Let*

$$\Phi = \{\pm\mathbf{e}_i \pm \mathbf{e}_j \mid i < j\} = \Phi^\vee,$$
$$\Phi^+ = \{\mathbf{e}_i \pm \mathbf{e}_j \mid i < j\}.$$

The weight lattice Λ is the same as the $\operatorname{spin}(2r+1)$ weight lattice: $\mathbb{Z}^r$ plus those vectors of the form $(n_1/2, \ldots, n_r/2)$ where the n_i are odd integers. A weight $\lambda = (\lambda_1, \ldots, \lambda_r)$ is dominant if and only if $\lambda_1 \geqslant \lambda_2 \geqslant \cdots \geqslant \lambda_{r-1} \geqslant |\lambda_r|$. This example is simply-laced and of simply-connected type. We call this the $\operatorname{spin}(2r+1)$ weight lattice and root system. See Figure 2.1 for the simple roots and fundamental weights.

Example 2.11. (Cartan Type D_r, $\mathrm{SO}(2r)$) *As a variant of the last example, we could take Λ to be just the lattice $\mathbb{Z}^r$. In this case, we would speak of the (orthogonal) $\mathrm{SO}(2r)$ root system and weight lattice. Again, we will call a weight orthogonal if it is in the adjoint lattice $\mathbb{Z}^r$, and we will call it a spin weight if it is not. Thus the spin weights are of the form $(n_1/2, \ldots, n_r/2)$ where the n_i are odd integers.*

For types A_r through D_r, Figure 2.1 gives the simple roots and fundamental weights.

Example 2.12. (Exceptional Types) *The Cartan types $G_2, F_4, E_6, E_7,$ and E_8 are described in the appendices to [Bourbaki (2002)].*

We discuss Dynkin diagrams corresponding to a root system in Section 2.8.

Cartan Type	Simple Roots	Fundamental Weights
A_r	$\alpha_1 = \mathbf{e}_1 - \mathbf{e}_2$ $\alpha_2 = \mathbf{e}_2 - \mathbf{e}_3$ $\vdots$ $\alpha_r = \mathbf{e}_r - \mathbf{e}_{r+1}$	$\varpi_1 = \mathbf{e}_1$ $\varpi_2 = \mathbf{e}_1 + \mathbf{e}_2$ $\vdots$ $\varpi_r = \mathbf{e}_1 + \mathbf{e}_2 + \cdots + \mathbf{e}_r$
B_r	$\alpha_1 \quad = \mathbf{e}_1 - \mathbf{e}_2$ $\alpha_2 \quad = \mathbf{e}_2 - \mathbf{e}_3$ $\vdots$ $\alpha_{r-1} = \mathbf{e}_{r-1} - \mathbf{e}_r$ $\alpha_r \quad = \mathbf{e}_r$	$\varpi_1 \quad = \mathbf{e}_1$ $\varpi_2 \quad = \mathbf{e}_1 + \mathbf{e}_2$ $\vdots$ $\varpi_{r-1} = \mathbf{e}_1 + \cdots + \mathbf{e}_{r-1}$ $\varpi_r \quad = \frac{1}{2}(\mathbf{e}_1 + \mathbf{e}_2 + \cdots + \mathbf{e}_r)$
C_r	$\alpha_1 \quad = \mathbf{e}_1 - \mathbf{e}_2$ $\alpha_2 \quad = \mathbf{e}_2 - \mathbf{e}_3$ $\vdots$ $\alpha_{r-1} = \mathbf{e}_{r-1} - \mathbf{e}_r$ $\alpha_r \quad = 2\mathbf{e}_r$	$\varpi_1 \quad = \mathbf{e}_1$ $\varpi_2 \quad = \mathbf{e}_1 + \mathbf{e}_2$ $\vdots$ $\varpi_{r-1} = \mathbf{e}_1 + \mathbf{e}_2 + \cdots + \mathbf{e}_{r-1}$ $\varpi_r \quad = \mathbf{e}_1 + \mathbf{e}_2 + \cdots + \mathbf{e}_r$
D_r	$\alpha_1 \quad = \mathbf{e}_1 - \mathbf{e}_2$ $\alpha_2 \quad = \mathbf{e}_2 - \mathbf{e}_3$ $\vdots$ $\alpha_{r-1} = \mathbf{e}_{r-1} - \mathbf{e}_r$ $\alpha_r \quad = \mathbf{e}_{r-1} + \mathbf{e}_r$	$\varpi_1 \quad = \mathbf{e}_1$ $\varpi_2 \quad = \mathbf{e}_1 + \mathbf{e}_2$ $\vdots$ $\varpi_{r-1} = \frac{1}{2}(\mathbf{e}_1 + \cdots + \mathbf{e}_{r-1} - \mathbf{e}_r)$ $\varpi_r \quad = \frac{1}{2}(\mathbf{e}_1 + \cdots + \mathbf{e}_{r-1} + \mathbf{e}_r)$

Fig. 2.1   Simple roots and fundamental weights for the classical Cartan types.

2.2   Kashiwara crystals

Originally crystal bases were introduced as certain limits of representations of quantum groups. In this section we define crystals in an axiomatic fashion. Each crystal is associated to a root system Φ with index set I and weight lattice Λ.

For us, the set $\mathbb{Z} \sqcup \{-\infty\}$ is ordered with $-\infty < n$ for all $n \in \mathbb{Z}$. We also define $-\infty + n = -\infty$ for all $n \in \mathbb{Z}$.

Definition 2.13. Fix a root system Φ with index set I and weight lattice Λ. A *Kashiwara crystal* (or *crystal* for short) of type Φ is a nonempty set $\mathcal{B}$ together with maps

$$e_i, f_i \colon \mathcal{B} \longrightarrow \mathcal{B} \sqcup \{0\}, \tag{2.5a}$$

$$\varepsilon_i, \varphi_i \colon \mathcal{B} \longrightarrow \mathbb{Z} \sqcup \{-\infty\}, \tag{2.5b}$$

$$\mathrm{wt} \colon \mathcal{B} \longrightarrow \Lambda, \tag{2.5c}$$

where $i \in I$ and $0 \notin \mathcal{B}$ is an auxiliary element, satisfying the following conditions:

A1. If $x, y \in \mathcal{B}$ then $e_i(x) = y$ if and only if $f_i(y) = x$. In this case, it is assumed that

$$\mathrm{wt}(y) = \mathrm{wt}(x) + \alpha_i, \qquad \varepsilon_i(y) = \varepsilon_i(x) - 1, \qquad \varphi_i(y) = \varphi_i(x) + 1.$$

A2. We require that

$$\varphi_i(x) = \langle \mathrm{wt}(x), \alpha_i^\vee \rangle + \varepsilon_i(x)$$

for all $x \in \mathcal{B}$ and $i \in I$. In particular, if $\varphi_i(x) = -\infty$, then $\varepsilon_i(x) = -\infty$ also. If $\varphi_i(x) = -\infty$, then we require that $e_i(x) = f_i(x) = 0$.

The number of elements of a crystal is referred to as its *degree*. The map wt is called the *weight map*. The operators e_i and f_i are called *Kashiwara* or *crystal operators*. The maps φ_i and ε_i are sometimes referred to as *string lengths*.

If φ_i and ε_i do not take the value $-\infty$, then we say that $\mathcal{B}$ is of *finite type*. This does not necessarily imply that $\mathcal{B}$ is finite. If

$$\varphi_i(x) = \max\{k \in \mathbb{Z}_{\geqslant 0} \mid f_i^k(x) \neq 0\} \quad \text{and} \quad \varepsilon_i(x) = \max\{k \in \mathbb{Z}_{\geqslant 0} \mid e_i^k(x) \neq 0\},$$
$$(2.6)$$

for all i, then $\mathcal{B}$ is called *seminormal*. This implies, of course, that $\mathcal{B}$ is of finite type. It also implies that the values of φ_i and ε_i are nonnegative. If just the second condition is assumed, we say that $\mathcal{B}$ is *upper seminormal*.

Lemma 2.14. *If the root system is semisimple and $\mathcal{C}$ is a crystal of finite type, then*

$$\mathrm{wt}(x) = \sum_{i \in I} \big(\varphi_i(x) - \varepsilon_i(x)\big) \varpi_i. \qquad (2.7)$$

Proof. Consider the inner product

$$\left\langle \mathrm{wt}(x) - \sum_{i \in I} \big(\varphi_i(x) - \varepsilon_i(x)\big) \varpi_i, \alpha_j^\vee \right\rangle.$$

By (2.2) and Axiom A2, this equals

$$\langle \mathrm{wt}(x), \alpha_j^\vee \rangle - \varphi_j(x) + \varepsilon_j(x) = 0.$$

Since Φ is semisimple, any vector orthogonal to the simple coroots is zero. $\qquad \square$

We will mainly be interested in seminormal crystals, but it is convenient to allow more general crystals. For example, Kashiwara's crystal $\mathcal{B}_\infty$ is quite important in parts of the theory but it is not seminormal; see Example 2.27 below.

If $\mathcal{B}$ is a crystal, we associate with $\mathcal{B}$ a directed graph with vertices in $\mathcal{B}$ and edges labelled by $i \in I$. If $f_i(x) = y$ for $x, y \in \mathcal{B}$, then we draw an edge $x \xrightarrow{i} y$. This is the *crystal graph* of $\mathcal{B}$.

We may consider the equivalence on $\mathcal{B}$ generated by $x \sim y$ if $y = f_i(x)$ or $y = e_i(x)$ for some $i \in I$, which is precisely the condition that x and y are joined by an edge in the crystal graph. So the equivalence classes in this relation are the connected components of the crystal graph. If $\mathcal{B}$ consists of just one equivalence class, we say that $\mathcal{B}$ is *connected*. The equivalence classes of $\mathcal{B}$ are called the *connected components* of $\mathcal{B}$. Any subset $\mathcal{B}'$ of $\mathcal{B}$ that is a union of connected components inherits a crystal structure from $\mathcal{B}$, and we call $\mathcal{B}'$ a *full subcrystal*. Clearly $\mathcal{B}$ is a disjoint union of its maximal connected subcrystals. We call these full subcrystals its *connected components*.

An element $u \in \mathcal{B}$ such that $e_i(u) = 0$ for all $i \in I$ is called a *highest weight element*. The weight $\mathrm{wt}(u)$ of a highest weight element u is called a *highest weight*. The motivation for the term is the following fact.

Lemma 2.15. *If u is an element of $\mathcal{B}$ such that $\mathrm{wt}(u)$ is maximal with respect to the partial order $\succcurlyeq$ on the weight lattice Λ, then u is a highest weight element.*

Proof. This is obvious since if any $e_i(u) \neq 0$, then $\mathrm{wt}(e_i(u)) \succ \mathrm{wt}(u)$ contradicting the assumed maximality of $\mathrm{wt}(u)$. $\qquad\square$

For a converse, see Theorem 4.12.

Proposition 2.16. *Suppose that $\mathcal{B}$ is seminormal. Let u be a highest weight element. Then $\mathrm{wt}(u)$ is dominant.*

Proof. If $\mathrm{wt}(u)$ is not dominant, then $\langle \mathrm{wt}(u), \alpha_i^\vee \rangle < 0$ for some coroot $\alpha_i^\vee$. This equals $\varphi_i(u) - \varepsilon_i(u)$ and therefore $\varepsilon_i(u) > 0$. Since $\mathcal{B}$ is seminormal, $e_i(u) \neq 0$, contradicting our assumption that u is a highest weight element. $\qquad\square$

Proposition 2.17. *Suppose that $\mathcal{B}$ is seminormal. Let μ, ν be elements of Λ such that $w(\mu) = \nu$ for some Weyl group element w. Then the number of elements of $\mathcal{B}$ of weight μ equals the number of elements of weight ν.*

Proof. It is sufficient to show this if $\nu = s_i(\mu)$, where s_i is the simple reflection. We may also assume that $\mu \neq \nu$. Since $\nu = \mu - \langle \mu, \alpha_i^\vee \rangle \alpha_i$, this means that $k = \langle \mu, \alpha_i^\vee \rangle \neq 0$. Interchanging μ and ν if necessary, we may assume that $k > 0$. Now let X and Y be the sets of elements of $\mathcal{B}$ of weight μ and ν respectively. If $x \in X$ then since $k = \varphi_i(x) - \varepsilon_i(x)$ by Axiom A2, we have $\varphi_i(x) \geqslant k$ and since $\mathcal{B}$ is seminormal it follows that $f_i^k(x)$ is nonzero. Thus f_i^k is a map from X to Y and similarly e_i^k is a map from Y to X. By Axiom A1 these maps are inverses, and hence X and Y are in bijection. $\qquad\square$

Example 2.18. *Let Φ and Λ be the GL(2) root system and weight lattice. For each integer $k > 0$, there is a crystal of cardinality $k+1$ that is seminormal. The weight lattice is $\mathbb{Z}^2$, and there is one element v_ℓ for each $\ell \in \{-k, -k+2, \ldots, k-4, k-2, k\}$. The element v_ℓ has weight $\frac{1}{2}(k+\ell, k-\ell)$. We have $e_1(v_\ell) = v_{\ell+2}$ (except $e_1(v_k) = 0$)*

while $f_1(v_\ell) = v_{\ell-2}$ *(except $f_1(v_{-k}) = 0$). We have $\varphi_1(v_\ell) = \frac{1}{2}(k + \ell)$, $\varepsilon_1(v_\ell) = \frac{1}{2}(k - \ell)$. Here is the crystal graph:*

$$v_k \xrightarrow{\;1\;} v_{k-2} \xrightarrow{\;1\;} \cdots \xrightarrow{\;1\;} v_{-k+2} \xrightarrow{\;1\;} v_{-k} \;.$$

We will call this crystal $\mathcal{B}_{(k)}$. The highest weight element is v_k of weight $k\mathbf{e}_1$.

In this example, (2.7) is not satisfied with the fundamental weight (2.4). This does not contradict Lemma 2.14 since the GL(2) weight lattice is not semisimple.

Example 2.19. *For type A_r, there is a standard crystal whose crystal graph looks like this:*

$$\boxed{1} \xrightarrow{\;1\;} \boxed{2} \xrightarrow{\;2\;} \cdots \xrightarrow{\;r\;} \boxed{r+1} \;.$$

We use the GL$(r + 1)$ weight lattice as in Example 2.5, so we will refer to this (or any crystal with this weight lattice) as a GL$(r+1)$-crystal. The weight $\mathrm{wt}\!\left(\boxed{i}\right) = \mathbf{e}_i$. We must define the φ_i and ε_i. We do this by requiring it to be seminormal, from which the reader may infer the values of these crystal operators. We will usually denote this crystal as $\mathcal{B}_{(1)}$ or (in some chapters) as $\mathbb{B}$.

Definition 2.20. Let $\mathcal{C}$ be a crystal of finite type. We may define a new crystal as follows. Let $\mathcal{C}^\vee$ be a set that is in bijection with $\mathcal{C}$. Denote the bijection $\mathcal{C} \to \mathcal{C}^\vee$ by $x \mapsto x^\vee$. Then if we define $\mathrm{wt}(x^\vee) = -\mathrm{wt}(x)$

$$\varepsilon_i(x^\vee) = \varphi_i(x), \quad \varphi_i(x^\vee) = \varepsilon_i(x), \quad e_i(x^\vee) = f_i(x)^\vee, \quad f_i(x^\vee) = e_i(x)^\vee,$$

we may check (Exercise 2.6) that $\mathcal{C}^\vee$ is a crystal, called the *dual* or *contragredient* of $\mathcal{C}$.

Example 2.21. *For type A_r, if $\mathbb{B}$ is the standard crystal of Example 2.19, the dual $\mathbb{B}^\vee$ is the* dual standard crystal. *Its crystal graph looks like this:*

$$\boxed{\overline{r+1}} \xrightarrow{\;r\;} \boxed{\overline{r}} \xrightarrow{\;r-1\;} \cdots \xrightarrow{\;2\;} \boxed{\overline{2}} \xrightarrow{\;1\;} \boxed{\overline{1}} \;.$$

The weight $\mathrm{wt}\!\left(\boxed{\overline{i}}\right) = -\mathbf{e}_i$. Here we are using $\boxed{\overline{i}}$ instead of $\boxed{i}^\vee$ for reasons of legibility.

For the other classical Cartan types there are also standard crystals, which we now describe. Unlike the case of type A_r, for type B_r, C_r or D_r the standard crystals are self-dual. In these crystals $\mathrm{wt}\!\left(\boxed{i}\right) = \mathbf{e}_i$ unless $i = 0$, in which case $\mathrm{wt}\!\left(\boxed{0}\right) = 0$, and $\mathrm{wt}\!\left(\boxed{\overline{i}}\right) = -\mathbf{e}_i$. Note that here $\boxed{0}$ is an element of the crystal and not the auxiliary element 0. The crystals are seminormal, from which the values of φ_i and ε_i may be inferred.

Example 2.22. *For type B_r, the standard crystal is the seminormal crystal with crystal graph:*

$$\boxed{1} \xrightarrow{\;1\;} \boxed{2} \xrightarrow{\;2\;} \cdots \xrightarrow{\;r-1\;} \boxed{r} \xrightarrow{\;r\;} \boxed{0} \xrightarrow{\;r\;} \boxed{\overline{r}} \xrightarrow{\;r-1\;} \cdots \xrightarrow{\;1\;} \boxed{\overline{1}} \;.$$

Example 2.23. *For type C_r, the standard crystal is the seminormal crystal with crystal graph:*

$$\boxed{1} \xrightarrow{1} \boxed{2} \xrightarrow{2} \cdots \xrightarrow{r-1} \boxed{r} \xrightarrow{r} \boxed{\bar{r}} \xrightarrow{r-1} \cdots \xrightarrow{1} \boxed{\bar{1}}.$$

Example 2.24. *For type D_r, the standard crystal is the seminormal crystal with crystal graph:*

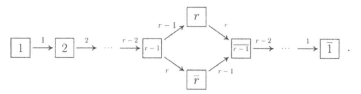

Let us give a couple of examples of $\mathrm{GL}(n)$ crystals that are special cases of the *crystals of tableaux* that will occupy us in Chapter 3.

Example 2.25. *Let $\Lambda = \mathbb{Z}^n$ be the $\mathrm{GL}(n)$ weight lattice, so $n = r + 1$ in our previous notation. By a* row *of length k we mean a row*

$$R = \boxed{j_1 \, j_2 \, \cdots \, j_k}$$

of k boxes filled with integers $j_1 \leqslant \cdots \leqslant j_k$ taken from the alphabet $[n] = \{1, 2, 3, \ldots, n\}$. Let $\mathcal{B}_{(k)}$ be the set of rows of length k.[1] We define $\varphi_i(R)$ to be the number of i's that occur among the entries $j_1, \ldots, j_k$, and $\varepsilon_i(R)$ to be the number of $i+1$'s. We define $\mathrm{wt}(R) = (\mu_1, \ldots, \mu_n)$ where μ_i is the number of i's in R. If $\varphi_i(R) > 0$, we define $f_i(R)$ to be the row obtained by changing the rightmost i to an $i + 1$; otherwise $f_i(R) = 0$. If $\varepsilon_i(R) > 0$, we define $e_i(R)$ to be the row obtained by changing the leftmost $i + 1$ to an i; otherwise $e_i(R) = 0$. It is easy to see that with these definitions, $\mathcal{B}_{(k)}$ is a seminormal crystal. For example, if $n = 3$, the $\mathrm{GL}(3)$ crystal $\mathcal{B}_{(3)}$ is shown in Figure 2.2.

Example 2.26. *Similary to the crystal of rows, there is the* crystal of columns. *This differs from the crystal of rows in that the entries in a column must be strictly increasing. Thus let $\mathcal{B}_{(1^k)}$ be the set of*

$$C = \begin{array}{|c|} \hline j_1 \\ \hline j_2 \\ \hline \vdots \\ \hline j_k \\ \hline \end{array}, \qquad j_1 < \cdots < j_k,$$

with $j_1, \ldots, j_k \in [n]$.[2] We define $\varphi_i(C)$ to be 1 if there is an entry i but no entry $i + 1$; in this case $f_i(C)$ is obtained from C by changing the i to $i + 1$. Otherwise, $\varphi_i(C) = 0$. In other words, $\varphi_i(C) = 1$ if it is possible to change one i to $i + 1$ and still obtain a column that is strictly increasing. Similarly, $\varepsilon_i(C) = 1$ if an $i + 1$ can

---

[1] The subscript (k) denotes the partition of k with just one part; we will define more general $\mathcal{B}_\lambda$ for all partitions λ of k in Chapter 3.

[2] The subscript (1^k) denotes the partition $(1, 1, \ldots, 1)$ with k parts equal to 1.

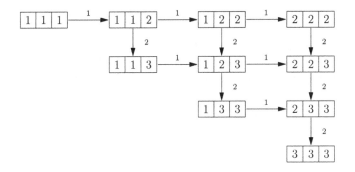

Fig. 2.2   The GL(3) crystal $\mathcal{B}_{(3)}$ of rows of length 3.

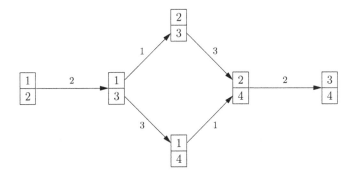

Fig. 2.3   The GL(4) crystal $\mathcal{B}_{(1,1)}$ of columns of length 2.

be changed to an i. Again, it is easy to check that this crystal is seminormal. The crystal of columns of length 2 when $n = 4$ is illustrated in Figure 2.3.

The next two examples are crystals that are *not* seminormal.

Example 2.27. *For type A_1, here is a crystal called $\mathcal{B}_\infty$ that is of finite type but not seminormal. Although it is of finite type, the crystal is infinite:*

$$v_0 \xrightarrow{1} v_{-2} \xrightarrow{1} v_{-4} \xrightarrow{1} \cdots$$

We have $\mathrm{wt}(v_{-2k}) = (-k, k)$, $\varphi_1(v_{-2k}) = -k$ *and* $\varepsilon_1(v_{-2k}) = k$.

Example 2.28. *Here is a crystal that is finite, but not of finite type. Let λ be any weight, and let $\mathcal{T}_\lambda$ be the crystal that has one element t_λ such that $f_i(t_\lambda) = e_i(t_\lambda) = 0$ for all $i \in I$, $\mathrm{wt}(t_\lambda) = \lambda$ and $\varphi_i(t_\lambda) = \varepsilon_i(t_\lambda) = -\infty$.*

These two crystals will be studied in more detail and for general root systems in Chapter 12.

2.3   Tensor products of crystals

If $\mathcal{B}$ and $\mathcal{C}$ are two crystals associated to the same root system Φ, then we define the *tensor product* $\mathcal{B} \otimes \mathcal{C}$. As a set, it is the Cartesian product, but we denote the ordered pair (x, y) with $x \in \mathcal{B}$ and $y \in \mathcal{C}$ by $x \otimes y$. We define $\mathrm{wt}(x \otimes y) = \mathrm{wt}(x) + \mathrm{wt}(y)$,

$$f_i(x \otimes y) = \begin{cases} f_i(x) \otimes y & \text{if } \varphi_i(y) \leqslant \varepsilon_i(x), \\ x \otimes f_i(y) & \text{if } \varphi_i(y) > \varepsilon_i(x), \end{cases}$$

and

$$e_i(x \otimes y) = \begin{cases} e_i(x) \otimes y & \text{if } \varphi_i(y) < \varepsilon_i(x), \\ x \otimes e_i(y) & \text{if } \varphi_i(y) \geqslant \varepsilon_i(x). \end{cases}$$

It is understood that $x \otimes 0 = 0 \otimes x = 0$. We also let

$$\varphi_i(x \otimes y) = \max\big(\varphi_i(x), \varphi_i(y) + \langle \mathrm{wt}(x), \alpha_i^\vee \rangle\big) \tag{2.8}$$

and

$$\varepsilon_i(x \otimes y) = \max\big(\varepsilon_i(y), \varepsilon_i(x) - \langle \mathrm{wt}(y), \alpha_i^\vee \rangle\big). \tag{2.9}$$

If $\mathcal{B}$ and $\mathcal{C}$ are of finite type, we may write these as

$$\begin{aligned} \varphi_i(x \otimes y) &= \varphi_i(x) + \max\big(0, \varphi_i(y) - \varepsilon_i(x)\big), \\ \varepsilon_i(x \otimes y) &= \varepsilon_i(y) + \max\big(0, \varepsilon_i(x) - \varphi_i(y)\big). \end{aligned} \tag{2.10}$$

Note that our convention for tensor products is opposite to that in Kashiwara's original papers [Kashiwara (1990, 1991, 1994)]. Our convention is better suited in conjunction with combinatorial properties such as the Robinson–Schenstend correspondence.

Proposition 2.29. *The tensor product $\mathcal{B} \otimes \mathcal{C}$ is a crystal. If $\mathcal{B}$ and $\mathcal{C}$ are seminormal, so is $\mathcal{B} \otimes \mathcal{C}$.*

Proof. We first show that $f_i(x \otimes y) = z \otimes w$ if and only if $e_i(z \otimes w) = x \otimes y$. There are two cases. If $\varphi_i(y) \leqslant \varepsilon_i(x)$, then $z = f_i(x)$ and $w = y$. Then $\varepsilon_i(z) = \varepsilon_i(x) + 1$ and $\varphi_i(w) = \varphi_i(y)$ so $\varphi_i(w) < \varepsilon_i(z)$ and $e_i(z \otimes w) = e_i(z) \otimes w = x \otimes y$. On the other hand, if $\varphi_i(y) > \varepsilon_i(x)$ then $z = x$ and $w = f_i(y)$. Then $\varphi_i(w) = \varphi_i(y) - 1 \geqslant \varepsilon_i(x) = \varepsilon_i(z)$ and so $e_i(w \otimes z) = w \otimes e_i(z) = x \otimes y$. This proves that $f_i(x \otimes y) = z \otimes w$ implies $e_i(z \otimes w) = x \otimes y$. We leave the other direction to the reader.

The next thing to check is that if $f_i(x \otimes y) = z \otimes w$ then $\varphi_i(z \otimes w) = \varphi_i(x \otimes y) - 1$. Again there are two cases. If $\varphi_i(y) \leqslant \varepsilon_i(x)$ then $z = f_i(x)$ and $w = y$. We have

$$\varphi_i(x \otimes y) = \max\big(\varphi_i(x), \varphi_i(y) + \langle \mathrm{wt}(x), \alpha_i^\vee \rangle\big) = \varphi_i(x),$$

while

$$\varphi_i(z \otimes w) = \max\big(\varphi_i(x) - 1, \varphi_i(y) + \langle \mathrm{wt}(x), \alpha_i^\vee \rangle - 2\big) = \varphi_i(x) - 1,$$

as required. On the other hand, if $\varphi_i(y) > \varepsilon_i(x)$ then $z = x$ and $w = f_i(y)$. In this case

$$\varphi_i(x \otimes y) = \varphi_i(y) + \langle \text{wt}(x), \alpha_i^\vee \rangle$$

and

$$\varphi_i(z \otimes w) = \max\big(\varphi_i(x), \varphi_i(y) - 1 + \langle \text{wt}(x), \alpha_i^\vee \rangle\big) = \varphi_i(y) - 1 + \langle \text{wt}(x), \alpha_i^\vee \rangle$$
$$= \varphi_i(x \otimes y) - 1,$$

as desired.

We leave it to the reader to check that $\varepsilon_i(z \otimes w) = \varepsilon_i(x \otimes y) + 1$. It is straightforward to see that $\text{wt}(z \otimes w) = \text{wt}(x \otimes y) - \alpha_i$ since this is true whether $z = f_i(x)$ and $w = y$, or $z = x$ and $w = f_i(y)$. To prove Axiom 2 note that, using (2.8) and (2.9),

$$\langle \text{wt}(x \otimes y), \alpha_i^\vee \rangle + \varepsilon_i(x \otimes y)$$
$$= \langle \text{wt}(x), \alpha_i^\vee \rangle + \langle \text{wt}(y), \alpha_i^\vee \rangle + \max\big(\varepsilon_i(y), \varepsilon_i(x) - \langle \text{wt}(y), \alpha_i^\vee \rangle\big)$$
$$= \max\big(\langle \text{wt}(x), \alpha_i^\vee \rangle + \varphi_i(y), \varphi_i(x)\big) = \varphi_i(x \otimes y).$$

This proves that $\mathcal{B} \otimes \mathcal{C}$ is a crystal.

It remains to be shown that if $\mathcal{B}$ and $\mathcal{C}$ are seminormal, so is $\mathcal{B} \otimes \mathcal{C}$. We must show that $\varphi_i(x \otimes y)$ is the number of times f_i can be applied to $x \otimes y$ before we get 0. First suppose that $\varphi_i(y) \leqslant \varepsilon_i(x)$. Then $f_i^{\varphi_i(x)}(x \otimes y) = f_i^{\varphi_i(x)}(x) \otimes y$, and the next application gives 0, so $\varphi_i(x) = \varphi_i(x \otimes y)$ is the exact number of times f_i may be applied to $x \otimes y$ in this case. The case $\varphi_i(y) > \varepsilon_i(x)$ is more complicated. In this case $\varphi_i(x \otimes y) = \varphi_i(x) + \varphi_i(y) - \varepsilon_i(x)$. First apply $f_i^{\varphi_i(y) - \varepsilon_i(x)}$ to $x \otimes y$ to obtain $x \otimes f_i^{\varphi_i(y) - \varepsilon_i(x)}(y)$. Now $\varphi_i\big(f_i^{\varphi_i(y) - \varepsilon_i(x)}(y)\big) = \varepsilon_i(x)$, so the next time we apply f_i it goes on the x. We can apply it $\varphi_i(x)$ times before we obtain

$$f_i^{\varphi_i(y) - \varepsilon_i(x) + \varphi_i(x)}(x \otimes y) = f_i^{\varphi_i(x)}(x) \otimes f_i^{\varphi_i(y) - \varepsilon_i(x)}(y).$$

The next time we apply f_i we get zero, so the number of times f_i can be applied is $\varphi_i(y) - \varepsilon_i(x) + \varphi_i(x) = \varphi_i(x \otimes y)$. We leave the reader to check that $\varepsilon_i(x \otimes y)$ is the number of times that e_i can be applied. $\qquad\square$

Example 2.30. *Let $\mathbb{B}$ be the GL(3) crystal from Example 2.19. Then $\mathbb{B} \otimes \mathbb{B}$ has two connected components, one with three elements and one with six, as shown in Figure 2.4. See also Exercise 2.1.*

Let $\mathcal{B}$ and $\mathcal{C}$ be two crystals associated to the root system Φ and index set I. A *crystal morphism* is a map $\psi \colon \mathcal{B} \longrightarrow \mathcal{C} \sqcup \{0\}$ such that

(1) if $b \in \mathcal{B}$ and $\psi(b) \in \mathcal{C}$, then

    (a) $\text{wt}\big(\psi(b)\big) = \text{wt}(b)$,
    (b) $\varepsilon_i\big(\psi(b)\big) = \varepsilon_i(b)$ for all $i \in I$, and
    (c) $\varphi_i\big(\psi(b)\big) = \varphi_i(b)$ for all $i \in I$;

(2) if $b, e_i b \in \mathcal{B}$ such that $\psi(b), \psi(e_i b) \in \mathcal{C}$, then we have $\psi(e_i b) = e_i \psi(b)$;

(3) if $b, f_i b \in \mathcal{B}$ such that $\psi(b), \psi(f_i b) \in \mathcal{C}$, then we have $\psi(f_i b) = f_i \psi(b)$.

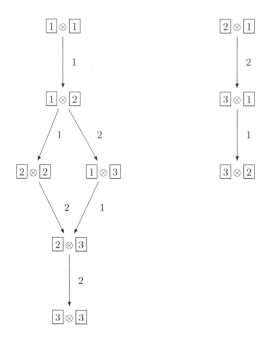

Fig. 2.4   The tensor product of the standard GL(3) crystal with itself.

A morphism ψ is called *strict* if ψ commutes with e_i and f_i for all $i \in I$. Moreover, a crystal morphism $\psi \colon \mathcal{B} \longrightarrow \mathcal{C} \sqcup \{0\}$ is called a *crystal isomorphism* if the induced map $\psi \colon \mathcal{B} \sqcup \{0\} \longrightarrow \mathcal{C} \sqcup \{0\}$ with $\psi(0) = 0$ is a bijection.

With this definition, crystals become a category. If ψ is an isomorphism in this category, then ψ is a bijection and $\psi \circ f_i = f_i \circ \psi$, $\psi \circ e_i = e_i \circ \psi$. However, for a general morphism, what we have assumed is weaker since it would be possible for $f_i(\psi(b)) \neq \psi(f_i(b))$ if one or the other is zero (where we assume that $\psi(0) = 0$).

Example 2.31. *Let $\mathcal{B}_{(k)}$ be the $(k+1)$-dimensional crystal of type A_1 of Example 2.18 with highest weight element $u := v_k$, $\mathcal{B}_\infty$ the crystal of Example 2.27 with highest weight element $u_\infty := v_0$, and $\mathcal{T}_{(k)}$ the singleton crystal of Example 2.28. Then*

$$\psi \colon \mathcal{B}_{(k)} \longrightarrow \mathcal{T}_{(k)} \otimes \mathcal{B}_\infty$$
$$f_1^j u \longmapsto t_{(k)} \otimes f_1^j u_\infty \qquad \text{for } 0 \leqslant j \leqslant k$$

is a crystal morphism, which is not strict. Indeed, by (2.8) and (2.9) we have $\varepsilon_1(t_{(k)} \otimes f_1^j u_\infty) = \varepsilon_1(f_1^j u_\infty) = j = \varepsilon_1(f_1^j u)$ and $\varphi_1(t_{(k)} \otimes f_1^j u_\infty) = \varphi_1(f_1^j u_\infty) + \langle \mathrm{wt}(t_{(k)}), \alpha_1^\vee \rangle = -j + k = \varphi_1(f_1^j u)$. The other conditions can be checked similarly.

This embedding will be studied in more detail in Chapter 12.

Proposition 2.32. *Let $\mathcal{B}$, $\mathcal{C}$ and $\mathcal{D}$ be crystals. Then the bijection $(\mathcal{B} \otimes \mathcal{C}) \otimes \mathcal{D} \longrightarrow \mathcal{B} \otimes (\mathcal{C} \otimes \mathcal{D})$ in which $(x \otimes y) \otimes z \longmapsto x \otimes (y \otimes z)$ is an isomorphism.*

Proof. Let $x \otimes y \otimes z$ denote either $(x \otimes y) \otimes z$ or $x \otimes (y \otimes z)$. It follows directly from the definitions that

$$\varphi_i(x \otimes y \otimes z) = \max\left(\varphi_i(x), \varphi_i(y) + \langle \mathrm{wt}(x), \alpha_i^\vee \rangle, \varphi_i(z) + \langle \mathrm{wt}(y), \alpha_i^\vee \rangle + \langle \mathrm{wt}(x), \alpha_i^\vee \rangle\right).$$

Moreover $f_i(x \otimes y \otimes z)$ is given by the following table:

$\varphi_i(x \otimes y \otimes z)$	$f_i(x \otimes y \otimes z)$ at first maximum
$\varphi_i(x)$	$f_i(x) \otimes y \otimes z$
$\varphi_i(y) + \langle \mathrm{wt}(x), \alpha_i^\vee \rangle$	$x \otimes f_i(y) \otimes z$
$\varphi_i(z) + \langle \mathrm{wt}(y), \alpha_i^\vee \rangle + \langle \mathrm{wt}(x), \alpha_i^\vee \rangle$	$x \otimes y \otimes f_i(z)$

The meaning of "at first maximum" in this table is that $\varphi_i(x \otimes y \otimes z)$ is the maximum of the values in the first column—but it could appear more than once. We take $f_i(x \otimes y \otimes z)$ to be the second column value from the *first time* the maximum is attained. Similarly, $e_i(x \otimes y \otimes z)$ is given from the following table:

$\varepsilon_i(x \otimes y \otimes z)$	$e_i(x \otimes y \otimes z)$ at first maximum
$\varepsilon_i(z)$	$x \otimes y \otimes e_i(z)$
$\varepsilon_i(y) - \langle \mathrm{wt}(z), \alpha_i^\vee \rangle$	$x \otimes e_i(y) \otimes z$
$\varepsilon_i(x) - \langle \mathrm{wt}(y), \alpha_i^\vee \rangle - \langle \mathrm{wt}(z), \alpha_i^\vee \rangle$	$e_i(x) \otimes y \otimes z$

This description is the same whether we parse $x \otimes y \otimes z$ as $(x \otimes y) \otimes z$ or $x \otimes (y \otimes z)$, so the two crystals are isomorphic. $\qquad\square$

The formulas in this proof have an obvious generalization to k-fold tensor products.

Lemma 2.33. *Let $x_1 \otimes \cdots \otimes x_k \in \mathcal{B}_1 \otimes \cdots \otimes \mathcal{B}_k$ for crystals $\mathcal{B}_1, \ldots, \mathcal{B}_k$. Then*

$$\varphi_i(x_1 \otimes x_2 \otimes \cdots \otimes x_k) = \max_{j=1}^{k}\left(\varphi_i(x_j) + \sum_{h=1}^{j-1} \langle \mathrm{wt}(x_h), \alpha_i^\vee \rangle\right), \tag{2.11}$$

and if j is the first value where the maximum is attained, then

$$f_i(x_1 \otimes x_2 \otimes \cdots \otimes x_k) = x_1 \otimes \cdots \otimes f_i(x_j) \otimes \cdots \otimes x_k. \tag{2.12}$$

Similarly, for $x_k \otimes \cdots \otimes x_2 \otimes x_1 \in \mathcal{B}_k \otimes \cdots \otimes \mathcal{B}_2 \otimes \mathcal{B}_1$,

$$\varepsilon_i(x_k \otimes \cdots \otimes x_2 \otimes x_1) = \max_{j=1}^{k}\left(\varepsilon_i(x_j) - \sum_{h=1}^{j-1} \langle \mathrm{wt}(x_h), \alpha_i^\vee \rangle\right), \tag{2.13}$$

and when e_i is applied to the tensor, it applies to x_j where j is the first value in (2.13) where the maximum is attained.

Proof. The reader will easily prove this by induction. $\qquad\square$

If $\mathcal{B}$ is of finite type, we may write (2.11) in the form

$$\varphi_i(x_1 \otimes x_2 \otimes \cdots \otimes x_k) = \max_{j=1}^{k} \left(\sum_{h=1}^{j} \varphi_i(x_h) - \sum_{h=1}^{j-1} \varepsilon_i(x_h) \right). \qquad (2.14)$$

Similarly, for $\mathcal{B}$ of finite type:

$$\varepsilon_i(x_k \otimes \cdots \otimes x_2 \otimes x_1) = \max_{j=1}^{k} \left(\sum_{h=1}^{j} \varepsilon_i(x_h) - \sum_{h=1}^{j-1} \varphi_i(x_h) \right). \qquad (2.15)$$

2.4 The signature rule

There is an easy combinatorial interpretation of which tensor factor x_j the crystal operator f_i acts on in (2.12). We begin by assuming that $x_1, \ldots, x_k \in \mathbb{B}$ are elements of the standard crystal of type A_r of Example 2.19. (The same idea easily generalizes to general finite type crystals.) We consider

$$f_i \left(\boxed{x_1} \otimes \boxed{x_2} \otimes \cdots \otimes \boxed{x_k} \right),$$

where $x_1, \ldots, x_k \in \{1, 2, \ldots, r+1\}$. It is immediately clear from Lemma 2.33 that if every i among the x_j is to the left of every $i+1$, then the x_j chosen in (2.12) is the rightmost i.

If this is not the case, then we can find some x_a equal to $i+1$ to the left of some x_b equal to i, and we may assume that no intervening entry is equal to either i or $i+1$. Now by Lemma 2.33,

$$\varphi_i(\xi) = \varepsilon_i(\xi) = 0, \qquad \text{where} \qquad \xi = \boxed{x_a} \otimes \cdots \otimes \boxed{x_b}.$$

This means that, invoking the associativity of $\otimes$, we may treat ξ as a unit, which is "invisible" to e_i and f_i. We will indicate this by "bracketing" x_a and x_b thus:

$$f_i \left(\boxed{x_1} \otimes \cdots \otimes \overbrace{\boxed{x_a} \otimes \cdots \otimes \boxed{x_b}} \otimes \cdots \otimes \boxed{x_k} \right).$$

We repeat this process until we reach our original situation in which every unbracketed entry equal to i is to the left of every unbracketed entry equal to $i+1$, and then select the rightmost i. (If no i's remain, then the result is 0.)

This process is sometimes called the *signature rule* since in practice it may be helpful to write $-$ by every i and $+$ over every $i+1$ before doing the bracketing.

Example 2.34. *To compute*

$$f_2 \left(\boxed{1} \otimes \boxed{3} \otimes \boxed{3} \otimes \boxed{2} \otimes \boxed{2} \otimes \boxed{1} \otimes \boxed{2} \otimes \boxed{3} \right)$$

we first mark the entries 2 with a $-$ and 3 with a $+$

$$f_2(\boxed{1} \otimes \underset{+}{\boxed{3}} \otimes \underset{+}{\boxed{3}} \otimes \underset{-}{\boxed{2}} \otimes \underset{-}{\boxed{2}} \otimes \boxed{1} \otimes \underset{-}{\boxed{2}} \otimes \underset{+}{\boxed{3}}).$$

We now bracket the signs in two steps:

$$\text{first: } + \overset{\frown}{+ -} - - + \qquad\qquad \text{then: } \overset{\frown}{+ \overset{\frown}{+ -}} - - +$$

We could also represent this as:

$$\text{first: } + (+-) - - + \qquad\qquad \text{second: } (+ (+-) -) - +.$$

Either way, the rightmost unbracketed − *is the third, and so the third* 2 *is the one that* f_2 *acts on in the tensor product, yielding*

$$f_2\left(\boxed{1} \otimes \boxed{3} \otimes \boxed{3} \otimes \boxed{2} \otimes \boxed{2} \otimes \boxed{1} \otimes \boxed{2} \otimes \boxed{3}\right) = \boxed{1} \otimes \boxed{3} \otimes \boxed{3} \otimes \boxed{2} \otimes \boxed{2} \otimes \boxed{1} \otimes \boxed{3} \otimes \boxed{3}.$$

For a general tensor product of crystals $\mathcal{B}_1 \otimes \cdots \otimes \mathcal{B}_k$ of arbitrary type, we decorate each tensor factor x_j in (2.12) by $\varphi_i(x_j)$ times − followed by $\varepsilon_i(x_j)$ times +:

$$x_1 \otimes \qquad \cdots \qquad \otimes x_j \otimes \qquad \cdots \qquad \otimes x_k$$

$$\underbrace{- \cdots -}_{\varphi_i(x_1)} \underbrace{+ \cdots +}_{\varepsilon_i(x_1)} \qquad \underbrace{- \cdots -}_{\varphi_i(x_j)} \underbrace{+ \cdots +}_{\varepsilon_i(x_j)} \qquad \underbrace{- \cdots -}_{\varphi_i(x_k)} \underbrace{+ \cdots +}_{\varepsilon_i(x_k)}.$$

Then successively bracket any pair +− until all unbracketed symbols are of the form

$$-^a +^b.$$

Then $\varphi_i(x_1 \otimes \cdots \otimes x_k) = a$ and $\varepsilon_i(x_1 \otimes \cdots \otimes x_k) = b$ and f_i acts on the tensor factor associated to the rightmost unbracketed −, whereas e_i acts on the tensor factor associated to the leftmost unbracketed +.

2.5 Root strings

Let $\mathcal{C}$ be a seminormal crystal with weight lattice Λ and root system Φ. For each $i \in I$ we may partition $\mathcal{C}$ into *root strings*, which may be defined as follows. We impose an equivalence relation on $\mathcal{C}$ in which $x \sim y$ if $x = e_i^k(y)$ or $x = f_i^k(y)$ for some k. The equivalence classes are the *i-root strings*.

It follows clearly from Axiom A1 that the weights of the elements of a root string all differ by integer multiples of α_i, and may be arranged in a sequence

$$u_1, u_2, \ldots, u_m,$$

where u_m is the unique element of the string such that $e_i(u_m) = 0$, and where if $\mathrm{wt}(u_m) = \mu$ we have

$$\mathrm{wt}(u_{m-t}) = \mu - t\alpha_i, \qquad 0 \leqslant t \leqslant m - 1.$$

The crystal graph of the root string (obtained by discarding the rest of the crystal) looks like this:

$$\overset{u_1}{\bullet} \leftarrow \overset{u_2}{\bullet} \leftarrow \overset{u_3}{\bullet} \quad - - - - \quad \leftarrow \overset{u_m}{\bullet}$$

Let s_i be the simple reflection of Λ corresponding to i, so that

$$s_i(\mu) = \mu - \langle \mu, \alpha_i^\vee \rangle \alpha_i. \tag{2.16}$$

We will define a symmetry σ_i of the crystal that "covers" the simple reflection.

Definition 2.35. Let $x \in \mathcal{C}$. Let $k = \langle \mathrm{wt}(x), \alpha_i^\vee \rangle$. Define

$$\sigma_i(x) = \begin{cases} f_i^k(x) & \text{if } k > 0, \\ x & \text{if } k = 0, \\ e_i^{-k}(x) & \text{if } k < 0. \end{cases}$$

Proposition 2.36. *The map σ_i takes $\mathcal{C}$ to itself. It has order two, and*

$$\mathrm{wt}\,(\sigma_i(x)) = s_i\,(\mathrm{wt}(x)). \tag{2.17}$$

The map σ_i takes every i-root string to itself.

Proof. By Axiom A2, $k = \varphi_i(x) - \varepsilon_i(x)$. Since $\mathcal{C}$ is seminormal, this means that we may apply f_i to x for k times if $k > 0$, and e_i for $-k$ times if $k < 0$. Hence σ_i is defined. Whether $k \geqslant 0$ or $k < 0$ we have

$$\mathrm{wt}\big(\sigma_i(x)\big) = \mathrm{wt}(x) - k\alpha_i = s_i\big(\mathrm{wt}(x)\big).$$

Now the i-root string through x has a unique element of weight $s_i\big(\mathrm{wt}(x)\big)$, so we see that σ_i reverses the root string end to end, that is, it sends u_t to u_{m+1-t} in our previous notation. Thus it has order 2. □

See Section 11.3 for more information about the σ_i.

2.6 The character

Like finite-dimensional representations of Lie groups, crystals have characters, which we now introduce. Let $\mathcal{C}$ be a crystal with weight lattice Λ and root system Φ. Let $\mathcal{E}$ be the free abelian group on Λ, with basis elements t^μ ($\mu \in \Lambda$). Define the *character*

$$\chi_{\mathcal{C}}(t) = \sum_{v \in \mathcal{C}} t^{\mathrm{wt}(v)}. \tag{2.18}$$

Let W be the *Weyl group* of Φ, which is the group of transformations of Λ generated by the s_i.

Proposition 2.37. *The character of the crystal is invariant under the action of the Weyl group.*

Proof. Since the transformation σ_i of the crystal satisfies (2.17), the character is unchanged by s_i. □

For example, let $\mathcal{C}$ be the $\mathrm{GL}(n)$ crystal of rows (see Example 2.25). Since we identify the weight lattice Λ with $\mathbb{Z}^n$, we may identify $\mathcal{E}$ with the Laurent polynomial ring in variables $t_1, \ldots, t_n$, with $t^\mu = \prod_i t_i^{\mu_i}$. The Weyl group may be identified with the symmetric group S_n, and so Proposition 2.37 means that the character is a symmetric polynomial.

For example, since

$$t^{\mathrm{wt}\left(\boxed{j_1}\boxed{j_2}\boxed{\cdots}\boxed{j_k}\right)} = \prod_i t_{j_i}$$

we have

$$\chi_{\mathcal{B}_{(k)}}(t) = \sum_{j_1 \leqslant j_2 \leqslant \cdots \leqslant j_k} t_{j_1} \cdots t_{j_k} = h_k(t_1, \ldots, t_n). \tag{2.19}$$

This is the k-th *complete symmetric polynomial*. Similarly the character of the crystal of columns of length k is the k-th *elementary symmetric function*:

$$\chi_{\mathcal{B}_{(1^k)}}(t) = \sum_{j_1 < \cdots < j_k} t_{j_1} \cdots t_{j_k} = e_k(t_1, \ldots, t_n). \tag{2.20}$$

2.7 Related crystals and twisting

We have seen that there may be different but closely related weight lattices. Taking type A as an example, the group $\mathrm{GL}(n)$ has as a subgroup $\mathrm{SL}(n)$, and correspondingly there is a relationship between their weight lattices. The weight lattice $\Lambda_{\mathrm{GL}(n)}$ of $\mathrm{GL}(n)$ is $\mathbb{Z}^n$ (see Example 2.5), while the weight lattice $\Lambda_{\mathrm{SL}(n)}$ of $\mathrm{SL}(n)$ is $\mathbb{Z}^n$ modulo the diagonal subgroup (see Example 2.6). Dual to the embedding of $\mathrm{SL}(n)$ we have a surjective homomorphism $\Lambda_{\mathrm{GL}(n)} \to \Lambda_{\mathrm{SL}(n)}$. Similarly we have a surjective homomorphism from $\mathrm{spin}(2r + 1)$ to $\mathrm{SO}(2r + 1)$ and this corresponds to the embedding of the orthogonal weight lattice (see Example 2.8) into the spin weight lattice (see Example 2.7).

So maps of weight lattices correspond (contravariantly) to maps of Lie groups. The simplest case is that of an *isogeny*, in which the weight lattice map induces a bijection of the roots. The maps $\mathrm{SL}(n) \to \mathrm{GL}(n)$ and $\mathrm{spin}(2r + 1) \to \mathrm{SO}(2r)$ that we have just considered are isogenies. Moreover, given an isogeny (or indeed any Lie group homomorphism) $G \to H$, a representation of H may be pulled back to give a representation of G.

Let us formulate crystal analogs of these facts. We consider a category in which the objects are pairs (Φ, Λ), where Φ is a root system in a lattice Λ. A morphism $(\Phi, \Lambda) \to (\Phi', \Lambda')$ in this category will be a homomorphism $m\colon \Lambda \to \Lambda'$ of lattices that maps Φ bijectively onto Φ'. We call such a map an *isogeny* of weight lattices.

Now if $\mathcal{C}$ is a crystal for the root system Φ with weight lattice Λ, then we can compose the weight function $\mathrm{wt}\colon \mathcal{C} \to \Lambda$ with m and obtain a crystal $\mathcal{C}'$ for the second weight lattice. Since the weight lattice is part of the data that specifies the crystal, the crystals $\mathcal{C}$ and $\mathcal{C}'$ are distinct, but we say they are *related* by the isogeny m. More generally, we say that two crystals are *related* if they are equivalent under the equivalence relation generated by this isogeny relationship. Crystals that are related this way have isomorphic crystal graphs.

Another construction that we can make with representations of Lie groups is *twisting*. If χ is a character of G, then we may tensor any representation of G with

χ. Let us describe the corresponding operation on crystals. Let $\mathcal{C}$ be a crystal with wt taking values in the weight lattice Λ. Let θ be any element of Λ such that $\langle \theta, \alpha^\vee \rangle = 0$ for every coroot $\alpha^\vee$. Then we may modify the crystal $\mathcal{C}$ by changing its weight function to a new one wt$': \mathcal{C} \to \Lambda$, in which wt$'(u) = $ wt$(u) + \theta$. The meaning of e_i, f_i, ε_i and φ_i are unchanged in this definition. We call the crystal $\mathcal{C}'$ a *twist* of $\mathcal{C}$. If $\mathcal{C}$ is seminormal, so is $\mathcal{C}'$.

This operation is equivalent to tensoring the crystal $\mathcal{C}$ with the crystal $\mathcal{T}_\theta$ (see Example 2.28). The condition that $\langle \theta, \alpha^\vee \rangle = 0$ can be relaxed, but if $\langle \theta, \alpha_i^\vee \rangle \neq 0$ the meaning of either φ_i or ε_i must be changed, and seminormality will not be preserved.

2.8   Dynkin diagrams and Levi branching

Certain subgroups of a Lie group G are called *Levi subgroups*. For example, $GL(r) \times GL(s)$ is a Levi subgroup of $GL(r+s)$, $GL(r) \times O(s)$ is a Levi subgroup of $O(s+2r)$ and $GL(r) \times Sp(2s)$ is a Levi subgroup of $Sp(2(r+s))$. The root system of a Levi subgroup is naturally contained in the root system of G. The term *branching* refers to the decomposition of representations of G into irreducibles of a subgroup when restricted to H. In this section we will describe branching of crystals to Levi subtypes.

Let Φ_1 and Φ_2 be root systems, with ambient spaces V_1 and V_2 and weight lattices Λ_1 and Λ_2. Let $V = V_1 \oplus V_2$ be the orthogonal direct sum of the Euclidean spaces V_1 and V_2, and let $\Lambda = \Lambda_1 \oplus \Lambda_2$. Let Φ be the disjoint union of Φ_1 and Φ_2. Then Φ is a root system. We will denote this root system $\Phi_1 \oplus \Phi_2$, and call it the *direct sum* of the root systems Φ_1 and Φ_2. If X and Y are the Cartan types of Φ_1 and Φ_2 we will use the notation $X \times Y$ for the Cartan type of $\Phi_1 \oplus \Phi_2$; for example we will write $A_1 \times A_1$ for the Cartan type of $GL(2) \times GL(2)$.

Let Φ be a root system with weight lattice Λ and index set I. If J is any subset of I, we may produce a root system Φ_J, called the *Levi root system* as follows. Let $\{\alpha_i \mid i \in I\}$ be the simple roots in Φ, and let Φ_J be the set of $\alpha \in \Phi$ such that when α is written as a linear combination of the α_i, only α_i with $i \in J$ appear with nonzero coefficients. Alternatively, we may describe Φ_J as follows. The reflections $\{s_j \mid j \in J\}$ generate a subgroup W_J of the Weyl group W, and Φ_J is the set of roots equivalent to some α_j with $j \in J$ under the action of W_J. We may take the ambient space and weight lattices to be the same as Φ.

The Levi root system may be read off from the Dynkin diagram of Φ. The Dynkin diagram is a graph with vertices labeled by elements of I. There is an edge joining i and j if and only if α_i and α_j are not orthogonal. This is equivalent to the simple reflections s_i and s_j not commuting. Let $n(i,j)$ be the order of $s_i s_j$.

- If $n(i,j) = 2$, then α_i and α_j are orthogonal and s_i and s_j commute. In this case vertex i and vertex j of the Dynkin diagram are not joined by an edge.
- If $n(i,j) = 3$, then α_i and α_j have the same length and the angle between them

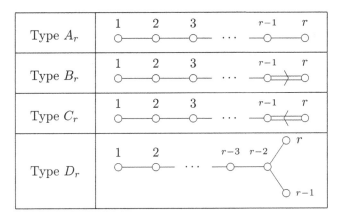

Fig. 2.5 The classical Dynkin diagrams.

is $\frac{2\pi}{3}$. We join them by a simple edge: $\overset{i}{\circ}\!\!-\!\!\overset{j}{\circ}$
- If $n(i,j) = 4$, then α_i and α_j do not have the same length. If α_i is the longer root then $|\alpha_j| = \sqrt{2}|\alpha_j|$. The angle between them is $\frac{3\pi}{4}$. In the Dynkin diagram they are joined by a double bond with an arrow from the vertex corresponding to the longer root to the shorter: $\overset{i}{\circ}\!\!\Longleftarrow\!\!\overset{j}{\circ}$
- If $n(i,j) = 6$, then α_i and α_j do not have the same length. If α_i is the longer root then $|\alpha_j| = \sqrt{3}|\alpha_j|$. The angle between them is $\frac{5\pi}{6}$. Among the Dynkin diagrams of semisimple Lie algebras, this situation only occurs in type G_2. In the Dynkin diagram the vertices are joined by a triple bond with an arrow from the vertex of the longer root to the shorter.

The Dynkin diagrams of the classical Cartan types may be found in Figure 2.5, whereas those for the exceptional types are listed in Figure 2.6. Note that we use the Bourbaki labeling [Bourbaki (2002)].

To obtain the Dynkin diagram of Φ_J from the Dynkin diagram of Φ, we simply eliminate all the nodes whose labels are not in J. For example, if Φ is the GL(4) root system, then the Dynkin diagram of Φ is of type A_3:

$$\overset{1}{\circ}\!\!-\!\!\overset{2}{\circ}\!\!-\!\!\overset{3}{\circ}$$

Taking $J = \{1, 3\}$, eliminating the central node, gives the Dynkin diagram:

$$\overset{1}{\circ}\qquad\overset{3}{\circ}$$

Thus the Levi root system is of type $A_1 \times A_1$. This Levi embedding corresponds to the inclusion of GL(2) × GL(2) as a Levi subgroup of GL(4).

Now suppose that $\mathcal{B}$ is a crystal for the root system Φ. Then we may obtain a crystal for the root system Φ_J by simply forgetting the maps f_i, e_i, φ_i and ε_i for

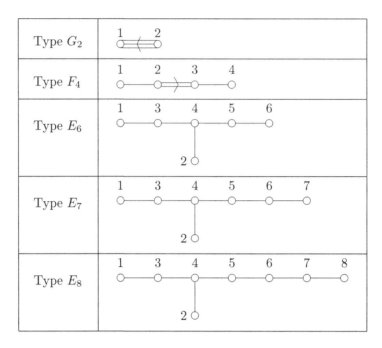

Fig. 2.6 Dynkin diagrams for exceptional types in the Bourbaki labeling.

$i \notin J$. The resulting Φ_J crystal is called the (Levi) *branched crystal* $\mathcal{B}_J$. Thus the crystal graph of $\mathcal{B}_J$ is obtained from the crystal graph of $\mathcal{B}$ by erasing those edges labeled by indices not in J.

Exercises

To gain familiarity with crystals it would be good to learn SAGE's capabilities. These are explained in the Thematic Tutorial [Bump, Schilling and Salisbury (2015)]. You should be able to do the following exercises by hand, but you may find SAGE useful for checking your work.

Exercise 2.1. Let $\mathbb{B}$ be the GL(3) crystal from Example 2.19. Compute $f_i(x \otimes y)$ by hand for every x and y in $\mathbb{B}$ and thus compute by hand the tensor product $\mathbb{B} \otimes \mathbb{B}$. Check that it decomposes as a disjoint union of two connected crystals, one with three elements and one with six.

Exercise 2.2. Show that one of the two connected subcrystals of $\mathbb{B} \otimes \mathbb{B}$ that you found in Exercise 2.1 is a twist of the dual standard crystal in Example 2.21. Explain why it is necessary to twist.

Exercise 2.3. Generalize Exercise 2.2 by showing that the dual standard crystal of type $GL(n)$ is a twist of a subcrystal of $\mathbb{B}^{\otimes n-1}$.
Hint: The n elements of this subcrystal have the shape

$$\boxed{n} \otimes \cdots \otimes \widehat{\boxed{j}} \otimes \cdots \otimes \boxed{1}$$

with the entries in decreasing order, where the notation means that the factor $\boxed{j}$ is omitted.

Exercise 2.4. Let $\mathbb{B}$ be the $GL(3)$ crystal from Example 2.19 and let $\mathbb{B}^\vee$ be the dual standard crystal from Example 2.21. Compute the tensor product $\mathbb{B} \otimes \mathbb{B}^\vee$. You should find that this crystal has two components, one of degree one and one of degree 8.

Exercise 2.5. Show that the row and column crystals of Examples 2.25 and 2.26 are seminormal.

Exercise 2.6.

(i) Prove that the dual crystal $\mathcal{C}^\vee$ of a crystal $\mathcal{C}$ of finite type is a crystal.
(ii) Assume that $\mathcal{C}$ is seminormal. Prove that if λ is the highest weight of $\mathcal{C}$, then $-w_0(\lambda)$ is the highest weight of $\mathcal{C}^\vee$, where w_0 is the long Weyl group element.
Hint: Use Proposition 2.37.

Exercise 2.7. If $\mathcal{C}$ and $\mathcal{D}$ are crystals and $\mathcal{C}^\vee$, $\mathcal{D}^\vee$ are their duals, prove that the dual of $\mathcal{C} \otimes \mathcal{D}$ is isomorphic to $\mathcal{D}^\vee \otimes \mathcal{C}^\vee$.

Exercise 2.8. Let Λ be the C_2 weight lattice, and let $\mathbb{B}$ be the standard crystal of degree 4. Show that $\mathbb{B} \otimes \mathbb{B}$ has three connected components, of degrees 1, 5 and 10. What are their highest weight elements?

Exercise 2.9. To generalize the last exercise, let $\mathbb{B}$ be the C_r standard crystal. In order to look for connected components of $\mathbb{B} \otimes \mathbb{B}$, a shortcut is to look for highest weight elements. That is, assume that

$$\varepsilon_i\left(\boxed{x} \otimes \boxed{y}\right) = 0$$

for all i and try to deduce what the possibilities are for x and y using (2.10). You should find that there are exactly three highest weight elements. Can you conjecture the degrees of the crystals containing them?

Exercise 2.10. Construct a connected $GL(4)$ crystal $\mathcal{C}$ with highest weight $(1, 1, 0, 0)$ having six elements. Compute its Levi branching to a $GL(2) \times GL(2)$ crystal.

The next exercise assumes some familiarity with Lie group representations, and with a theme that will be developed in later chapters; namely, the analogy between crystal bases and representations of Lie groups.

Exercise 2.11. Prove that if $\mathcal{B}$ and $\mathcal{C}$ are crystals then the character of $\mathcal{B} \otimes \mathcal{C}$ is the product of the characters of $\mathcal{B}$ and $\mathcal{C}$.

Exercise 2.12. Every one of these Exercises is the analog of some computation involving irreducible representations of Lie groups. Describe these.

For example, regarding Exercise 2.1, the analog of the standard GL(3) crystal $\mathbb{B}$ is the standard three-dimensional module $V = \mathbb{C}^3$ of $GL(3, \mathbb{C})$. If we decompose this representation into irreducibles, $V \otimes V$ decomposes into the direct sum of the exterior square module $\wedge^2 V$ and the symmetric square module $\mathrm{Sym}^2(V)$. These have dimensions 3 and 6. So the two subcrystals of $\mathbb{B} \otimes \mathbb{B}$ found in Exercise 2.1 can be thought of as the crystals of these two representations of $GL(3, \mathbb{C})$.

Exercise 2.13.

(i) Let ρ be the Weyl vector defined in (2.3), and let ϖ_i be the fundamental weights. Show that $\rho - \sum_i \varpi_i$ is orthogonal to all the coroots.
(ii) If the weight lattice is semisimple, show that $\rho = \sum_i \varpi_i$.

Chapter 3

Crystals of Tableaux

[Littlewood (1938)] proved that Schur polynomials, which are essentially the characters of irreducible representations of $\mathrm{GL}(n, \mathbb{C})$, can be expressed as sums over semistandard Young tableaux, see (3.3). [Kashiwara and Nakashima (1994)] reinterpreted this as a reflection of the fact that the tableaux can be arranged into crystals, and extended this construction to other Cartan types. In this chapter, we will define these crystals of tableaux of type A. The extension to other types is discussed in Chapter 6.

3.1 Type A crystals of tableaux

In this section we will construct many $\mathrm{GL}(n)$ crystals, where n is a fixed integer.

A *partition* of the nonnegative integer k is a sequence $\lambda = (\lambda_1, \lambda_2, \ldots)$ of nonnegative integers such that $\lambda_1 \geqslant \lambda_2 \geqslant \cdots \geqslant 0$ and $\sum_{i \geqslant 1} \lambda_i = k$. This implies that λ_i is eventually zero. It will be convenient to suppress trailing zeros from the notation; thus $(5,3)$ and $(5,3,0)$ both represent the same partition $(5,3,0,0,0,\ldots)$. The largest ℓ such that $\lambda_\ell \neq 0$ is called the *length* of λ. The elements λ_i for $1 \leqslant i \leqslant \ell$ are called the *parts* of λ. We write $\lambda \vdash k$ or $k = |\lambda|$ and $\ell(\lambda) = \ell$ if λ is a partition of k of length ℓ. We will sometimes use an abbreviated notation for partitions, where a^j means j parts of length a; thus $(3^2 2 1^3)$ is short for $(3, 3, 2, 1, 1, 1)$.

Given a partition λ, the *Young diagram* $\mathrm{YD}(\lambda)$ (also known as the *Ferrer's diagram*) is an array of boxes arranged in the lower right quadrant of the plane. The i-th row has λ_i boxes. Given a partition λ, the *conjugate partition* λ' is characterized by the property that the Young diagrams of λ and λ' are transposes of each other. Thus $(4, 2, 1)$ and $(3, 2, 1, 1)$ are conjugate partitions, whose Young diagrams are

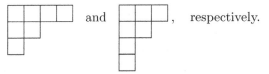

and , respectively.

A *tableau* is a filling of a Young diagram $\mathrm{YD}(\lambda)$ with elements of some alphabet. In this chapter the alphabet is $[n] := \{1, 2, 3, \ldots, n\}$. We call the tableau *semistan-*

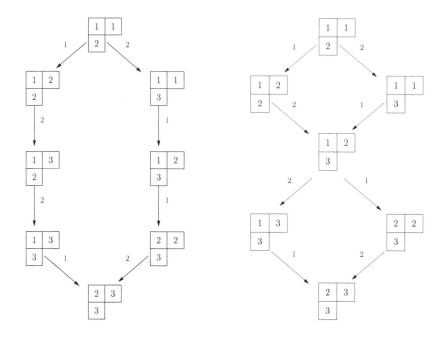

Fig. 3.1 Left: The GL(3) crystal $\mathcal{B}_{(2,1)}$ of shape $\lambda = (2,1)$. Right: A degeneration of $\mathcal{B}_{(2,1)}$.

dard if the rows are weakly increasing from left to right and the columns are strictly increasing from top to bottom. We call the partition $\lambda = (\lambda_1, \lambda_2, \ldots)$ the *shape* of the tableau.

Our goal is to assemble the semistandard Young tableaux of shape λ in the alphabet $[n]$ into a GL(n) crystal. For this, we need a map from the set of such tableaux into the GL(n) weight lattice Λ, which we identify with $\mathbb{Z}^n$. If T is a tableau, define wt(T) = $(\mu_1, \ldots, \mu_n)$ where μ_i is the number of i's in T.

For example, if $n = 3$, there are eight semistandard Young tableaux of shape $\lambda = (2,1)$. These appear in Figure 3.1 (left), where they have been arranged into a crystal graph. (The second crystal in Figure 3.1 is a degeneration of the first one that we will discuss in Chapter 4.)

We have not yet explained the general definition of the operators f_i and e_i (which you can read off from the crystal graph). We will explain that later in this chapter.

Our strategy for assembling the set $\mathcal{B}_\lambda$ of semistandard Young tableaux of shape λ into a crystal is to exhibit a map from $\mathcal{B}_\lambda$ into $\mathbb{B}^{\otimes k}$, where as before $\mathbb{B} := \mathcal{B}_{(1)}$ is the GL(n) standard crystal described in Example 2.19 with rank $r = n - 1$ and $k = |\lambda|$. We will show that the image of this map in $\mathbb{B}^{\otimes k}$ is closed under the operators e_i and f_i, and hence is a full subcrystal. There are different ways to do this, but in this chapter we will emphasize two. The first construction is based on *row reading*, and it builds the crystal up from crystals of rows. The alternative, which we will call the *column reading*, maps the set $\mathcal{B}_\lambda$ onto another, (usually) different subcrystal of

$\mathbb{B}^{\otimes k}$. The two subcrystals of $\mathbb{B}^{\otimes k}$ are isomorphic. We will study the column reading in Chapter 6.

The terms *Middle Eastern* and *Far Eastern* readings are used in some of the literature instead of row and column readings.

We call a semistandard tableau R a *row* of length k if its shape is (k). Since the tableau is semistandard, the entries in a row R are weakly increasing. This is consistent with the notation in Example 2.25.

Similarly, a *column* of length k is a semistandard tableau of shape (1^k), and since it is semistandard, the column is strictly increasing, consistent with Example 2.26.

Returning to the crystal $\mathcal{B}_{(k)}$ of rows, define a map $R \mapsto \mathrm{RR}(R)$ from $\mathcal{B}_{(k)}$ into $\mathbb{B}^{\otimes k}$ by sending the row $\boxed{i_1}\cdots\boxed{i_k}$ to $\boxed{i_1}\otimes\cdots\otimes\boxed{i_k}$.

Proposition 3.1. *The map* RR *is a morphism of crystals from* $\mathcal{B}_{(k)}$ *to* $\mathbb{B}^{\otimes k}$.

Proof. Let $R = \boxed{x_1}\cdots\boxed{x_k}$. It is clear that $\mathrm{wt}(R) = \mathrm{wt}\big(\mathrm{RR}(R)\big)$ since both equal $(\mu_1, \ldots, \mu_n)$ where μ_i is the number of x_j equal to i. It is enough to show that if R and R' are rows then $f_i(R) = R'$ if and only if $f_i\big(\mathrm{RR}(R)\big) = \mathrm{RR}(R')$. For this, we can appeal to the signature rule described in Section 2.4. Recall from Example 2.25 that f_i acts on R by changing the rightmost i into $i + 1$; if there is no i, then f_i annihilates R. Hence R' is obtained from R by replacing the rightmost i by $i + 1$. By the signature rule, the Kashiwara operator f_i acts on the rightmost unbracketed letter i in the tensor product $\mathrm{RR}(R)$. Since the sequence $x_1, \ldots, x_k$ is increasing, it is clear that no letter is bracketed and hence f_i also acts on the rightmost i in $\mathrm{RR}(R)$. This proves $f_i(R) = R'$ if and only if $f_i\big(\mathrm{RR}(R)\big) = \mathrm{RR}(R')$.

The analogous proof for e_i is left to the reader. Both $\mathcal{B}_{(k)}$ and $\mathbb{B}^{\otimes k}$ are seminormal. (For $\mathcal{B}_{(k)}$ this is easy to check and see Proposition 2.29 for $\mathbb{B}^{\otimes k}$.) Because of this, the fact that φ_i and ε_i commute with RR follows from the statements for f_i and e_i, which we already proved. $\qquad\square$

We now extend the map $R \mapsto \mathrm{RR}(R)$ that we have defined on rows to a map on all semistandard Young tableaux T of shape λ into $\mathbb{B}^{\otimes|\lambda|}$. This map, also denoted $T \mapsto \mathrm{RR}(T)$, is called the *row reading*. To define $\mathrm{RR}(T)$ we read each row of T in order, and we take the rows in order from the bottom to the top. Thus if

$$T = \begin{array}{|c|c|c|c|c|c|} \hline 1 & 1 & 2 & 2 & 2 & 4 \\ \hline \multicolumn{1}{|c|}{2} & \multicolumn{1}{c|}{3} & \multicolumn{1}{c|}{3} \\ \cline{1-3} \multicolumn{1}{|c|}{4} \\ \cline{1-1} \end{array}$$

then

$$\mathrm{RR}(T) = \mathrm{RR}(R_3) \otimes \mathrm{RR}(R_2) \otimes \mathrm{RR}(R_1)$$

$$= \boxed{4} \otimes \boxed{2} \otimes \boxed{3} \otimes \boxed{3} \otimes \boxed{1} \otimes \boxed{1} \otimes \boxed{2} \otimes \boxed{2} \otimes \boxed{2} \otimes \boxed{4},$$

where

$$R_1 = \boxed{1\;1\;2\;2\;2\;4}, \qquad R_2 = \boxed{2\;3\;3}, \qquad R_3 = \boxed{4}$$

are the three rows of T, and

$$\mathrm{RR}(R_1) = \boxed{1} \otimes \boxed{1} \otimes \boxed{2} \otimes \boxed{2} \otimes \boxed{2} \otimes \boxed{4},$$

$$\mathrm{RR}(R_2) = \boxed{2} \otimes \boxed{3} \otimes \boxed{3},$$

$$\mathrm{RR}(R_3) = \boxed{4}.$$

Let λ be a partition of k of length $\leqslant n$. Let $\mathcal{B}_\lambda$ be the set of semistandard Young tableaux of shape λ in the alphabet $[n]$. We will single out a particular tableau u_λ in $\mathcal{B}_\lambda$, which is the tableau of shape λ such that all entries in the i-th row are equal to i. Such a tableau is called a *Yamanouchi tableau.*

Theorem 3.2. *Let λ be a partition of k of length $\leqslant n$. Then $\mathrm{RR}(\mathcal{B}_\lambda)$ is a connected component of $\mathbb{B}^{\otimes k}$. It has a unique highest weight element, namely $\mathrm{RR}(u_\lambda)$.*

Proof. First we show that $\mathrm{RR}(\mathcal{B}_\lambda) \sqcup \{0\}$ is closed under the f_i and e_i. Thus if $f_i(\mathrm{RR}(T)) \neq 0$, our first task is to verify that $f_i(\mathrm{RR}(T)) = \mathrm{RR}(T')$, where T' is another tableau of shape λ.

Let $R_1, \ldots, R_m$ be the rows of T. Then $\mathrm{RR}(T) = \mathrm{RR}(R_m) \otimes \cdots \otimes \mathrm{RR}(R_1)$, and by Lemma 2.33

$$f_i(\mathrm{RR}(T)) = \mathrm{RR}(R_m) \otimes \cdots \otimes f_i(\mathrm{RR}(R_j)) \otimes \cdots \otimes \mathrm{RR}(R_1),$$

where j is the *largest* integer with $1 \leqslant j \leqslant m$ where

$$\sum_{h \geqslant j} \varphi_i(\mathrm{RR}(R_h)) - \sum_{h > j} \varepsilon_i(\mathrm{RR}(R_h))$$

attains its maximum value. Using the fact that the crystal $\mathbb{B}^{\otimes k}$ is seminormal and Proposition 3.1, this equals

$$\sum_{h \geqslant j} \varphi_i(R_h) - \sum_{h > j} \varepsilon_i(R_h). \tag{3.1}$$

Now let T' be the array with rows $R_1, \ldots, f_i(R_j), \ldots, R_m$. Thus T' is obtained from T by changing the last i in the j-th row to $i + 1$. We will show that T' is a semistandard tableau. The only way it could fail would be if the i that is changed to $i + 1$ lies above an entry (in the $(j + 1)$-st row) that is equal to $i + 1$. However in that case, the fact that T has strictly increasing columns and weakly increasing rows means that every entry in the $(j + 1)$-st row that is below an i in the j-th row is equal to $i + 1$. Therefore $\varepsilon_i(R_{j+1}) \geqslant \varphi_i(R_j)$, contradicting the assumption that j is the largest integer such that (3.1) attains its maximum.

We have shown that $\mathrm{RR}(\mathcal{B}_\lambda) \sqcup \{0\}$ is closed under the f_i. The analogous arguments to show that it is under the e_i are left to the reader. Thus $\mathrm{RR}(\mathcal{B}_\lambda)$ is a full subcrystal of $\mathbb{B}^{\otimes k}$.

Next we check that u_λ is the unique highest weight vector in $\mathrm{RR}(\mathcal{B}_\lambda)$. Suppose that T is a highest weight vector. Let $R_1, R_2, \ldots, R_m$ be the rows of T, so that $\mathrm{RR}(T) = R_m \otimes \cdots \otimes R_1$. Since $\mathbb{B}^{\otimes k}$ is seminormal by Proposition 2.29, $e_i(\mathrm{RR}(T)) = 0$ implies that $\varepsilon_i(\mathrm{RR}(T)) = 0$ and so by (2.15) we have, for all indices i

$$\varepsilon_i(R_1) = 0, \qquad \varepsilon_i(R_1) + \varepsilon_i(R_2) - \varphi_i(R_1) \leqslant 0, \qquad \ldots.$$

The first identity means that R_1 can contain only 1's. The second identity then means that

$$\varepsilon_i(R_2) - \varphi_i(R_1) \leqslant 0.$$

Hence $\varepsilon_i(R_2) = 0$ for $i > 1$, and therefore R_2 can contain only 1's and 2's. However it cannot contain any 1's since T is semistandard. Continuing this way, we eventually see that $T = u_\lambda$.

Every connected component of the crystal $\mathrm{RR}(\mathcal{B}_\lambda)$ contains a highest weight vector by Lemma 2.15. Since there is a unique highest weight vector, it follows that $\mathrm{RR}(\mathcal{B}_\lambda)$ is a connected component of $\mathbb{B}^{\otimes k}$. $\qquad\square$

By Theorem 3.2, the map RR gives a bijection of $\mathcal{B}_\lambda$ with a connected component of the seminormal crystal $\mathbb{B}^{\otimes k}$. By transportation of structure, $\mathcal{B}_\lambda$ may be given the structure of a connected seminormal crystal with a unique highest weight element. Then RR is a morphism. We call $\mathcal{B}_\lambda$ a *crystal of tableaux* (of type $\mathrm{GL}(n)$).

When convenient, we will *identify* the tableau T with its image $\mathrm{RR}(T)$ in $\mathbb{B}^{\otimes|\lambda|}$. Thus $\mathcal{B}_\lambda$ is identified with a specific subcrystal of $\mathbb{B}^{\otimes|\lambda|}$.

Let us contrast the case of crystals of rows with crystals of columns. Just as the map $R \mapsto \mathrm{RR}(R)$ takes a row to an increasing sequence of tensors, it takes a column to a (strictly) decreasing sequence of tensors. Thus if $i_1 \leqslant \cdots \leqslant i_k$ and $j_1 < \cdots < j_k$, then $T \mapsto \mathrm{RR}(T)$ sends:

$$\boxed{i_1\,\cdots\,i_k} \;\longmapsto\; \boxed{i_1} \otimes \cdots \otimes \boxed{i_k} \qquad \begin{array}{c}\boxed{j_1}\\ \vdots \\ \boxed{j_k}\end{array} \;\longmapsto\; \boxed{j_k} \otimes \cdots \otimes \boxed{j_1}. \qquad (3.2)$$

We have already given the columns $\mathcal{B}_{(1^k)}$ of length k a crystal structure. It gets another crystal structure by Theorem 3.2. These are the same. Indeed, analogous to Proposition 3.1, we have:

Proposition 3.3. *The map $C \longmapsto \mathrm{RR}(C)$ is a morphism of crystals from $\mathcal{B}_{(1^k)}$ to $\mathbb{B}^{\otimes k}$.*

Proof. We leave it to the reader to check this. $\qquad\square$

3.2 An example

We will prove later in Corollary 6.2 that $\mathbb{B}^{\otimes k}$ decomposes into a disjoint union of crystals each isomorphic to $\mathcal{B}_\lambda$ for some $\lambda \vdash k$. But some $\mathcal{B}_\lambda$ may occur more than once. To get some feeling for the general situation, let us consider the case where $n = 3$ and $k = 3$. Then $\mathbb{B}$ is the $\mathrm{GL}(3)$ standard crystal

$$\boxed{1} \xrightarrow{\;1\;} \boxed{2} \xrightarrow{\;2\;} \boxed{3}.$$

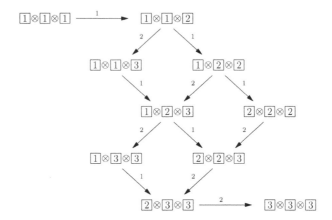

Fig. 3.2   The crystal of rows $\mathcal{B}_{(3)}$ as a subcrystal of $\mathbb{B}^{\otimes 3}$. Compare this with Figure 2.2.

The crystal $\mathbb{B}^{\otimes 3}$ is disconnected; its crystal graph has 4 connected components that we will now describe. First, there is the crystal $\mathcal{B}_{(1,1,1)}$ consisting of a single element

$$\boxed{3} \otimes \boxed{2} \otimes \boxed{1} = \begin{array}{c} \boxed{1} \\ \boxed{2} \\ \boxed{3} \end{array}.$$

Next there is the crystal $\mathcal{B}_{(3)}$, which is the image of $\mathcal{B}_{(3)}$ under $T \mapsto \mathrm{RR}(T)$. This crystal is shown as a crystal of rows in Figure 2.2, and as a subcrystal of $\mathbb{B}^{\otimes 3}$ in Figure 3.2.

Finally, there are two isomorphic crystals with eight elements. These are shown in Figure 3.3. The first of these two crystals has highest weight element $\boxed{2} \otimes \boxed{1} \otimes \boxed{1}$ and we recognize it as the image under $v \mapsto \mathrm{RR}(T)$ of $\mathcal{B}_{(2,1)}$. The second crystal has highest weight element $\boxed{1} \otimes \boxed{2} \otimes \boxed{1}$ and consists exclusively of elements *not* of the form $\mathrm{RR}(T)$ for tableaux T of any shape. However it is isomorphic to $\mathcal{B}_{(2,1)}$. We will denote this crystal $\mathcal{B}'_{(2,1)}$.

Exercises

Exercise 3.1. Let $\mathbb{B}$ be the standard $\mathrm{GL}(n)$ crystal. Show that for fixed k, the number of full connected subcrystals of $\mathbb{B}^{\otimes k}$ has an upper bound independent of n. **Hint**: Every connected subcrystal of $\mathbb{B}^{\otimes k}$ contains a highest weight element. Show that a highest weight element contains only entries $\leqslant k$, and deduce the conclusion.

Exercise 3.2. In the previous exercise, if $k = 3$ and $n \geqslant k$, there is one highest weight element of weight (3), namely $\boxed{1} \otimes \boxed{1} \otimes \boxed{1}$, and one of weight $(1, 1, 1)$, namely $\boxed{3} \otimes \boxed{2} \otimes \boxed{1}$. (Here we are interpreting the partition (3) as the dominant

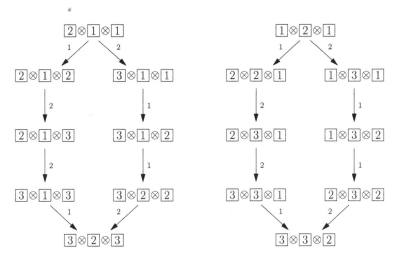

Fig. 3.3 Two subcrystals of $\mathbb{B}^{\otimes 3}$ isomorphic to $\mathcal{B}_{(2,1)}$. Left: the subcrystal $\mathcal{B}_{(2,1)}$. Right: the isomorphic subcrystal $\mathcal{B}'_{(2,1)}$.

weight $(3, 0, 0, \ldots)$ in $\Lambda = \mathbb{Z}^n$.) There are two of weight $(2, 1)$, namely $\boxed{2} \otimes \boxed{1} \otimes \boxed{1}$ and $\boxed{1} \otimes \boxed{2} \otimes \boxed{1}$. Repeat this experiment for $k = 4$ and $k = 5$ (with $n \geqslant k$). If you do this correctly, you should find that if λ is a partition of k and λ' is the conjugate partition, then the number of highest weight vectors with weight λ equals the number of highest weight vectors with weight λ'.

As explained in Section 2.6, if we identify the ring $\mathcal{E}$ defined there with the ring of Laurent polynomials in $t_1, \ldots, t_n$, then the character of $\mathcal{B}_\lambda$ is a polynomial s_λ in the t_i. These are the *Schur polynomials*. Thus by definition of the character

$$s_\lambda(t) = \sum_T t^{\mathrm{wt}(T)}, \tag{3.3}$$

where the sum is over all semistandard tableaux of shape λ in the alphabet $[n]$ and by Proposition 2.37 it is a symmetric polynomial. Equation (3.3), due to [Littlewood (1938)], is called the *combinatorial definition of the Schur polynomial*.

Exercise 3.3.

(i) Prove that $\mathcal{B}_{(1)} \otimes \mathcal{B}_{(2)}$ is the disjoint union of $\mathcal{B}_{(3)}$ and $\mathcal{B}_{(2,1)}$, embedded in $\mathcal{B}_{(1)}^{\otimes 3}$ by RR.

(ii) Similarly prove that $\mathcal{B}_{(1,1)} \otimes \mathcal{B}_{(1)}$ is the disjoint union of $\mathcal{B}_{(1,1,1)}$ and $\mathcal{B}_{(2,1)}$.

(iii) Deduce that

$$s_{(2,1)} = h_2 h_1 - h_3 = e_2 e_1 - e_3,$$

where h_i and e_i are the complete and elementary symmetric polynomials defined in Section 2.6.

Exercise 3.4. Let $\lambda = (\lambda_1, \ldots, \lambda_n)$ be a partition of k, and let μ be the partition $(\lambda_1 + 1, \ldots, \lambda_n + 1)$ of $n + k$. Show that the GL(n) crystal $\mathcal{B}_\mu$ is a twist of $\mathcal{B}_\lambda$.

Chapter 4

Stembridge Crystals

In this chapter, we characterize crystals associated to simply-laced root systems using the local Stembridge axioms [Stembridge (2003)]. We begin with some motivation and examples in Section 4.1, followed by the technical exposition of the axioms in Section 4.2, which give rise to weak Stembridge crystals or Stembridge crystals if the crystal is also seminormal. In Section 4.3 we show that Stembridge crystals form a monoidal category, that is, they are closed under taking tensor products. We conclude in Section 4.4 with further properties of Stembridge crystals.

4.1 Motivation and examples

Let $\mathcal{B}$ be a crystal as in Definition 2.13. Recall that if $u \in \mathcal{B}$ such that $e_i(u) = 0$ for all i in the index set I, then u is called a *highest weight element*. Let us define a partial order on $\mathcal{B}$ by $x \succcurlyeq y$ if there exists a sequence $i_1, \ldots, i_k$ of indices such that $x = e_{i_1} \cdots e_{i_k}(y)$. This implies that

$$\mathrm{wt}(x) = \mathrm{wt}(y) + \sum_{j=1}^{k} \alpha_{i_j},$$

so $\mathrm{wt}(x) \succcurlyeq \mathrm{wt}(y)$ with respect to the partial order on Λ defined in Definition 2.4. The converse is not true: $\mathrm{wt}(x) \succcurlyeq \mathrm{wt}(y)$ does not imply that $x \succcurlyeq y$ since for example we may have $\mathrm{wt}(x) = \mathrm{wt}(y)$ but $x \neq y$.

Lemma 4.1. *If $\mathcal{B}$ has a unique highest weight element u, then $\mathrm{wt}(u) \succcurlyeq \mathrm{wt}(x)$ for all $x \in \mathcal{B}$. In particular, $\mathcal{B}$ is connected.*

Proof. Let y be maximal such that $y \succcurlyeq x$ for $x \in \mathcal{B}$. Then y is clearly a highest weight element. Therefore $y = u$. $\square$

Problem 4.2. *Associate with every dominant weight $\lambda \in \Lambda^+$ a unique seminormal crystal $\mathcal{B}_\lambda$. The crystal $\mathcal{B}_\lambda$ shall be a connected crystal with a unique highest weight element u_λ such that $\mathrm{wt}(u_\lambda) = \lambda$. We further require that the set of such $\mathcal{B}_\lambda$ is closed under tensor product in the sense that every connected component of $\mathcal{B}_\lambda \otimes \mathcal{B}_\mu$ is $\mathcal{B}_\nu$ for some $\nu \in \Lambda^+$.*

A further objective is that the character of $\mathcal{B}_\lambda$ should coincide with the character of the irreducible representation with highest weight λ, but we will postpone discussion of this goal until later (see Corollary 13.9).

The class of all seminormal crystals is too large for this requirement. Namely, if we take the crystal $\mathcal{B}_{(2,1)}$ of type A_2 depicted in Figure 3.1 and identify the two elements of weight $(1,1,1)$ in the middle of the crystal, then the resulting "pinched crystals" is still a seminormal crystal. This degeneration of $\mathcal{B}_{(2,1)}$ is also illustrated in Figure 3.1. Both crystals have a highest weight element of the same weight $(2,1)$. If Φ is simply-laced, however, the problem of constructing the class of $\mathcal{B}_\lambda$ has a satisfactory solution in the *Stembridge axioms*, which we will explain in this chapter. We say that $\mathcal{C}$ is a *Stembridge crystal* if the axioms are satisfied. This will imply that $\mathcal{C}$ is a disjoint union of connected crystals, each $\mathcal{B}_\lambda$ for some highest weight λ. It turns out that the "pinched crystal" described above is not a Stembridge crystal.

The axioms themselves are a little technical, so we begin with an explanation of what they say. We will postpone the statement of the axioms until after some informal discussion and an example.

We recall the notions of Levi root system and Levi branching of crystals from Section 2.8. Let Φ be a root system of rank r, and let $I = \{1, 2, \ldots, r\}$ be its index set. If J is any subset of I then the Levi root system Φ_J is the root system "generated by" just the α_i with $i \in J$. Let $\mathcal{C}$ be a crystal with root system Φ. Then we may obtain a crystal for Φ_J by simply disregarding the e_i, f_i, φ_i and ε_i for $i \notin J$. We call this crystal the *branched* crystal corresponding to the Levi root system Φ_J.

We will assume in this chapter that Φ is simply-laced of rank r. If $1 \leqslant i < j \leqslant r$, then either the roots α_i and α_j are orthogonal, meaning that $\langle \alpha_i, \alpha_j^\vee \rangle = 0$, or else $\langle \alpha_i, \alpha_j^\vee \rangle = -1$. In the first case, the roots α_i and α_j may be embedded in a root system of type $A_1 \times A_1$, and in the second case, they may be embedded in a root system of type A_2. The ambient vector space of Φ may be larger than the ambient spaces of the A_2 or $A_1 \times A_1$ root systems from Example 2.5 or Example 2.6, but this is irrelevant for the following considerations.

The Stembridge axioms will only concern two roots at a time, and it is a consequence that $\mathcal{C}$ satisfies them if and only if its Levi branching to every rank two Levi root system is a Stembridge crystal. Therefore we must understand the Stembridge axioms for rank two crystals.

Now Stembridge crystals of $A_1 \times A_1$ are very easy to describe: such a crystal is a disjoint union of Cartesian products of two type A_1 crystals, each of which is a linear crystal like:

So a connected Stembridge $A_1 \times A_1$ crystal is a rectangle.

Thus the interesting simply-laced rank two crystals are of type A_2. We will now assume that $\Phi = A_2$. We will denote the crystal operators by f_i, e_i and f_j, e_j.

Suppose that $x \in \mathcal{C}$ and assume that $e_i x$ and $e_j x$ are both nonzero. Since $e_i x$ and $e_j x$ lie in the same connected component of $\mathcal{C}$, they lie in $\mathcal{B}_\lambda$ for some λ. Now it turns out that there is a unique y that is minimal with respect to the partial order $\succeq$ such that $y \succeq e_i x$ and $y \succeq e_j x$. Moreover, there are only two ways this can happen. Either:

- $y = e_j e_i x = e_i e_j x$; or
- $y = e_j e_i^2 e_j x = e_i e_j^2 e_i x$.

We will refer to this alternative as the *first dichotomy*.

Example 4.3. *We will illustrate this dichotomy with an example. In Figure 4.1 is the crystal graph $\mathcal{C}$ of the $GL(3)$ crystal of tableaux of shape $\lambda = (5, 2, 0)$. We have omitted labeling the elements with tableaux to keep the image uncluttered. We have also omitted labels on the edges, but solid arrows correspond to f_1 and dashed arrows correspond to f_2. The highest weight vector u_λ and a couple of others are marked. In particular, $u_{w_0\lambda}$ is the unique element of lowest weight $w_0\lambda$, where w_0 is the long Weyl group element.*

The reader will observe that if $x = e_2 u_{w_0\lambda}$ then $e_2 e_1 x = e_2 e_1 x$, while if $x = u_{w_0\lambda}$, then the other possibility occurs: $e_1 e_2^2 e_1 x = e_2 e_1^2 e_2 x$. Moreover, one or the other of these two possibilities is true for every x such that $e_1 x$ and $e_2 x$ are both nonzero.

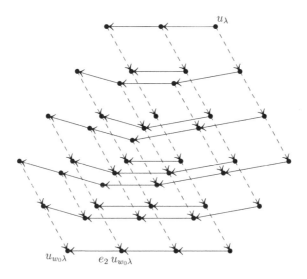

Fig. 4.1   The A_2 crystal with highest weight $\lambda = (5, 2, 0)$.

We have drawn this crystal graph in a way that f_2 always corresponds to a solid arrow pointing left, while f_1 corresponds to a dashed arrow pointing down and to the right. This way of drawing the crystal has the advantage that if $\mathrm{wt}(x) = \mathrm{wt}(y)$ then x and y are close to each other in the diagram. Therefore we may visualize the *character (2.18) as follows. In Figure 4.2 we show the set of weights $\mu \in \Lambda \cong \mathbb{Z}^3$ for which the coefficient $m(\mu)$ of t^μ in χ_C is nonzero. We have written the multiplicity of the weight in a circle, and we have labeled six weights around the edges to orient the reader. These weights are the orbit of λ under the Weyl group W.*

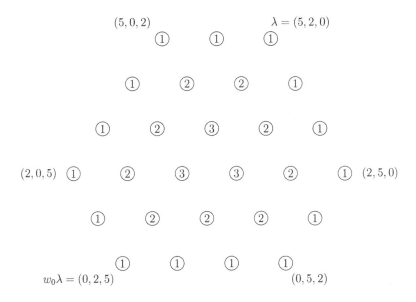

Fig. 4.2   The weight multiplicities for the crystal (or representation) with highest weight $\lambda = (5, 2, 0)$.

Now let us observe another property of simply-laced crystals.

Lemma 4.4. *Suppose that the roots α_i and α_j are not orthogonal. Suppose that x is an element of the crystal C such that $\varepsilon_i(x) > 0$. Then*

$$\varphi_j(e_i x) - \varphi_j(x) = \varepsilon_j(e_i x) - \varepsilon_j(x) - 1.$$

Proof. This follows from the definition of a crystal. We have $\mathrm{wt}(e_i x) = \mathrm{wt}(x) + \alpha_i$. Now $\langle \alpha_i, \alpha_j^\vee \rangle = -1$, so taking the inner product with $\alpha_j^\vee$ and remembering that $\langle \mathrm{wt}(x), \alpha_j^\vee \rangle = \varphi_j(x) - \varepsilon_j(x)$, the statement follows. $\qquad \square$

An examination of the crystal in Figure 4.1 shows that, at least in this example, a more precise statement than Lemma 4.4 holds. We have the following dichotomy. For every x such that $e_i x \neq 0$, either:

- $\varphi_j(e_i x) = \varphi_j(x)$ and $\varepsilon_j(e_i x) - \varepsilon_j(x) = 1$; or
- $\varphi_j(e_i x) - \varphi_j(x) = -1$ and $\varepsilon_j(e_i x) = \varepsilon_j(x)$.

Unlike the weaker statement in Lemma 4.4, this property does *not* follow from the definition of a crystal, but must be assumed, and indeed it will be part of the Stembridge axioms which will be stated in the next section. We may vizualize this dichotomy as in Figure 4.3. We will refer to this alternative as the *second dichotomy*.

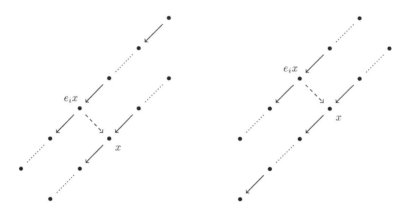

Fig. 4.3   Left: $\varphi_j(e_i x) = \varphi_j(x)$ and $\varepsilon_j(e_i x) - \varepsilon_j(x) = 1$.   Right: $\varphi_j(e_i x) - \varphi_j(x) = -1$ and $\varepsilon_j(e_i x) = \varepsilon_j(x)$. In this figure, we are depicting f_j by solid arrows and f_i by dashed ones.

As we will see in the next section, the second dichotomy determines which of the alternatives in the first dichotomy is followed. That is, given x such that $e_i x$ and $e_j x$ are nonzero, the first dichotomy tells us that there is a minimal y that succeeds both $e_i x$ and $e_j x$. The second dichotomy tells us that $\varepsilon_j(e_i x)$ equals either $\varepsilon_j(x)$ or $\varepsilon_j(x) + 1$, and similarly $\varepsilon_i(e_j x)$ equals either $\varepsilon_i(x)$ or $\varepsilon_i(x) + 1$. The connection between the two dichotomies is that if *either* $\varepsilon_j(e_i x) = \varepsilon_j(x)$ or $\varepsilon_i(e_j x) = \varepsilon_i(x)$ then $y = e_i e_j(x) = e_j e_i(x)$; and if *both* $\varepsilon_j(e_i x) = \varepsilon_j(x) + 1$ and $\varepsilon_i(e_j x) = \varepsilon_i(x) + 1$ then $y = e_i e_j^2 e_i(x) = e_j e_i^2 e_j(x)$. The reader is invited to confirm this for the crystal in Figure 4.1.

4.2   Stembridge axioms

We turn now to the statement of the Stembridge axioms. Let $\mathcal{C}$ be a crystal of finite type for a simply-laced root system Φ with index set $I = \{1, 2, \ldots, r\}$. We assume the following axioms:

Axiom S0. If $e_i(x) = 0$, then $\varepsilon_i(x) = 0$.

Axiom S0′. If $f_i(x) = 0$, then $\varphi_i(x) = 0$.

Note that a seminormal crystal clearly satisfies these axioms.

We say that $\mathcal{C}$ is a *weak Stembridge crystal* if in addition to Axioms S0 and S0′ the remaining *Stembridge axioms* (stated below) are satisfied. If in addition $\mathcal{C}$ is seminormal, we call it a *Stembridge crystal*.

Recall that if $\mathcal{C}$ is a crystal and $x \in \mathcal{C}$, then $\varphi_i(x) - \varepsilon_i(x) = \langle \mathrm{wt}(x), \alpha_i^\vee \rangle$ for all $i \in I$. We will often use this in the following form:

$$\varphi_i(y) - \varphi_i(x) = \varepsilon_i(y) - \varepsilon_i(x) + \langle \mathrm{wt}(y) - \mathrm{wt}(x), \alpha_i^\vee \rangle. \tag{4.1}$$

Axiom S1. *When $i, j \in I$ and $i \neq j$, if $x, y \in \mathcal{C}$ and $y = e_i x$, then $\varepsilon_j(y)$ equals either $\varepsilon_j(x)$ or $\varepsilon_j(x) + 1$. The second case where $\varepsilon_j(y) = \varepsilon_j(x) + 1$ is possible only if α_i and α_j are not orthogonal roots.*

This has an immediate consequence which we note before proceeding to the other axioms.

Proposition 4.5. *Suppose that $\mathcal{C}$ satisfies Axiom S1. If $\varepsilon_i(x) > 0$, then exactly one of the following three possibilities is true:*

$$
\begin{array}{llll}
\text{(i)} & \varepsilon_j(e_i x) = \varepsilon_j(x), & \varphi_j(e_i x) = \varphi_j(x) - 1, & \text{where } \langle \alpha_i, \alpha_j^\vee \rangle = -1, \\
\text{(ii)} & \varepsilon_j(e_i x) = \varepsilon_j(x) + 1, & \varphi_j(e_i x) = \varphi_j(x), & \text{where } \langle \alpha_i, \alpha_j^\vee \rangle = -1, \quad (4.2) \\
\text{(iii)} & \varepsilon_j(e_i x) = \varepsilon_j(x), & \varphi_j(e_i x) = \varphi_j(x), & \text{where } \langle \alpha_i, \alpha_j^\vee \rangle = 0.
\end{array}
$$

Proof. We note that if $y = e_i x$ as in Axiom S1, then $\varphi_j(y) - \varepsilon_j(y) = \langle \mathrm{wt}(y), \alpha_j^\vee \rangle = \langle \mathrm{wt}(x), \alpha_j^\vee \rangle + \langle \alpha_i, \alpha_j^\vee \rangle = \varphi_j(x) - \varepsilon_j(x) + \langle \alpha_i, \alpha_j^\vee \rangle$. The inner product $\langle \alpha_i, \alpha_j^\vee \rangle = 0$ or -1 since Φ is simply-laced. Thus if α_i and α_j are orthogonal, the statement that $\varepsilon_j(x) = \varepsilon_j(y)$ implies that $\varphi_j(x) = \varphi_j(y)$. On the other hand, if α_i and α_j are not orthogonal, then the axiom implies that $\varphi_j(y) = \varphi_j(x)$ or $\varphi_j(x) - 1$. $\qquad \square$

Axiom S2. *Assume that $i, j \in I$ and $i \neq j$. If $x \in \mathcal{C}$ with $\varepsilon_i(x) > 0$ and $\varepsilon_j(e_i x) = \varepsilon_j(x) > 0$, then $e_i e_j x = e_j e_i x$ and $\varphi_i(e_j x) = \varphi_i(x)$.*

Note that the requirement that $\varepsilon_j(e_i x) = \varepsilon_j(x) > 0$ implies that $e_i e_j x \neq 0$.

Axiom S3. *Assume that $i, j \in I$ and $i \neq j$. If $x \in \mathcal{C}$ with $\varepsilon_j(e_i x) = \varepsilon_j(x) + 1 > 1$ and $\varepsilon_i(e_j x) = \varepsilon_i(x) + 1 > 1$, then*

$$e_j e_i^2 e_j x = e_i e_j^2 e_i x \neq 0,$$

$\varphi_i(e_j x) = \varphi_i(e_j^2 e_i x)$ *and* $\varphi_j(e_i x) = \varphi_j(e_i^2 e_j x)$.

Axioms S2 and S3 are illustrated in Figure 4.4. Note that the condition $e_j e_i^2 e_j x = e_i e_j^2 e_i x \neq 0$ implies in particular that all eight elements $x, e_i x, e_j x, e_j e_i x, e_i e_j x, e_j^2 e_i x, e_i^2 e_j x, e_i e_j^2 e_i x$ are nonzero. In fact, they are all distinct (as indicated in Figure 4.4), which is proved in the next lemma.

Lemma 4.6. *For x, i, j satisfying the conditions of Axiom S3, all eight elements $x, e_i x, e_j x, e_j e_i x, e_i e_j x, e_j^2 e_i x, e_i^2 e_j x, e_i e_j^2 e_i x = e_j e_i^2 e_j x$ are distinct.*

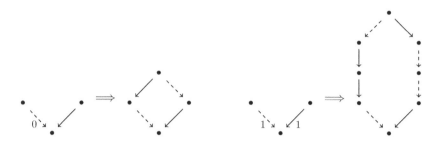

Fig. 4.4   Illustration of Axioms S2 and S3. Dashed arrows denote f_i, whereas solid arrows stand for f_j. The 0 below the dashed arrow in the left picture indicates the condition $\varepsilon_j(e_i x) = \varepsilon_j(x)$ (that is, ε_j changes by 0 along the edge). Similarly, the 1 below the two arrows in the right picture indicate the conditions $\varepsilon_j(e_i x) = \varepsilon_j(x) + 1$ and $\varepsilon_i(e_j x) = \varepsilon_i(x) + 1$ (that is, ε_j and ε_i change by 1 along the respective edges).

Proof. Since $i \neq j$ all elements are distinct by weight considerations except possibly $e_i e_j x$ and $e_j e_i x$.

Let us assume that $e_i e_j x = e_j e_i x$ and show that this leads to a contradiction. This yields $\varepsilon_j(e_i e_j x) = \varepsilon_j(e_j e_i x) = \varepsilon_j(e_i x) - 1 = \varepsilon_j(x)$, where the last equality comes from the assumptions for Axiom S3. On the other hand, $\varepsilon_j(x) = \varepsilon_j(e_j x) + 1 = \varepsilon_j(e_i e_j x)$. Hence setting $z = e_j x$, we have by Proposition 4.5 that $\varphi_j(e_i z) = \varphi_j(z)$ or equivalently $\varphi_j(e_i e_j x) = \varphi_j(x) + 1$. Now Axiom S3 and Proposition 4.5 imply that $\varphi_j(e_i^2 e_j x) = \varphi_j(e_i x) = \varphi_j(x)$, so that in particular $\varphi_j(e_i^2 e_j x) = \varphi_j(e_i e_j x) - 1$. By Proposition 4.5 this yields $\varepsilon_j(e_i^2 e_j x) = \varepsilon_j(e_i e_j x)$. Hence Axiom S2 applies and $\varphi_i(e_j e_i e_j x) = \varphi_i(e_i e_j x)$ or, since $e_i e_j x = e_j e_i x$, also $\varphi_i(e_j^2 e_i x) = \varphi_i(e_j e_i x)$. All arguments so far have been symmetric in i and j, so that we also obtain $\varphi_j(e_i^2 e_j x) = \varphi_j(e_i e_j x)$, which contradicts $\varphi_j(e_i^2 e_j x) = \varphi_j(e_i e_j x) - 1$. $\qquad\square$

Axioms S1-S3 have "dual" statements which we also assume.

Axiom S1'. *When $i, j \in I$ and $i \neq j$, if $x, y \in C$ and $y = f_i x$, then $\varphi_j(y)$ equals either $\varphi_j(x)$ or $\varphi_j(x) + 1$. The second case is possible only if α_i and α_j are not orthogonal roots.*

By Proposition 4.5, Axiom S1' is actually equivalent to Axiom S1, but we are stating it for completeness. The other two dual Axioms need to be assumed.

Axiom S2'. *Assume that $i, j \in I$ and $i \neq j$. If $x \in C$ with $\varphi_i(x) > 0$ and $\varphi_j(f_i x) = \varphi_j(x) > 0$, then $f_i f_j x = f_j f_i x$ and $\varepsilon_i(f_j x) = \varepsilon_i(x)$.*

Axiom S3'. *Assume that $i, j \in I$ and $i \neq j$. If $x \in C$ with $\varphi_j(f_i x) = \varphi_j(x) + 1 > 1$ and $\varphi_i(f_j x) = \varphi_i(x) + 1 > 1$, then*

$$f_j f_i^2 f_j x = f_i f_j^2 f_i x \neq 0,$$

$\varepsilon_i(f_j x) = \varepsilon_i(f_j^2 f_i x)$ *and* $\varepsilon_j(f_i x) = \varepsilon_j(f_i^2 f_j x)$.

Proposition 4.7. *Assume that C is a weak Stembridge crystal for a simply-laced root system. Let i and j be (distinct) indices such that the roots α_i and α_j are not orthogonal. Let $x \in C$ satisfy $\varepsilon_j(x) > 0$ and $\varepsilon_i(e_j x) = \varepsilon_i(x) + 1$. Then $\varepsilon_j(e_i e_j x) = \varepsilon_j(x) - 1$.*

Proof. By Proposition 4.5 we have either $\varepsilon_j(e_i e_j x) = \varepsilon_j(e_j x)$ or $\varepsilon_j(e_j x) + 1$, that is, either $\varepsilon_j(x) - 1$ or $\varepsilon_j(x)$. We will assume $\varepsilon_j(e_i e_j x) = \varepsilon_j(x)$ and deduce a contradiction. By (4.1)

$$\varphi_j(e_i e_j x) - \varphi_j(x) = \langle \alpha_i + \alpha_j, \alpha_j^\vee \rangle = 1,$$

which we can write in the form $\varphi_j(e_i e_j x) = \varphi_j(e_j x)$. Let $z = e_i e_j x$. This equation can also be written $\varphi_j(z) = \varphi_j(f_i z)$. Since $f_j f_i z = x$ we have $\varphi_j(f_i z) > 0$ and so both $f_i z$ and $f_j z$ are nonzero. Thus by Axiom S2' we have $f_i f_j z = f_j f_i z = x$, so that $e_i e_j x = e_j e_i x$. In particular, we have $e_i x \neq 0$.

Now we have $\varepsilon_j(x) = \varepsilon_j(e_i e_j x) = \varepsilon_j(e_j e_i x) = \varepsilon_j(e_i x) - 1$, or $\varepsilon_j(e_i x) = \varepsilon_j(e_j e_i x) + 1 = \varepsilon_j(e_i e_j x) + 1 = \varepsilon_j(x) + 1$. Since we also have $\varepsilon_i(e_j x) = \varepsilon_i(x) + 1$ Axiom S3 applies. But by Lemma 4.6, we have $e_i e_j x \neq e_j e_i x$, which is a contradiction. $\square$

Remark 4.8. Let C be a crystal of finite type, and let $C^\vee$ be the dual crystal (Definition 2.20). Then it is straightforward to check that C satisfies Axioms S1, S2, and S3 if and only if $C^\vee$ satisfies Axioms S1', S2', and S3'.

Proposition 4.9. *Suppose that C is a weak Stembridge crystal. Assume that $x \in C$ such that $e_j x, e_i x, e_j e_i x, e_j^2 e_i x \neq 0$ and $\varphi_i(e_j x) < \varphi_i(x)$. Then $\varphi_i(e_j^2 e_i x) < \varphi_i(e_j e_i x)$.*

Proof. By Proposition 4.5 we have $\varepsilon_i(e_j x) = \varepsilon_i(x)$, and the roots α_i and α_j are not orthogonal. Thus Axiom S2 applies, so that $e_i e_j x = e_j e_i x$ and moreover $\varphi_j(e_i x) = \varphi_j(x)$. Now suppose that it is not true that $\varphi_i(e_j^2 e_i x) < \varphi_i(e_j e_i x)$. Then $\varphi_i(e_j^2 e_i x) = \varphi_i(e_j e_i x)$. We have $\varphi_i(e_j e_i x) > 0$ since $f_i(e_j e_i x) = e_j x \neq 0$, so writing $z = e_j^2 e_i x$ we see that $f_i(z) \neq 0$, and also $f_j(z) = e_j e_i x \neq 0$, and $\varphi_i(z) = \varphi_i(f_j z)$. Thus by Axiom S2' we have $f_i f_j z = f_j f_i z$ and $\varepsilon_j(f_i z) = \varepsilon_j(z)$. Now $f_i z = e_j f_j f_i z = e_j f_i f_j z = e_j f_i e_j e_i x = e_j f_i e_i e_j x = e_j^2 x$, so this proves that $\varepsilon_j(e_j^2 x) = \varepsilon_j(e_j^2 e_i x)$, which implies $\varepsilon_j(x) = \varepsilon_j(e_i x)$. But $\varphi_j(x) = \varphi_j(e_i x)$, and since the roots α_i and α_j are not orthogonal, this contradicts Proposition 4.5. $\square$

4.3 Stembridge crystals as a monoidal category

A *monoidal category* is a category having a bifunctor $\otimes$ that is associative in a suitable sense. That is, there is a natural isomorphism

$$\phi_{A,B,C} \colon A \otimes (B \otimes C) \longrightarrow (A \otimes B) \otimes C$$

such that all "obvious" diagrams commute, for example the pentagon:

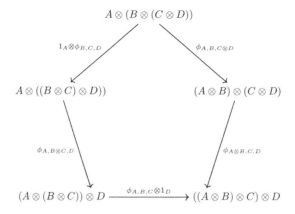

A monoidal category must also have a *unit element* I with isomorphisms $A \otimes I \cong I \otimes A \cong I$ that are compatible with the associativity morphisms in the sense that the diagram

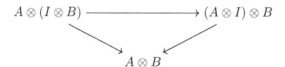

is commutative.

Mac Lane proved the *coherence theorem* for monoidal categories, which implies that all such diagrams commute if we assume the commutativity of the above pentagon. See the Corollary on page 165 of [Mac Lane (1971)]. The term *tensor category* is sometimes used, but this term is also used to mean different things, so we will use the unambiguous term "monoidal".

By Proposition 2.32, crystals form a monoidal category. The unit element is the crystal $\mathcal{T}_0$ defined in Example 2.28. By Proposition 2.29, seminormal crystals also form a monoidal category. We now use the Stembridge axioms to show that Stembridge crystals are closed under tensor products, so Stembridge crystals also form a monoidal category.

Theorem 4.10. *If $\mathcal{C}$ and $\mathcal{D}$ are Stembridge crystals, then so is $\mathcal{C} \otimes \mathcal{D}$.*

Proof. By Proposition 2.29 $\mathcal{C} \otimes \mathcal{D}$ is seminormal, and hence Axioms S0 and S0' are satisfied.

Let us confirm Axiom S1 for $\mathcal{C} \otimes \mathcal{D}$. We show that if $e_j(x \otimes y) = z \otimes w$, then $\varepsilon_i(z \otimes w) = \varepsilon_i(x \otimes y)$ or $\varepsilon_i(x \otimes y) + 1$, and that the second possibility can occur only if α_i and α_j are not orthogonal. Indeed

$$\varepsilon_i(x \otimes y) = \max\big(\varepsilon_i(y), \varepsilon_i(x) + \varepsilon_i(y) - \varphi_i(y)\big),$$
$$\varepsilon_i(z \otimes w) = \max\big(\varepsilon_i(w), \varepsilon_i(z) + \varepsilon_i(w) - \varphi_i(w)\big).$$

We have either $z = e_j(x)$ and $w = y$, or $z = x$ and $w = e_i(y)$. In view of (4.2) it follows that $\varepsilon_i(z \otimes w) = \varepsilon_i(x \otimes y)$ or $\varepsilon_i(x \otimes y) + 1$, and that the second possibility can only happen if α_i and α_j are not orthogonal.

Let us confirm Axiom S2 for $\mathcal{C} \otimes \mathcal{D}$. We assume that both $\varepsilon_i(x \otimes y) > 0$ and $\varepsilon_j\big(e_i(x \otimes y)\big) = \varepsilon_j(x \otimes y) > 0$. We must show that $e_i e_j(x \otimes y) = e_j e_i(x \otimes y)$ and that $\varphi_i\big(e_j(x \otimes y)\big) = \varphi_i(x \otimes y)$. There are four cases.

Case 1. *Suppose that* $\varphi_i(y) \geqslant \varepsilon_i(x)$ *and that* $\varphi_j(y) \geqslant \varepsilon_j(x)$. We have $e_i(x \otimes y) = x \otimes e_i y$. Thus the assumption that $\varepsilon_j\big(e_i(x \otimes y)\big) = \varepsilon_j(x \otimes y)$ means

$$\max\big(\varepsilon_j(e_i y), \varepsilon_j(x) + \varepsilon_j(e_i y) - \varphi_j(e_i y)\big) = \max\big(\varepsilon_j(y), \varepsilon_j(x) + \varepsilon_j(y) - \varphi_j(y)\big).$$

Because $\varphi_j(y) \geqslant \varepsilon_j(x)$ this maximum is $\varepsilon_j(y)$. Because $\varepsilon_j(e_i y) \geqslant \varepsilon_j(y)$, this implies that

$$\varepsilon_j(e_i y) = \varepsilon_i(y), \qquad \varphi_j(e_i y) \geqslant \varepsilon_j(x).$$

We may apply Axiom S2 for $\mathcal{D}$ to conclude that

$$e_i e_j y = e_j e_i y \neq 0, \qquad \varphi_i(e_j y) = \varphi_i(y). \tag{4.3}$$

Now $\varphi_j(y) \geqslant \varepsilon_j(x)$ and $\varphi_i(e_j y) = \varphi_i(y) \geqslant \varepsilon_i(x)$ so

$$e_i e_j(x \otimes y) = e_i(x \otimes e_j y) = x \otimes e_i e_j y.$$

Also $\varphi_i(y) \geqslant \varepsilon_i(x)$ and $\varphi_j(e_i y) \geqslant \varepsilon_j(x)$ so

$$e_j e_i(x \otimes y) = e_j(x \otimes e_i y) = x \otimes e_j e_i y$$

which proves that $e_i e_j(x \otimes y) = e_j e_i(x \otimes y)$. Furthermore

$$\varphi_i\big(e_j(x \otimes y)\big) = \max\big(\varphi_i(x), \varphi_i(x) + \varphi_i(e_j y) - \varepsilon_i(x)\big)$$

which, in view of (4.3), equals $\varphi_i(x \otimes y)$.

Case 2. *Suppose* $\varphi_i(y) \geqslant \varepsilon_i(x)$ *and* $\varphi_j(y) < \varepsilon_j(x)$. We will leave the reader to check that in this case Axiom S2 is satisfied with

$$e_i e_j(x \otimes y) = e_i(e_j x \otimes y) = e_j x \otimes e_i y = e_j(x \otimes e_i y) = e_j e_i(x \otimes y).$$

Case 3. *Suppose that* $\varphi_i(y) < \varepsilon_i(x)$ *and* $\varphi_j(y) \geqslant \varepsilon_j(x)$. We will leave it to the reader to check that Axiom S2 is satisfied with

$$e_i e_j(x \otimes y) = e_i(x \otimes e_j y) = e_i x \otimes e_j y = e_j(e_i x \otimes y) = e_j e_i(x \otimes y).$$

Case 4. *Suppose that* $\varphi_i(y) < \varepsilon_i(x)$ *and* $\varphi_j(y) < \varepsilon_j(x)$. We will leave it to the reader to check that Axiom S2 is satisfied with

$$e_i e_j(x \otimes y) = e_i(e_j x \otimes y) = e_i e_j x \otimes y = e_j e_i x \otimes y = e_j(e_i x \otimes y) = e_j e_i(x \otimes y).$$

Now let us confirm Axiom S3 for $\mathcal{C} \otimes \mathcal{D}$. We assume that $x \in \mathcal{C}$ and $y \in \mathcal{D}$ are given such that $\varepsilon_i\big(e_j(x \otimes y)\big) = \varepsilon_i(x \otimes y) + 1 > 1$ and $\varepsilon_j\big(e_i(x \otimes y)\big) = \varepsilon_j(x \otimes y) + 1 > 1$. We note that by Proposition 4.5 the roots α_i and α_j are not orthogonal. (Proposition 4.5 depends on Axiom S1, but we have already proved this for $\mathcal{C} \otimes \mathcal{D}$.) There are several cases. We will verify in each that with $z = x \otimes y$ that $e_j e_i^2 e_j z = e_i e_j^2 e_i z$. We also need $\varphi_i(e_j z) = \varphi_i(e_j^2 e_i z)$ and $\varphi_j(e_i z) = \varphi_j(e_i^2 e_j z)$, but we will leave these verifications to the reader.

Case 1. *Assume that* $\varphi_i(y) \geqslant \varepsilon_i(x)$ *and that* $\varphi_j(y) \geqslant \varepsilon_j(x)$. We have $e_i(x \otimes y) = x \otimes e_i y$ and $e_j(x \otimes y) = x \otimes e_j y$. In particular $e_i y$ and $e_j y$ are nonzero, or equivalently $\varepsilon_i(y) > 0$ and $\varepsilon_j(y) > 0$. There are three subcases.

Case 1A. *Suppose that* $\varphi_i(e_j y) < \varepsilon_i(x)$. Since $\varphi_i(y) \geqslant \varepsilon_i(x)$ we have $\varphi_i(e_j y) < \varphi_i(y)$. By Proposition 4.5, this implies $\varphi_i(e_j y) = \varphi_i(y) - 1$ and so $\varphi_i(y) = \varepsilon_i(x)$. Also by Proposition 4.5 we have $\varepsilon_i(e_j y) = \varepsilon_i(y) > 0$. Since in addition $\varepsilon_j(y) > 0$, Axiom S2 applies (with i and j switched), giving us $e_i e_j y = e_j e_i y \neq 0$ and $\varphi_j(e_i y) = \varphi_j(y)$. Using Proposition 4.5 once again, we obtain $\varepsilon_j(e_i y) = \varepsilon_j(y) + 1$. Since $\varepsilon_j(y) > 0$, this means that $\varepsilon_j(e_i y) \geqslant 2$ so $e_j e_i y \neq 0$ and $e_j^2 e_i y \neq 0$. Also we have $\varphi_j(e_i e_j y) = \varphi_j(e_j e_i y) = \varphi_j(e_i y) + 1 = \varphi_j(y) + 1 \geqslant \varepsilon_j(x) + 1 \geqslant \varepsilon_j(e_i x)$. Proposition 4.9 gives us $\varphi_i(e_j^2 e_i y) < \varphi_i(e_j e_i y) = \varphi_i(e_i e_j y) = \varphi_i(e_j y) + 1 = \varphi_i(y) = \varepsilon_i(x)$.

Since by assumption $\varphi_i(e_j y) < \varepsilon_i(x)$, we conclude that $e_i(x \otimes e_j y) = e_i x \otimes e_j y$. Now $\varphi_i(e_j y) = \varphi_i(y) - 1 = \varepsilon_i(x) - 1 = \varepsilon_i(e_i x)$ and so $e_i(e_i x \otimes e_j y) = e_i x \otimes e_i e_j y = e_i x \otimes e_j e_i y$. Furthermore $\varphi_j(e_i e_j y) \geqslant \varepsilon_j(e_i x)$ implies that $e_j(e_i x \otimes e_j e_i y) = e_i x \otimes e_j^2 e_i y$. In addition $\varphi_i(y) \geqslant \varepsilon_i(x)$ yields $e_i(x \otimes y) = x \otimes e_i y$, and $\varphi_j(e_i y) = \varphi_j(y) \geqslant \varepsilon_j(x)$ implies $e_j(x \otimes e_i y) = x \otimes e_j e_i y$. Also $\varphi_j(e_j e_i y) = \varphi_j(e_i y) + 1 > \varepsilon_j(x)$ so $e_j(x \otimes e_j e_i y) = x \otimes e_j^2 e_i y$. Finally, since $\varphi_i(e_j^2 e_i y) < \varepsilon_i(x)$ we have $e_i(x \otimes e_j^2 e_i y) = e_i x \otimes e_j^2 e_i y$. We have confirmed the following scenario, where as usual $z \xrightarrow{\;i\;} w$ means $f_i z = w$ and $e_i w = z$:

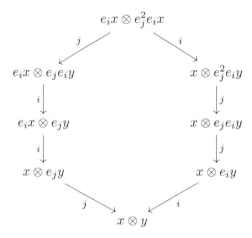

In particular, $e_i x \neq 0$ and $e_j^2 e_i x \neq 0$ so that $e_i e_j^2 e_i(x \otimes y) = e_j e_i^2 e_j(x \otimes y) \neq 0$.

Case 1B. *Suppose that* $\varphi_j(e_i y) < \varepsilon_j(x)$. This is similar to Case 1A with i and j reversed, and so we obtain the following scenario:

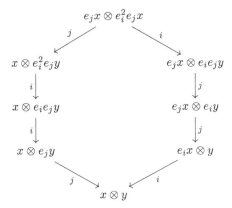

Case 1C. *Suppose that $\varphi_i(e_j y) \geqslant \varepsilon_i(x)$ and $\varphi_j(e_i y) \geqslant \varepsilon_j(x)$.* In this case we have the following scenario, which we will leave to the reader to confirm:

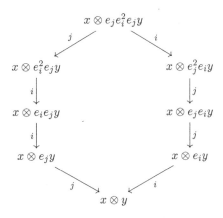

Case 2. *Assume that $\varphi_i(y) \geqslant \varepsilon_i(x)$ and that $\varphi_j(y) < \varepsilon_j(x)$.* This implies the following scenario, which we leave to the reader to check:

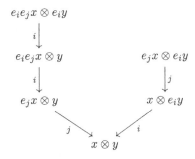

Now there are two possibilities.

Case 2A. *Suppose that $\varphi_j(e_i y) \geqslant \varepsilon_j(e_i e_j x)$.* In this case we find:

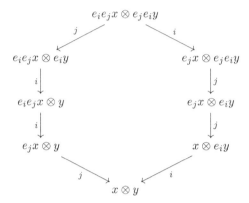

Case 2B. *Suppose that $\varphi_j(e_i y) < \varepsilon_j(e_i e_j x)$. In this case we find:*

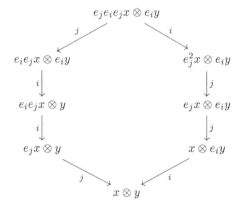

Case 3. *Assume that $\varphi_i(y) < \varepsilon_i(x)$ and that $\varphi_j(y) \geqslant \varepsilon_j(x)$. This is the same as Case 2 with i and j reversed, so we have already analyzed this situation.*

Case 4. *Suppose that $\varphi_i(y) < \varepsilon_i(x)$ and $\varphi_j(y) < \varepsilon_j(y)$. In this case, we have the following scenario:*

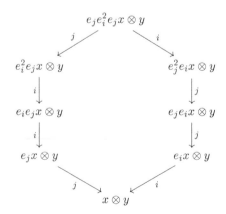

We have proved that Axioms S1, S2 and S3 are true for $\mathcal{C} \otimes \mathcal{D}$. Axioms S1′, S2′ and S3′ follow by dualizing. Thus Axioms S2′ and S3′ are assumed for $\mathcal{C}$ and $\mathcal{D}$, so Axioms S2 and S3 are true for $\mathcal{D}^\vee$ and $\mathcal{C}^\vee$, and hence for $\mathcal{D}^\vee \otimes \mathcal{C}^\vee$. By Remark 4.8 this means that Axioms S2 and S3 are true for $(\mathcal{C} \otimes \mathcal{D})^\vee$, which implies that Axioms S2′ and S3′ are true for $\mathcal{C} \otimes \mathcal{D}$.   □

4.4   Properties of Stembridge crystals

In this section, we investigate some further properties of Stembridge crystals. We will assume throughout that the Cartan type is simply-laced.

Let $\mathcal{C}$ be a Stembridge crystal. Recall that any subset $\mathcal{C}'$ of $\mathcal{C}$ that is a union of connected components inherits a crystal structure from $\mathcal{C}$, and we call $\mathcal{C}'$ a *full subcrystal*.

Theorem 4.11. *Let* $\mathbb{B}$ *be the standard crystal of type* A_r *or* D_r. *Then any full subcrystal of* $\mathbb{B}^{\otimes k}$ *is a Stembridge crystal. In particular, a crystal of tableaux is a Stembridge crystal.*

Proof. Indeed the standard crystal $\mathbb{B}$ is easily checked to satisfy the Stembridge axioms and so does $\mathbb{B}^{\otimes k}$ by Theorem 4.10. Since $\mathcal{B}_\lambda$ is a full subcrystal of $\mathbb{B}^{\otimes k}$ it satisfies the axioms.   □

Stembridge's axioms have potent implications. We will prove in Theorem 4.13 below that a connected crystal is characterized by its highest weight. In combination with this fact, Theorem 4.11 becomes a powerful tool, as we will see in Chapters 8, 9 and 11.

In this section, we are mainly interested in seminormal crystals. We recall this means that $\varepsilon_i(x)$ is the largest k such that $e_i^k x \neq 0$ and $\varphi_i(x)$ is the largest ℓ such that $f_i^\ell x \neq 0$. However, for our first result a slightly weaker condition is sufficient, so let us define a crystal to be *upper seminormal* if $\varepsilon_i(x)$ is the largest k such that $e_i^k x \neq 0$, with no corresponding assumption on $\varphi_i(x)$. We recall that an element x of a crystal will be called a *highest weight element* $e_i(x) = 0$ for all $i \in I$. If the crystal is seminormal, this is equivalent to $\varepsilon_i(x) = 0$ for all $i \in I$.

We consider the partial order on $\mathcal{C}$ generated by the condition that $x \prec y$ if $y = e_i x$ for some i in the index set I. Thus x is maximal with respect to this partial order if and only if it is a highest weight element. We will say that $\mathcal{C}$ is *bounded above* if for every $x \in \mathcal{C}$ there exists a highest weight element y such that $x \prec y$. For example if $\mathcal{C}$ is finite, it is obviously bounded above.

Theorem 4.12 ([Stembridge (2003)]). *Assume that the root system is simply-laced. Let* $\mathcal{C}$ *be a connected weak Stembridge crystal that is nonempty, upper seminormal and bounded above. Then* $\mathcal{C}$ *has a unique highest weight element.*

Proof. Since $\mathcal{C}$ is nonempty and bounded above, it has a maximal element x. We will show that $y \preccurlyeq x$ for all y. Let $\Omega = \{y \in \mathcal{C} \mid y \preccurlyeq x\}$. Let S be the set of all

$y \in \mathcal{C}$ such that $y \in \Omega$ but $0 \neq e_i y \notin \Omega$ for some i.

We show S is empty. If not, let y be a maximal element of S. By assumption $0 \neq e_i y \notin \Omega$ for some i. But since $y \in \Omega$ and at least e_i does not annihilate y, we have $y \prec x$ and so there exists j such that $0 \neq e_j y \in \Omega$. Since $e_i y$ and $e_j y$ are both nonzero, either Axiom S2 or Axiom S3 applies. Thus either $e_i e_j y = e_j e_i y \neq 0$ or $e_i e_j^2 e_i y = e_j e_i^2 e_j y \neq 0$. In both cases, $e_j y \succ y$ so $e_j y \notin S$ by the maximality of y in S. So we have $e_i e_j y \in \Omega$. If Axiom S2 applies, $e_i y \prec e_j e_i y = e_i e_j y \preccurlyeq x$ which is a contradiction. If Axiom S3 applies, then $e_i e_j y \succ y$ so $e_i e_j y \notin S$ by the maximality of y in S and therefore $e_i^2 e_j y \in \Omega$. Then $e_i^2 e_j y \succ y$ so $e_i^2 e_j y \notin S$ by the maximality of y in S and therefore $e_i^2 e_j y \in \Omega$. And finally, by the same reasoning $e_j e_i^2 e_j y \in \Omega$. Now $e_i y \prec e_i e_j^2 e_i y = e_j e_i^2 e_j y \preccurlyeq x$ which is a contradiction. This proves that S is empty.

Now we may show that $\Omega = \mathcal{C}$. If not, since $\mathcal{C}$ is connected and Ω is nonempty $(x \in \Omega)$ there exists a $y \in \Omega$ that is adjacent to some element $z \in \mathcal{C} \setminus \Omega$. We have either $z = e_i y$ or $z = f_i y$. The second is impossible since it would imply that $z \prec y \preccurlyeq x$ but $z \notin \Omega$. The first is also impossible since it implies $y \in S = \varnothing$. □

If $u \in \mathcal{C}$ is maximal with respect to the partial order $\prec$, then $\mathrm{wt}(u)$ is maximal among $\mathrm{wt}(x)$ for all other $x \in \mathcal{C}$ with respect to the partial order on the weight lattice in which $\mu \preccurlyeq \lambda$ if $\lambda - \mu = \sum k_i \alpha_i$ with $k_i \geqslant 0$. This is clear since $\mathrm{wt}(e_i x) = \mathrm{wt}(x) + \alpha_i$, so if the maximal element u is obtained from x by repeated applications of e_i, then $\mathrm{wt}(u)$ is a linear combination with nonnegative integer coefficients of the simple roots α_i. Evidently the highest weight element u of a connected crystal is uniquely characterized by the property that $e_i u = 0$ for all $i \in I$.

If $x \in \mathcal{C}$, then we define the *rank* of x to be $\mathrm{rank}(x) = \langle \mathrm{wt}(u) - \mathrm{wt}(x), \rho^\vee \rangle$, where $\rho^\vee$ is a vector such that $\langle \alpha_i, \rho^\vee \rangle = 1$ for all $i \in I$; we may take $\rho^\vee$ to be half the sum of the positive coroots. Write a sequence $x, e_{i_1} x, e_{i_2} e_{i_1} x, \ldots$ of nonzero elements of $\mathcal{C}$ that is as long as possible. If $y = e_{i_N} \cdots e_{i_1} x$ is the last element, then $e_i y = 0$ for all $i \in I$, so $y = u$ is the unique highest weight vector. Now $\mathrm{wt}(u) = \mathrm{wt}(x) + \alpha_{i_1} + \cdots + \alpha_{i_N}$ and so $\mathrm{rank}(x) = N$.

Theorem 4.13 ([Stembridge (2003)]). *Let $\mathcal{C}$ and $\mathcal{C}'$ be connected Stembridge crystals. Let $u \in \mathcal{C}$ and $u' \in \mathcal{C}'$ be the highest weight vectors. If $\mathrm{wt}(u) = \mathrm{wt}(u')$, then $\mathcal{C}$ and $\mathcal{C}'$ are isomorphic.*

Proof. Let Ω be the set of all subsets S of $\mathcal{C}$ with the following properties. First $u \in S$. Second, if $x \in S$ and $e_i x \in \mathcal{C}$ then $e_i x \in S$. Finally, there exists a subset S' of $\mathcal{C}'$ and a bijection $x \longmapsto x'$ from $S \longrightarrow S'$ mapping $u \longmapsto u'$ such that if $x \in S$ then $e_i x \neq 0$ if and only if $e_i(x') \neq 0$ and $(e_i x)' = e_i x'$. The set Ω is nonempty since $\{u\} \in \Omega$. Let us assume that S is an element of Ω that is maximal with respect to the inclusion of sets. We will prove that $S = \mathcal{C}$.

First let us show that if $x \in S$, then $\mathrm{wt}(x') = \mathrm{wt}(x)$ and for each i we have $\varepsilon_i(x') = \varepsilon_i(x)$ and $\varphi_i(x') = \varphi_i(x)$. Indeed, if $N = \mathrm{rank}(x)$ we may find a sequence $x, e_{i_1} x, e_{i_2} e_{i_1} x, \ldots, e_{i_N} \cdots e_{i_1} x = u$ of nonzero elements of $\mathcal{C}$, the last of which is

u. By the defining properties of S, $e_{i_k} \cdots e_{i_1} x' \neq 0$ for all k and $e_{i_k} \cdots e_{i_1} x' = (e_{i_k} \cdots e_{i_1} x)'$. Thus $e_{i_N} \cdots e_{i_1} x' = u'$ and $\text{wt}(x') = \text{wt}(u') - \sum_{j=1}^{N} \alpha_{i_j} = \text{wt}(x)$. Next $\varepsilon_i(x)$ is the largest k such that $e_i^k x \neq 0$, and since Ω is closed under the e_i, the sequence $x, e_i x, \ldots, e_i^k x$ is contained in S. From this, $\varepsilon_i(x')$ is the largest k' such that $e_i^{k'} x' = (e_i^{k'} x)'$ is nonzero, and evidently $k = k'$, so $\varepsilon_i(x') = \varepsilon_i(x)$. Finally, $\varphi_i(x) = \langle \text{wt}(x), \alpha_i^\vee \rangle + \varepsilon_i(x)$, from which we deduce that $\varphi_i(x') = \varphi_i(x)$.

If $S \neq C$, then let $z \in C \setminus S$ be an element of minimal rank. Since $z \neq u$ there exists an i such that $e_i z \neq 0$. Because $\text{rank}(e_i z) < \text{rank}(z)$ we have $e_i z \in S$. We have $\varphi_i((e_i z)') = \varphi_i(e_i z) > 0$ so there exists a $z' \in C'$ such that $e_i z' = (e_i z)'$. We claim that z' does not depend on the choice of i. In other words, if j is another index such that $e_j z \neq 0$, then by the same reasoning there exists $z'' \in C'$ such that $e_j z'' = (e_j z)'$ and we will show that $z'' = z'$. Since $e_i z, e_j z \neq 0$ either Axiom S2 or Axiom S3 applies. This means that either $e_i e_j z = e_j e_i z$ or $e_j e_i^2 e_j z = e_i e_j^2 e_i z \neq 0$. Let $w = e_i e_j z$ in the first case and $w = e_j e_i^2 e_j z$ in the second. Note that $w \in S$ since S is closed under the e_i. By Axioms S2' and S3' in the first case we have $\varphi_j(f_i w) = \varphi_j(w)$ or $\varphi_i(f_j w) = \varphi_i(w)$, and in the second case we have $\varphi_j(f_i w) = \varphi_j(w) + 1$ and $\varphi_i(f_j w) = \varphi_i(w) + 1$. Note that $f_i w$ and $f_j w \in S$ in both cases. We have $\varphi_j(w') = \varphi_j(w)$ by the last paragraph and also $\varphi_j(f_i w') = \varphi_j((f_i w)') = \varphi_j(f_i w)$ and $\varphi_i(f_j w') = \varphi_i(f_j w)$. So we can apply Axiom S2' or Axiom S3' in C' and deduce that $f_i f_j w' = f_j f_i w'$ in the first case while $f_i f_j^2 f_i w' = f_j f_i^2 f_j w'$ in the second.

In the first case since $f_i f_j w = z$ we have $f_j w = e_i z$ and so $e_i z' = (f_j w)' = f_j w'$ and therefore $w' = e_j e_i z'$. Similarly $w' = e_j e_i z''$. Therefore $z' = f_i f_j w' = f_j f_i w' = z''$. In the second case since $f_i f_j^2 f_i w = z$ we have $f_j^2 f_i w = e_i z$ and so $e_i z' = (e_i z)' = f_j^2 f_i w'$ and similarly $e_j z'' = f_i^2 f_j w$, so $z' = f_i f_j^2 f_i w' = f_j f_i^2 f_j w' = z''$.

We now see that if $S \neq C$ then we can extend the map $x \mapsto x'$ to $S \cup \{z\}$ by mapping $z \mapsto z'$ and that this is well-defined. This contradicts the maximality of S. Therefore $S = C$ and the map $x \mapsto x'$ is a morphism of crystals. Its inverse map $C' \longrightarrow C$ exists by the same reasoning. $\qquad \square$

Exercises

Exercise 4.1. Consider the crystal on the right in Figure 3.1. Does it satisfy the Stembridge Axioms? If not, which ones are not satisfied?

Exercise 4.2. Recall that $\rho^\vee$ is defined as the vector such that $\langle \alpha_i, \rho^\vee \rangle = 1$ for all $i \in I$. Show that $\rho^\vee$ can be taken to be half the sum of the positive coroots.

Exercise 4.3. Fill in all details in the proof of Theorem 4.10.

Exercise 4.4. Prove the statement of Remark 4.8.

Chapter 5

Virtual, Fundamental, and Normal Crystals

Chapter 4 on (weak) Stembridge crystals focused on the local characterization via axioms for highest-weight crystals of simply-laced types. In this chapter, we provide a characterization that extends to non-simply-laced types. We begin in Section 5.1 by describing natural inclusions of non-simply-laced Lie algebras into simply-laced ones. These embeddings are used in Section 5.2 to define virtual crystals. In Section 5.3 we then prove properties of the class of virtual crystals analogous to ones we proved for the simply-laced types. In Sections 5.4 through 5.6 we provide crystals for each fundamental weight, including the spin weights for types B and D and the fundamental weights for the exceptional types. For the simply-laced cases the fundamental crystals are Stembridge and for the non-simply-laced cases they are virtual crystals. Next we combine the results of Chapter 4 on Stembridge crystals and of Section 5.3 on virtual crystals to give a definition of *normal crystals* in Section 5.7. Normal crystals turn out to be crystals corresponding to representations of quantum groups as originally defined by [Kashiwara (1994)]. We show that there is a unique connected normal crystal for every dominant weight, and that the class of normal crystals is closed under tensor product. Finally, we introduce reducible Cartan types and similarity of crystals in Sections 5.8 and 5.9 in order to show in Section 5.10 that the class of normal crystals is preserved by Levi branching.

5.1 Embeddings of root systems

There exist natural inclusions of semisimple complex Lie algebras

$$C_r \hookrightarrow A_{2r-1} \tag{5.1a}$$

$$B_r \hookrightarrow D_{r+1} \tag{5.1b}$$

$$F_4 \hookrightarrow E_6 \tag{5.1c}$$

$$G_2 \hookrightarrow D_4 \tag{5.1d}$$

which we describe in this section. They are illustrated in Figures 5.1 and 5.2. Note that these are all embeddings of non-simply-laced Lie algebras (or equivalently, the corresponding simply-connected Lie groups) into simply-laced ones. Given such an embedding $X \hookrightarrow Y$ for one of these pairs, there is a natural restriction map from the

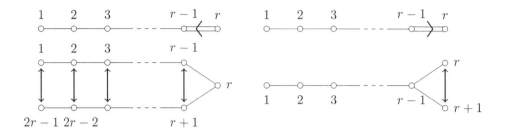

Fig. 5.1   Dynkin diagram folding $C_r \hookrightarrow A_{2r-1}$ (left) and $B_r \hookrightarrow D_{r+1}$ (right).

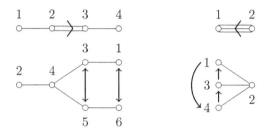

Fig. 5.2   Dynkin diagram folding $F_4 \hookrightarrow E_6$ (left) and $G_2 \hookrightarrow D_4$ (right).

weight lattice Λ_Y of Y to the weight lattice Λ_X of X. The adjoint of this mapping, with respect to Weyl group-invariant inner products on these weight lattices, is the key to the virtual crystal construction. We will describe this embedding of Λ_X into Λ_Y concretely by telling what it does to the fundamental weights.

Thus, consider an embedding of the Lie algebra X with Dynkin diagram $DD(X)$ into the Lie algebra Y with diagram $DD(Y)$. To describe the embedding, let aut be one of the following graph automorphisms of $DD(Y)$. We define aut: $I^Y \to I^Y$ on the index set I^Y for the diagram $DD(Y)$. For type $Y = A_{2r-1}$, aut$(i) = 2r - i$ for $i \in I^Y = \{1, 2, \ldots, 2r - 1\}$. For type D_{r+1}, the automorphism interchanges the nodes r and $r+1$ and fixes all other nodes. There is an additional automorphism for type D_4 called *triality*, namely, the cyclic permutation of the nodes 1, 3 and 4. For type E_6, the automorphism exchanges nodes 1 and 6 and nodes 3 and 5. In Figures 5.1 and 5.2 the various automorphisms aut are illustrated pictorially by arrows.

Let I^X and I^Y be the vertex sets of the diagrams X and Y respectively, $I^Y/$aut the set of orbits of the action of aut on I^Y, and $\sigma \colon I^X \to I^Y/$aut a bijection which preserves edges. In our cases:

(1) If $X = C_r$ and $Y = A_{2r-1}$, then $\sigma(i) = \{i, 2r - i\}$ for $1 \leqslant i < r$ and $\sigma(r) = \{r\}$.
(2) If $X = B_r$ and $Y = D_{r+1}$, then $\sigma(i) = \{i\}$ for $1 \leqslant i < r$ and $\sigma(r) = \{r, r+1\}$.
(3) If $X = F_4$ and $Y = E_6$, then $\sigma(1) = \{2\}$, $\sigma(2) = \{4\}$, $\sigma(3) = \{3, 5\}$ and $\sigma(4) = \{1, 6\}$.

(4) If $X = G_2$ and $Y = D_4$, then $\sigma(1) = \{1, 3, 4\}$ and $\sigma(2) = \{2\}$.

To describe the embedding of weight lattices, we endow the bijection σ with additional data. For each $i \in I^X$, we shall define a *multiplication factor* γ_i that depends on the location of i with respect to the arrow (multiple bond) in X. Removing the arrow leaves two connected components. The factor γ_i is defined as follows:

(1) If i is in the component that the arrow points towards, then $\gamma_i = 1$.
(2) If i is in the component that the arrow points away from, then γ_i is equal to the size of the bond (2 for a double arrow and 3 for a triple arrow).

If we normalize the roots so that the short roots have length 1, then γ_i is precisely the square of the length of α_i. The values of γ_i in the various cases are summarized in Table 5.1.

Table 5.1 Summary of the multiplication factors.

Cartan type X	Multiplication factors	
C_r	$\gamma_i = 1$	for $1 \leqslant i < r$
	$\gamma_r = 2$	
B_r	$\gamma_i = 2$	for $1 \leqslant i < r$
	$\gamma_r = 1$	
F_4	$\gamma_1 = \gamma_2 = 2$	
	$\gamma_3 = \gamma_4 = 1$	
G_2	$\gamma_1 = 1,$	$\gamma_2 = 3$

The Lie algebra embeddings yield natural embeddings $\Psi \colon \Lambda^X \longrightarrow \Lambda^Y$ of the corresponding semisimple, simply-connected weight lattices as

$$\varpi_i^X \longmapsto \gamma_i \sum_{j \in \sigma(i)} \varpi_j^Y,$$

$$\alpha_i^X \longmapsto \gamma_i \sum_{j \in \sigma(i)} \alpha_j^Y.$$

Recall that ϖ_i is the i-th fundamental weight and α_i is the i-th simple root. We have decorated them with X and Y to indicate to which Dynkin diagram they correspond.

Example 5.1. *Take the embedding $C_2 \hookrightarrow A_3$. Then*

$$\Psi(\varpi_1^C) = \varpi_1^A + \varpi_3^A, \qquad \Psi(\varpi_2^C) = 2\varpi_2^A.$$

5.2 Virtual crystals

We now use the embeddings of Lie algebras of the last section to construct crystals of non-simply-laced types. This was first done by [Kashiwara (1996)] (called similarity of crystals in the reference) and then later extended by [Baker (2000)]. In [Okado, Schilling and Shimozono (2003a,b); Schilling and Shimozono (2006)] this was used to construct Kirillov–Reshetikhin crystals for non-simply-laced types and termed the *virtual crystal* construction.

We are assuming that the weight lattices Λ^X and Λ^Y of X and Y are semisimple, that is, spanned by their fundamental weights. This condition will be relaxed in Section 5.7, where we introduce the notion of *virtualizable crystals*, whose weight lattices are not required to be semisimple.

Consider the embedding $X \hookrightarrow Y$ of Lie algebras. Let $\widehat{V}$ be a crystal of type Y with crystal operators $\widehat{e}_i, \widehat{f}_i$ for $i \in I^Y$, which we all the *ambient crystal*. We denote the weight and string lengths for an element $b \in \widehat{V}$ by $\widehat{\mathrm{wt}}(b)$ and $\widehat{\varphi}_i(b), \widehat{\varepsilon}_i(b)$, respectively. The *virtual crystal operators* (of type X) for $i \in I^X$ are defined as

$$e_i := \prod_{j \in \sigma(i)} \widehat{e}_j^{\,\gamma_i} \quad \text{and} \quad f_i := \prod_{j \in \sigma(i)} \widehat{f}_j^{\,\gamma_i}, \tag{5.2}$$

where $\sigma \colon I^X \longrightarrow I^Y/\mathrm{aut}$ and γ_i are the multiplication factors of Section 5.1.

Remark 5.2. The order in which the operators $\widehat{e}_j$ and $\widehat{f}_j$ in (5.2) are applied does not matter since any $j, j' \in \sigma(i)$ are not connected in the Dynkin diagram for Y and hence the corresponding crystal operators commute.

A *virtual crystal* $V \subseteq \widehat{V}$ is a subset of the elements in the ambient crystal $\widehat{V}$ with crystal operators defined as in (5.2) if Axioms V1-V3 are satisfied.

Axiom V1. The crystal $\widehat{V}$ is Stembridge.

Axiom V2. If $b \in V$ and $i \in I^X$, then $\widehat{\varepsilon}_j(b)$ has the same value for all $j \in \sigma(i)$ and that value is a multiple of γ_i. Similarly, $\widehat{\varphi}_j(b)$ is also constant on $\sigma(i)$ and is a multiple of γ_i.

Assuming Axiom V2, we may define, for $b \in V$, $i \in I^X$ and any $j \in \sigma(i)$:

$$\varepsilon_i(b) := \frac{1}{\gamma_i} \widehat{\varepsilon}_j(b), \tag{5.3a}$$

$$\varphi_i(b) := \frac{1}{\gamma_i} \widehat{\varphi}_j(b). \tag{5.3b}$$

Note that by Axiom V2, $\varepsilon_i(b)$ and $\varphi_i(b)$ are integers.

Axiom V3. The subset $V \sqcup \{0\}$ of $\widehat{V} \sqcup \{0\}$ is closed under the virtual operators e_i and f_i and

$$\varepsilon_i(b) = \max\{k \mid e_i^k(b) \neq 0\} \quad \text{and} \quad \varphi_i(b) = \max\{k \mid f_i^k(b) \neq 0\}. \tag{5.4}$$

We define a *weak virtual crystal* in the same manner, except for replacing the condition in Axiom V1 which states that $\widehat{V}$ is Stembridge by the condition that $\widehat{V}$ is weak Stembridge and except for dropping (5.4).

Remark 5.3. The conditions in Axiom V2 for the string lengths are called *aligned* in [Okado, Schilling and Shimozono (2003a,b)]. In fact, they imply that the weight defined by

$$\text{wt}(b) := \sum_{i \in I^X} \left(\varphi_i(b) - \varepsilon_i(b) \right) \varpi_i^X \tag{5.5}$$

satisfies $\Psi\left(\text{wt}(b)\right) = \widehat{\text{wt}}(b)$.

Proposition 5.4. *Let $\widehat{\mathcal{V}}$ be a seminormal crystal of type Y. A virtual crystal $\mathcal{V} \subseteq \widehat{\mathcal{V}}$ for the embedding $X \hookrightarrow Y$ is a seminormal crystal of type X according to Definition 2.13.*

Proof. To check Axiom A1 of Definition 2.13 for $\mathcal{V}$, suppose that $e_i x = \prod_{j \in \sigma(i)} \widehat{e}_j^{\gamma_i} x = y$ for $x, y \in \mathcal{V}$. By Axiom A1 for $\widehat{\mathcal{V}}$, this is equivalent to $\prod_{j \in \sigma(i)} \widehat{f}_j^{\gamma_i} y = x$ (which can be rewritten as $f_i y = x$). In this case, we have for all $j \in \sigma(i)$

$$\varepsilon_i(y) = \frac{1}{\gamma_i} \widehat{\varepsilon}_j(y) = \frac{1}{\gamma_i} \left(\widehat{\varepsilon}_j(x) - \gamma_i \right) = \varepsilon_i(x) - 1$$

and similarly $\varphi_i(y) = \varphi_i(x) + 1$. Again by A1 for $\widehat{\mathcal{V}}$, we have

$$\widehat{\text{wt}}(y) = \widehat{\text{wt}}(x) + \gamma_i \sum_{j \in \sigma(i)} \alpha_j^Y = \widehat{\text{wt}}(x) + \Psi(\alpha_i^X)$$

which, using $\Psi(\text{wt}(x)) = \widehat{\text{wt}}(x)$, implies $\text{wt}(y) = \text{wt}(x) + \alpha_i^X$. This proves that Axiom A1 holds for $\mathcal{V}$.

By definition, $\varphi_i(x)$ and $\varepsilon_i(x)$ are given by (5.4) for all $i \in I^X$. Hence $\mathcal{V}$ is seminormal. Moreover, the weight defined by in (5.5) satisfies Axiom A2 since the fundamental weights are the orthonormal basis to the simple coroots. $\square$

Example 5.5. *Let us continue Example 5.1. We take $\widehat{\mathcal{V}} = \mathcal{B}_{\varpi_1} \otimes \mathcal{B}_{\varpi_3}$ of type A_3. Then $\mathcal{B}_{\varpi_1}$ of type C_2 (on the left in Figure 5.3) can be modeled by $\mathcal{V} \subseteq \widehat{\mathcal{V}}$ (in the middle and on the right in Figure 5.3) generated by $u_{\varpi_1} \otimes u_{\varpi_3}$ with crystal operators $f_1 = \widehat{f}_1 \widehat{f}_3$ and $f_2 = \widehat{f}_2^2$ according to (5.2) (and similarly for e_1 and e_2).*

Take for example

$$b = \boxed{1} \otimes \begin{array}{c} \boxed{1} \\ \boxed{2} \\ \boxed{3} \end{array} .$$

Then $\varphi_1(b) = \widehat{\varphi}_1(b) = \widehat{\varphi}_3(b) = 1$ and $\varepsilon_1(b) = \widehat{\varepsilon}_1(b) = \widehat{\varepsilon}_3(b) = 0$. Conditions (5.3a) and (5.3b) for all $b \in \mathcal{V}$ can be checked in a similar fashion. Hence $\mathcal{V}$ is a virtual crystal.

More generally for the embedding $C_r \hookrightarrow A_{2r-1}$, take

$$\mathcal{V} \subseteq \widehat{\mathcal{V}} = \mathcal{B}_{\varpi_1} \otimes \mathcal{B}_{\varpi_{2r-1}}$$

inside $\widehat{\mathcal{V}}$ of type A_{2r-1} generated by $u_{\varpi_1} \otimes u_{\varpi_{2r-1}} \in \widehat{\mathcal{V}}$ under the crystal operators $f_i = \widehat{f}_i \widehat{f}_{2r-i}$ for $1 \leqslant i < r$ and $f_r = \widehat{f}_r^2$ (and similarly for e_i). Then $\mathcal{V}$ is a virtual crystal and models the standard crystal $\mathcal{B}_{\varpi_1}$ of type C_r.

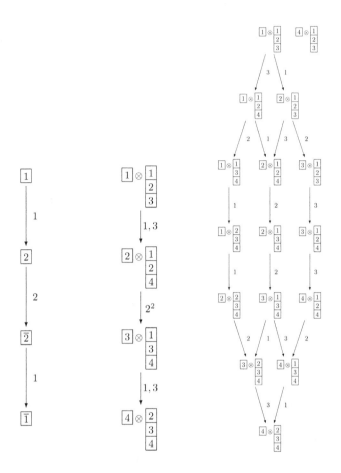

Fig. 5.3   The standard crystal of C_2 as a virtual crystal in A_3. Left: the C_2 crystal $\mathcal{B}_{\varpi_1}$. Right: the A_3 crystal $\mathcal{B}_{\varpi_1} \otimes \mathcal{B}_{\varpi_3}$, with two components. Middle: the virtual crystal.

Example 5.6. *Let us now consider the Lie algebra embedding* $B_2 \hookrightarrow D_3$*. Take* $\widehat{\mathcal{V}} = \mathcal{B}_{\varpi_1} \otimes \mathcal{B}_{\varpi_1}$ *of type* D_3*. Then* $\mathcal{B}_{\varpi_1}$ *of type* B_2 *(on the left in Figure 5.4) can be modeled by* $\mathcal{V} \subseteq \widehat{\mathcal{V}}$ *(in the middle and on the right in Figure 5.4) generated by* $u_{\varpi_1} \otimes u_{\varpi_1} \in \widehat{\mathcal{V}}$ *with crystal operators* $f_1 = \widehat{f}_1^2$ *and* $f_2 = \widehat{f}_2 \widehat{f}_3$ *according to (5.2) (and similarly for* e_1 *and* e_2*).*

In this case $\gamma_1 = 2$ *and* $\gamma_2 = 1$*. It is easily checked that* $\varphi_1(b) = \frac{1}{2}\widehat{\varphi}_1(b)$ *and* $\varphi_2(b) = \widehat{\varphi}_2(b) = \widehat{\varphi}_3(b)$ *(and similarly for* ε_i*) for all* $b \in \mathcal{V}$*. Hence* $\mathcal{V}$ *is a virtual crystal.*

In general for the embedding $B_r \hookrightarrow D_{r+1}$, take

$$\mathcal{V} \subseteq \widehat{\mathcal{V}} = \mathcal{B}_{\varpi_1} \otimes \mathcal{B}_{\varpi_1}$$

inside $\widehat{\mathcal{V}}$ of type D_{r+1} generated by $u_{\varpi_1} \otimes u_{\varpi_1} \in \widehat{\mathcal{V}}$ under the crystal operators $f_i = \widehat{f}_i^2$ for $1 \leqslant i < r$ and $f_r = \widehat{f}_r \widehat{f}_{r+1}$ (and similarly for e_i). Then $\mathcal{V}$ is a virtual

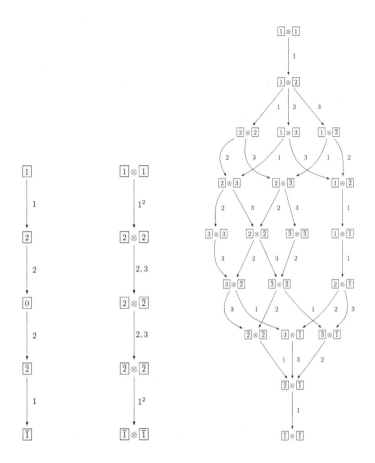

Fig. 5.4   The standard crystal of B_2 as a virtual crystal in D_3. Left: the B_2 crystal $\mathcal{B}_{\varpi_1}$. Right: one of three components in the D_3 crystal $\mathcal{B}_{\varpi_1} \otimes \mathcal{B}_{\varpi_1}$. Middle: the virtual crystal.

crystal and models the standard crystal B_{ϖ_1} of type B_r.

5.3   Properties of virtual crystals

In this section, we prove the analogues of Theorem 4.10 (monoidal category), Theorem 4.12 (unique maximal element in connected components), and Theorem 4.13 (connected components are characterized by the weights of their highest weight elements) that we established for simply-laced types for Stembridge crystals.

Proposition 5.7. *Suppose that $V \subseteq \widehat{V}$ is a virtual crystal for the embedding of Lie algebras $X \hookrightarrow Y$. Then a highest weight element u of V is a highest weight element of $\widehat{V}$. Moreover, if $\lambda^Y = \widehat{\mathrm{wt}}(u) \in \Lambda^Y$ for Y, then λ^Y is invariant under the automorphism* aut *described in Section 5.1. If V and $\widehat{V}$ are connected, then there exists a map* aut$: \widehat{V} \to \widehat{V}$ *that is an automorphism of the crystal graph, permuting* *the labels on the edges by the automorphism* aut *of the Dynkin diagram, and every element of V is fixed by this map.*

Proof. If u_λ is a highest weight element of V, then $\varepsilon_i(u_\lambda) = 0$ for all $i \in I^X$ and it follows from (5.3a) that $\widehat{\varepsilon}_j(u_\lambda) = 0$ for all $j \in I^Y$. Thus u_λ is also a highest weight element of $\widehat{V}$. Now by Remark 5.3, the highest weight λ^Y of $\widehat{V}$ equals $\Psi(\lambda)$. Since the image of Ψ consists of weights that are invariant under aut, the first statement follows.

We prove the last statement, with V and $\widehat{V}$ connected. Since $\widehat{V}$ has a unique highest weight element, so does V and therefore V is connected also. Since the highest weight λ^Y of $\widehat{V}$ is stable under aut, we may obtain a new Stembridge crystal with the same highest weight by using aut to permute the labels on the crystal graph. By Theorem 4.13, this relabeled crystal is isomorphic to the original one, and so the relabeling is induced by an automorphism of the crystal graph $\widehat{V}$.

Since the highest weight element u_λ of V is unique, it is fixed by aut and it follows from Remark 5.2 that any element derived from u_λ is also fixed by aut. But this is all of V. $\qquad\square$

Theorem 5.8. *Suppose that $V \subseteq \widehat{V}$ and $W \subseteq \widehat{W}$ are virtual crystals for the embedding of Lie algebras $X \hookrightarrow Y$. Then so is $V \otimes W \subseteq \widehat{V} \otimes \widehat{W}$.*

Proof. The crystal $\widehat{V} \otimes \widehat{W}$ is Stembridge by Theorem 4.10, so Axiom V1 is clear.

Consider $v \otimes w \in V \otimes W$. We consider the crystal graph automorphisms of $\widehat{V}$ and $\widehat{W}$ from Proposition 5.7. (We may assume that $\widehat{V}$ and $\widehat{W}$ are connected.) These automorphisms fix v and w and permute the labels in $\sigma(i)$ transitively, so they induce an automorphism of the crystal graph of $\widehat{V} \otimes \widehat{W}$ also permuting the labels. It follows that $\varepsilon_j(v \otimes w)$ and $\varphi_j(v \otimes w)$ are constant as functions of $j \in \sigma(i)$, as needed for Axiom V2.

We have two different formulas for ε_i and φ_i on $\widehat{V} \otimes \widehat{W}$, namely, the definitions (2.8) and (2.9) for the tensor product crystal and the definitions (5.3a) and (5.3b) for the virtual crystal. We will take (2.8) and (2.9) to be the definitions of ε_i and φ_i, and verify that (5.3a) and (5.3b) are satisfied on $V \otimes W$.

Since the ambient crystal $\widehat{V} \otimes \widehat{W}$ is a crystal, we have

$$\widehat{\varphi}_j(v \otimes w) = \widehat{\varphi}_j(v) + \max\{0, \widehat{\varphi}_j(w) - \widehat{\varepsilon}_j(v)\}$$

for any $j \in I^Y$ by (2.8). But by (5.3a) and (5.3b) for V and W, respectively, this implies

$$\widehat{\varphi}_j(v \otimes w) = \gamma_i\big(\varphi_i(v) + \max\{0, \varphi_i(w) - \varepsilon_i(v)\}\big) = \gamma_i \varphi_i(v \otimes w)$$

and similarly $\widehat{\varepsilon}_j(v \otimes w) = \gamma_i \varepsilon_i(v \otimes w)$ for all $j \in \sigma(i)$. This proves (5.3a) and (5.3b) for $V \otimes W$, and it also completes the proof of Axiom V2 because it shows that $\varepsilon_j(v \otimes w)$ and $\varphi_j(v \otimes w)$ are multiples of γ_i.

Finally, we must prove Axiom V3. We will show that $f_i(v \otimes w)$ equals either $f_i(v) \otimes w$ or $v \otimes f_i(w)$. This will show that $(V \otimes W) \sqcup \{0\}$ is stable under f_i (the proof for e_i is similar). There are two cases, depending on whether $\gamma_i > 1$, in which case $\sigma(i)$ consists of a single element j, or $\gamma_i = 1$. The second case is easy in view of Remark 5.2 and we leave it to the reader.

Thus assume that $\gamma_i > 1$ and $\sigma(i)$ consists of a single element j. Applying (2.10) γ_i times for each, we see that

$$f_i(v \otimes w) = \widehat{f}_j^{\gamma_i}(v \otimes w) = \begin{cases} \widehat{f}_j^{\gamma_i} v \otimes w & \text{if } \widehat{\varphi}_j(w) \leqslant \widehat{\varepsilon}_j(v), \\ v \otimes \widehat{f}_j^{\gamma_i} w & \text{if } \widehat{\varphi}_j(w) \geqslant \widehat{\varepsilon}_j(v) + \gamma_i. \end{cases}$$

The remaining cases $\widehat{\varepsilon}_j(v) < \widehat{\varphi}_j(w) < \widehat{\varepsilon}_j(v) + \gamma_i$ cannot occur, because $\widehat{\varepsilon}_j(v)$ and $\widehat{\varphi}_j(w)$ are multiples of γ_i by Axiom V2. $\qquad\square$

Theorem 5.9. *Let $V \subseteq \widehat{V}$ be a connected virtual crystal for the Lie algebra embedding $X \hookrightarrow Y$. Then V has a unique element of maximal weight with respect to the partial order $\prec$.*

Proof. Since V is connected, it must be contained in a single connected component of $\widehat{V}$. Thus we may assume that $\widehat{V}$ is connected. In view of Proposition 5.7, this result now follows from the corresponding property Theorem 4.12 of Stembridge crystals. $\qquad\square$

Theorem 5.10. *Let $V, V' \subseteq \widehat{V}$ be connected virtual crystals corresponding to the Lie algebra embedding $X \hookrightarrow Y$. Let $u \in V$ and $u' \in V'$ be the highest weight elements. If $\mathrm{wt}(u) = \mathrm{wt}(u')$, then V and V' are isomorphic.*

Proof. By Remark 5.3, we have that $\widehat{\mathrm{wt}}(u) = \widehat{\mathrm{wt}}(u')$ inside $\widehat{V}$. As we argued in the proof of Theorem 5.9, u and u' must be highest weight elements in $\widehat{V}$ and hence by Theorems 4.12 and 4.13 they generate isomorphic components. Now V and V' are generated by elements of equal weight in isomorphic components with the same crystal operators (5.2), so they must be isomorphic. $\qquad\square$

5.4   Fundamental crystals

As basic building blocks for crystals, we wish to construct a "fundamental crystal" for every fundamental weight. We can do this immediately for the classical Cartan types A_r, B_r, C_r, and D_r.

Let Φ and Λ be a root system and weight lattice, respectively. We assume that Λ is of simply-connected type by which we mean that Λ contains the fundamental weights (see Section 2.1). Denote by $\mathbb{B} := \mathcal{B}_{\varpi_1}$ the standard crystal described in Examples 2.19 through 2.24.

We recall that an element u of a crystal is called a *highest weight element* if $e_i(u) = 0$ for all $i \in I$.

Proposition 5.11. *For $1 \leqslant k \leqslant r$, the crystal $\mathbb{B}^{\otimes k}$ contains a full connected subcrystal with highest weight element of weight ϖ_k, with the following exceptions:*

(i) *If Φ is of type B_r, then there is no subcrystal of $\mathbb{B}^{\otimes r}$ with highest weight ϖ_r. However there is a subcrystal with highest weight $2\varpi_r$.*

(ii) *If Φ is of type D_r, then there is no subcrystal of $\mathbb{B}^{\otimes k}$ with highest weight ϖ_k when $k = r - 1$ or $k = r$. However there are subcrystals with highest weight $\varpi_{r-1} + \varpi_r$, $2\varpi_{r-1}$, and $2\varpi_r$.*

These full subcrystals of $\mathbb{B}^{\otimes k}$ have unique highest weight elements. For simply-laced types A_r and D_r, this follows from Theorem 4.12, whereas for types B_r and C_r uniqueness follows from Theorem 5.9 since we constructed the fundamental crystals of types B_r and C_r as virtual crystals.

Proof. Consulting Figure 2.1, we are claiming that $\mathbb{B}^{\otimes k}$ contains a subcrystal with an element of weight $\mathbf{e}_1 + \cdots + \mathbf{e}_k$. It is sufficient to exhibit an element u of $\mathbb{B}^{\otimes k}$ with that weight such that $e_i(u) = 0$ for all $i \in I$, since then we may take the maximal connected subcrystal containing u. Let

$$u = \boxed{k} \otimes \cdots \otimes \boxed{2} \otimes \boxed{1}.$$

The crystal $\mathbb{B}^{\otimes k}$ is seminormal by Proposition 2.29, and so it is enough to prove the vanishing of $\varepsilon_i(u)$ for all $i \in I$, which equals

$$\max_{j=1}^{k} \left(\sum_{h=1}^{j} \varepsilon_i\left(\boxed{h}\right) - \sum_{h=1}^{j-1} \varphi_i\left(\boxed{h}\right) \right)$$

by (2.15). We may discard $\varepsilon_i\left(\boxed{1}\right)$ which vanishes for all $i \in I$. Then

$$\varepsilon_i(u) = \max_{j=1}^{k} \left(\sum_{h=1}^{j-1} \varepsilon_i\left(\boxed{h+1}\right) - \varphi_i\left(\boxed{h}\right) \right),$$

but $\varepsilon_i\left(\boxed{h+1}\right) - \varphi_i\left(\boxed{h}\right)$ is always zero.

In type D_r, there is one additional highest weight vector

$$u = \boxed{\bar{r}} \otimes \boxed{r-1} \otimes \cdots \otimes \boxed{2} \otimes \boxed{1}$$

of weight $\mathbf{e}_1 + \cdots + \mathbf{e}_{r-1} - \mathbf{e}_r = 2\varpi_{r-1}$.

It remains to be explained why in type B_r there is no subcrystal of $\mathbb{B}^{\otimes k}$ with highest weight equal to the spin fundamental weight ϖ_r, and in type D_r there is no subcrystal with highest weight either of the spin fundamental weights ϖ_{r-1} or ϖ_r. This is because the highest weight of $\mathbb{B}$ lies in the orthogonal weight lattice $\mathbb{Z}^r$, while the spin fundamental weights do not. (See Example 2.8 for a definition of the orthogonal weight lattice.) So $\mathrm{wt}(\mathbb{B}^{\otimes k})$ does not contain the spin fundamental weights, and therefore $\mathbb{B}^{\otimes k}$ cannot have any component where the spin weights have nonzero multiplicitiy. $\qquad\square$

To complete the story for the missing Cartan types, we would like to define three "spin" crystals for the missing fundamental weights ϖ_r in the case of B_r and ϖ_{r-1},

ϖ_r for D_r. These cannot occur in $\mathbb{B}^{\otimes k}$, so we must construct them by a different method.

Let Φ be a root system with weight lattice Λ. Then $\lambda \in \Lambda$ is called *minuscule* if $\langle \lambda, \alpha^\vee \rangle = 0, 1$ or -1 for every coroot $\alpha^\vee$. In particular, this implies that the weights of the irreducible representation with highest weight $\lambda \in \Lambda^+$ consist of a single W-orbit, where W is the corresponding Weyl group (see Exercise 5.3). In type A_r, all fundamental weights are minuscule. In the other Cartan types, the minuscule fundamental weights are as follows:

- in type B_r the spin weight ϖ_r;
- in type C_r, ϖ_1;
- in type D_r, ϖ_1, and the spin weights ϖ_{r-1} and ϖ_r;
- in type E_6, ϖ_1 and ϖ_6;
- in type E_7, ϖ_7.

If $\lambda \in \Lambda^+$ is minuscule, we may define a crystal $\mathcal{M}_\lambda$ as follows. For every weight μ in the W-orbit of λ, we have an element v_μ. We write

$$f_i(v_\mu) = \begin{cases} v_{\mu-\alpha_i} & \text{if } \langle \mu, \alpha_i^\vee \rangle = 1, \\ 0 & \text{otherwise,} \end{cases}$$

and

$$e_i(v_\mu) = \begin{cases} v_{\mu+\alpha_i} & \text{if } \langle \mu, \alpha_i^\vee \rangle = -1, \\ 0 & \text{otherwise.} \end{cases}$$

It is easy to see that $\mathcal{M}_\lambda$ is a seminormal crystal with highest weight λ. Let us take a look at $\mathcal{M}_\lambda$ when λ is a spin weight of B_r or D_r. The weights that can appear in these crystals are

$$\frac{1}{2} \sum_{i=1}^r \epsilon_i \mathbf{e}_i \qquad \text{where } \epsilon_i = \pm.$$

We denote this weight μ as $\epsilon_1 \cdots \epsilon_r$ and the corresponding crystal element as $\boxed{\epsilon_1 \cdots \epsilon_r}$. For example, $\boxed{++-}$ denotes v_μ with $\mu = \frac{1}{2}(\mathbf{e}_1 + \mathbf{e}_2 - \mathbf{e}_3)$.

In Figure 5.5 we illustrate the spin crystal $\mathcal{M}_{\varpi_3}$ for type B_3, whereas in Figure 5.6 the spin crystals $\mathcal{M}_{\varpi_3}$ and $\mathcal{M}_{\varpi_4}$ for type D_4 are drawn.

For the classical Cartan types A_r, B_r, C_r, and D_r we have now constructed a crystal with highest weight element u_{ϖ_k} for every every fundamental weight ϖ_k.

Proposition 5.12. *Let $\mathcal{C}$ be a seminormal crystal with a highest weight element u such that $\mathrm{wt}(u) = \lambda$, where λ is a minuscule weight. Assume that $\mathcal{C}$ has the property that no two elements have the same weight. Then there is a unique morphism $\psi \colon \mathcal{M}_\lambda \longrightarrow \mathcal{C}$ such that $\psi(v_\lambda) = u$.*

Proof. By Proposition 2.17, $\mathcal{C}$ has an element of weight μ for every μ in the Weyl orbit of λ, and by our assumption on $\mathcal{C}$, this element is unique. Call it c_μ. The morphism ψ will be defined by $\psi(v_\mu) = c_\mu$ for every μ in the Weyl group orbit of λ. It is easily checked that this is a morphism using the seminormality of $\mathcal{C}$. □

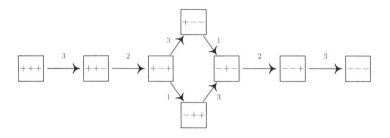

Fig. 5.5   The B_3 spin crystal $\mathcal{M}_{\varpi_3}$.

Proposition 5.13. *Suppose that Φ is a simply-laced root system with weight lattice Λ and that $\lambda \in \Lambda$ is minuscule. Then the minuscule crystal $\mathcal{M}_\lambda$ is a Stembridge crystal.*

Therefore the two spin crystals for type D_r are Stembridge crystals. We also obtain Stembridge crystals for the first fundamental representations of E_6 and E_7.

Proof. It is sufficient to show that the crystal obtained by Levi branching down to A_2 or $A_1 \times A_1$ is Stembridge. The branched crystal satisfies the conditions of Proposition 5.12. Each connected component is related (in the sense of Section 2.7) to a twist of a minuscule crystal for A_2 or $A_1 \times A_1$, and these crystals are easily identified. For type A_2, the connected component is related to a twist of $\mathcal{B}_{(1)}$ or $\mathcal{B}_{(1,1)}$, which are Stembridge, and similarly for $A_1 \times A_1$.   □

We can realize the spin crystal $\mathcal{B}_{\varpi_r}$ of type B_r as a virtual crystal inside $\widehat{\mathcal{V}} = \mathcal{B}_{\varpi_r} \otimes \mathcal{B}_{\varpi_{r+1}}$ of type D_{r+1}. The virtual crystal $\mathcal{V}$ is generated by $u_{\varpi_r} \otimes u_{\varpi_{r+1}}$ and the virtual crystal operators $f_i = \widehat{f_i}^2$ for $1 \leqslant i < r$ and $f_r = \widehat{f_r}\widehat{f_{r+1}}$.

Proposition 5.14. *$\mathcal{V}$ is a virtual crystal of type B_r and $\mathcal{V} \cong \mathcal{B}_{\varpi_r}$.*

Proof. The alignedness conditions (5.3a) and (5.3b) follow easily from the fact that $\mathcal{B}_{\varpi_r}$ and $\mathcal{B}_{\varpi_{r+1}}$ are minuscule, so that indeed for all $b \in \mathcal{V}$ we have $\varepsilon_i(b) = \frac{1}{2}\widehat{\varepsilon}_i(b)$ for $1 \leqslant i < r$ and $\varepsilon_r(b) = \widehat{\varepsilon}_r(b) = \widehat{\varepsilon}_{r+1}(b)$ and similarly for φ_i. The highest weight element $u_{\varpi_r} \otimes u_{\varpi_{r+1}}$ has weight $\Psi^{-1}(\varpi_r + \varpi_{r+1}) = \varpi_r$ in the type B_r weight lattice. Hence $\mathcal{V} \cong \mathcal{B}_{\varpi_r}$.   □

5.5   Adjoint crystals

Let Φ be any root system with weight lattice Λ. We define a crystal, the *adjoint crystal*, which is an analog of the adjoint representation. This has one element v_α for every root $\alpha \in \Phi$. In addition it has one element $\tilde{v}_i$ for every $i \in I$ (the index

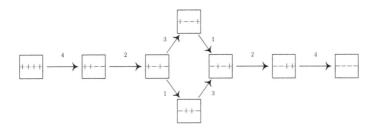

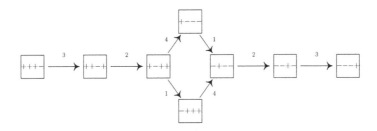

Fig. 5.6 The D_4 spin crystals $\mathcal{M}_{\varpi_3}$ and $\mathcal{M}_{\varpi_4}$. There is another D_4 crystal of degree 8, namely the standard crystal, and the three crystals have isomorphic crystal graphs modulo the triality automorphism $1 \to 3 \to 4 \to 1$ of the Dynkin diagram.

set). We define $\mathrm{wt}(v_\alpha) = \alpha$, $\mathrm{wt}(\tilde{v}_i) = 0$, and

$$f_i(v_\alpha) = \begin{cases} v_{\alpha-\alpha_i} & \text{if } \alpha - \alpha_i \in \Phi, \\ \tilde{v}_i & \text{if } \alpha = \alpha_i, \\ 0 & \text{otherwise,} \end{cases} \qquad e_i(v_\alpha) = \begin{cases} v_{\alpha+\alpha_i} & \text{if } \alpha + \alpha_i \in \Phi, \\ \tilde{v}_i & \text{if } \alpha = -\alpha_i, \\ 0 & \text{otherwise,} \end{cases}$$

$$f_i(\tilde{v}_j) = \begin{cases} v_{-\alpha_i} & \text{if } i = j, \\ 0 & \text{otherwise,} \end{cases} \qquad e_i(\tilde{v}_j) = \begin{cases} v_{\alpha_i} & \text{if } i = j, \\ 0 & \text{otherwise.} \end{cases}$$

The maps ε_i and φ_i are defined by (2.6). The crystal is thus seminormal.

For example, let α_1 and α_2 be the short and long roots of G_2, respectively. Then the adjoint crystal is illustrated in Figure 5.7.

Lemma 5.15. *Let Φ be a simply-laced root system. Let α and β be roots. If $\alpha + k\beta$ is a root and $k > 1$, then $k = 2$ and $\alpha = -\beta$.*

Proof. Let us consider the length of a root string

$$\alpha, \alpha + \beta, \alpha + 2\beta, \ldots, \alpha + k\beta \tag{5.6}$$

which is maximal in that it cannot be extended in either direction; that is, $\alpha - \beta$ is not a root and $\alpha + (k+1)\beta$ is not a root. Then if r_β is the reflection in the hyperplane orthogonal to $\beta^\vee$, we have $r_\beta(\alpha) = \alpha + k\beta$ and so $k = \langle \alpha, \beta^\vee \rangle$.

We observe that the intersection of Φ with the vector space spanned by α and β is a root system, and since α and β have the same length, this intersection is of type

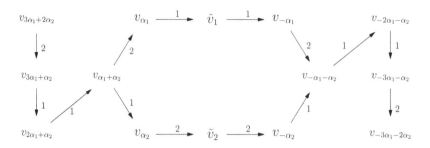

Fig. 5.7   The adjoint crystal of type G_2.

A_2 or $A_1 \times A_1$, or A_1 if $\alpha = \pm\beta$. Then we can see from inspection that $\langle \alpha, \beta^\vee \rangle \leqslant 2$, with equality only when $\alpha = -\beta$.   □

Proposition 5.16. *Let Φ be a simply-laced root system. Then the adjoint crystal is Stembridge.*

Proof. By Lemma 5.15 the length k of the root string (5.6) is bounded by 1, except when $\alpha = -\beta$ when it is 2. Thus $\varphi_i(v_\alpha) + \varepsilon_i(v_\alpha) \leqslant 1$ except when $\alpha = \pm\alpha_i$. On the other hand

$$\varphi_i(v_\alpha) - \varepsilon_i(v_\alpha) = \langle \alpha, \alpha_i^\vee \rangle. \tag{5.7}$$

We deduce that

$$\varphi_i(v_\alpha) = \begin{cases} 2 & \text{if } \alpha = \alpha_i, \\ 1 & \text{if } \langle \alpha, \alpha_i^\vee \rangle = 1, \\ 0 & \text{otherwise,} \end{cases} \qquad \varepsilon_i(v_\alpha) = \begin{cases} 2 & \text{if } \alpha = -\alpha_i, \\ 1 & \text{if } \langle \alpha, \alpha_i^\vee \rangle = -1, \\ 0 & \text{otherwise.} \end{cases}$$

Now the Stembridge axioms (Section 4.2) are easy to verify. With (5.7) in mind, (4.2) is easy to check, and this implies Axiom S1. The situation in Axiom S2 cannot occur. The situation in Axiom S3 can occur but only if $x = v_{-\alpha_i - \alpha_j}$. Regarding $\tilde{v}_i$, this cannot be x in either Axiom S2 or Axiom S3. It is straightforward to check Axiom S1 when either x or y equals $\tilde{v}_i$.

The dual axioms S1′, S2′ and S3′ are checked similarly.   □

5.6   Fundamental crystals: The exceptional cases

In Section 5.4 we exhibited specific crystals whose highest weights were the fundamental weights in the classical Cartan types A_r, B_r, C_r, and D_r. Now we do the same in the exceptional cases.

We begin with type E_r where $r = 6, 7$ or 8. In type E_6, the fundamental crystals $\mathcal{B}_{\varpi_1}$ and $\mathcal{B}_{\varpi_6}$ are minuscule both of dimension 27, whereas $\mathcal{B}_{\varpi_7}$ in type E_7 is minuscule of dimension 56. These crystals are Stembridge crystals by Proposition 5.13. On the other hand, for E_8 the highest root coincides with the fundamental weight

$$\varpi_8 = 2\alpha_1 + 3\alpha_2 + 4\alpha_3 + 6\alpha_4 + 5\alpha_5 + 4\alpha_6 + 3\alpha_7 + 2\alpha_8.$$

Table 5.2   Construction of the fundamental crystals for type E_6.

fundamental crystal	construction	dimension
$\mathcal{B}_{\varpi_1}$	minuscule	27
$\mathcal{B}_{\varpi_6}$	minuscule	27
$\mathcal{B}_{\varpi_2}$	$\mathcal{B}_{\varpi_1} \otimes \mathcal{B}_{\varpi_6}$	78
$\mathcal{B}_{\varpi_3}$	$\mathcal{B}_{\varpi_1} \otimes \mathcal{B}_{\varpi_1}$	351
$\mathcal{B}_{\varpi_5}$	$\mathcal{B}_{\varpi_6} \otimes \mathcal{B}_{\varpi_6}$	351
$\mathcal{B}_{\varpi_4}$	$\mathcal{B}_{\varpi_1} \otimes \mathcal{B}_{\varpi_1} \otimes \mathcal{B}_{\varpi_1}$	2925

Table 5.3   Construction of the fundamental crystals for type E_7.

fundamental crystal	construction	dimension
$\mathcal{B}_{\varpi_7}$	minuscule	56
$\mathcal{B}_{\varpi_1}$	$\mathcal{B}_{\varpi_7} \otimes \mathcal{B}_{\varpi_7}$	133
$\mathcal{B}_{\varpi_2}$	$\mathcal{B}_{\varpi_7} \otimes \mathcal{B}_{\varpi_7} \otimes \mathcal{B}_{\varpi_7}$	912
$\mathcal{B}_{\varpi_3}$	$\mathcal{B}_{\varpi_7} \otimes \mathcal{B}_{\varpi_7} \otimes \mathcal{B}_{\varpi_7} \otimes \mathcal{B}_{\varpi_7}$	8645
$\mathcal{B}_{\varpi_4}$	$\mathcal{B}_{\varpi_7} \otimes \mathcal{B}_{\varpi_7} \otimes \mathcal{B}_{\varpi_7} \otimes \mathcal{B}_{\varpi_7}$	365750
$\mathcal{B}_{\varpi_5}$	$\mathcal{B}_{\varpi_7} \otimes \mathcal{B}_{\varpi_7} \otimes \mathcal{B}_{\varpi_7}$	27664
$\mathcal{B}_{\varpi_6}$	$\mathcal{B}_{\varpi_7} \otimes \mathcal{B}_{\varpi_7}$	1539

Table 5.4   Certain useful roots for E_8.

α	notes
ϖ_8	$\varphi_8(v_\alpha) = 1$
$\varpi_7 - \varpi_8$	$\varphi_7(v_\alpha) = 1, \varepsilon_8(v_\alpha) = 1$
$\varpi_1 - \varpi_8$	$\varphi_1(v_\alpha) = 1, \varepsilon_8(v_\alpha) = 1$
$\varpi_2 - \varpi_1$	$\varphi_2(v_\alpha) = 1, \varepsilon_1(v_\alpha) = 1$
$\varpi_6 - \varpi_7$	$\varphi_6(v_\alpha) = 1, \varepsilon_7(v_\alpha) = 1$
$\varpi_3 - \varpi_2$	$\varphi_3(v_\alpha) = 1, \varepsilon_2(v_\alpha) = 1$
$\varpi_5 - \varpi_6$	$\varphi_5(v_\alpha) = 1, \varepsilon_6(v_\alpha) = 1$
$\varpi_4 - \varpi_5$	$\varphi_4(v_\alpha) = 1, \varepsilon_5(v_\alpha) = 1$

Hence we may define the fundamental crystal $\mathcal{B}_{\varpi_8}$ to be the adjoint crystal of dimension 248, which by Proposition 5.16 is a Stembridge crystal.

Theorem 5.17. *Let Φ be of type E_r where $r = 6, 7$ or 8, and let $\mathcal{B}$ be the Stembridge crystal $\mathcal{B}_{\varpi_1}$ or $\mathcal{B}_{\varpi_6}$ if $r = 6$, $\mathcal{B}_{\varpi_7}$ for $r = 7$, and $\mathcal{B}_{\varpi_8}$ if $r = 8$. Then, for each fundamental weight ϖ_i, some tensor power of the $\mathcal{B}$'s contains a connected component with highest weight ϖ_i for all $i \in I$.*

Proof. For types E_6 and E_7, the constructions are given in [Jones and Schilling (2010)], and are summarized in Tables 5.2 and 5.3.

For type E_8, the crystal $\mathcal{B}$ is the adjoint crystal. We will exploit the existence of roots that are the difference between two fundamental weights. In Table 5.4 we list the particular roots that we need. Any values of φ_i or ε_i that are not listed under "notes" are zero.

Now using (2.15), one may easily check that $\varepsilon_i(u) = 0$ for the tensors u in Table 5.5. These are therefore highest weight elements of subcrystals of $\mathcal{B}^{\otimes k}$ for the fundamental weights, and the case E_8 is proved. □

Table 5.5   Fundamental subcrystals of $\mathcal{B}_{\varpi_8}^{\otimes k}$ for E_8.

highest weight element u	$\mathrm{wt}(u)$	dimension
$v_{\varpi_1 - \varpi_8} \otimes v_{\varpi_8}$	ϖ_1	3875
$v_{\varpi_2 - \varpi_1} \otimes v_{\varpi_1 - \varpi_8} \otimes v_{\varpi_8}$	ϖ_2	147250
$v_{\varpi_3 - \varpi_2} \otimes v_{\varpi_2 - \varpi_1} \otimes v_{\varpi_1 - \varpi_8} \otimes v_{\varpi_8}$	ϖ_3	6696000
$v_{\varpi_4 - \varpi_5} \otimes v_{\varpi_5 - \varpi_6} \otimes v_{\varpi_6 - \varpi_7} \otimes v_{\varpi_7 - \varpi_8} \otimes v_{\varpi_8}$	ϖ_4	6899079264
$v_{\varpi_5 - \varpi_6} \otimes v_{\varpi_6 - \varpi_7} \otimes v_{\varpi_7 - \varpi_8} \otimes v_{\varpi_8}$	ϖ_5	146325270
$v_{\varpi_6 - \varpi_7} \otimes v_{\varpi_7 - \varpi_8} \otimes v_{\varpi_8}$	ϖ_6	2450240
$v_{\varpi_7 - \varpi_8} \otimes v_{\varpi_8}$	ϖ_7	30380
v_{ϖ_8}	ϖ_8	248

Table 5.6   Fundamental crystals for type F_4.

fundamental crystal	construction	dimension
$\mathcal{B}_{\varpi_1}$	$\mathcal{B}_{\varpi_4} \otimes \mathcal{B}_{\varpi_4}$	52
$\mathcal{B}_{\varpi_2}$	$\mathcal{B}_{\varpi_4} \otimes \mathcal{B}_{\varpi_4} \otimes \mathcal{B}_{\varpi_4}$	1274
$\mathcal{B}_{\varpi_3}$	$\mathcal{B}_{\varpi_4} \otimes \mathcal{B}_{\varpi_4}$	273
$\mathcal{B}_{\varpi_4}$	virtual	26

Next we turn our attention to type F_4. The crystal $\mathcal{B}_{\varpi_4}$ is depicted in Figure 5.8, where the vertices are labelled by the weights of the elements. This crystal can be realized as a virtual crystal under the folding $F_4 \hookrightarrow E_6$ given in Figure 5.2. Let $\mathcal{V}$ be the virtual crystal inside the ambient crystal $\widehat{\mathcal{V}} = \mathcal{B}_{\varpi_1} \otimes \mathcal{B}_{\varpi_6}$ of type E_6 generated by $u_{\varpi_1} \otimes u_{\varpi_6}$ and the crystal operators $f_1 = \widehat{f}_2^2$, $f_2 = \widehat{f}_4^2$, $f_3 = \widehat{f}_3 \widehat{f}_5$, and $f_4 = \widehat{f}_1 \widehat{f}_6$. Since all crystals involved are finite crystals, one can explicitly check that $\mathcal{V}$ is a virtual crystal and $\mathcal{V} \cong \mathcal{B}_{\varpi_4}$ of type F_4.

Theorem 5.18. *Let $\mathcal{B} := \mathcal{B}_{\varpi_4}$ of type F_4, which can be realized as a virtual crystal. Then, for each fundamental weight ϖ_i, some tensor power of $\mathcal{B}$ contains a connected component with highest weight ϖ_i for all $i \in I = \{1, 2, 3, 4\}$.*

Proof. The construction of each $\mathcal{B}_{\varpi_i}$ is summarized in Table 5.6.   □

Let us finally consider type G_2. We construct the fundamental crystal $\mathcal{B}_{\varpi_1}$ of type G_2 as a virtual crystal inside the ambient crystal $\widehat{\mathcal{V}} = \mathcal{B}_{\varpi_1} \otimes \mathcal{B}_{\varpi_3} \otimes \mathcal{B}_{\varpi_4}$ of type D_4 using the folding in Figure 5.2. Let $\mathcal{V} \subset \widehat{\mathcal{V}}$ be generated by $u_{\varpi_1} \otimes u_{\varpi_3} \otimes u_{\varpi_4}$ and crystal operators $f_1 = \widehat{f}_1 \widehat{f}_3 \widehat{f}_4$ and $f_2 = \widehat{f}_2^3$. One can explicitly check that $\mathcal{V}$ is a virtual crystal and that $\mathcal{V} \cong \mathcal{B}_{\varpi_1}$ of type G_2.

Theorem 5.19. *Let $\mathcal{B} := \mathcal{B}_{\varpi_1}$ of type G_2, which can be realized as a virtual crystal. Then $\mathcal{B}_{\varpi_2}$ is a connected component in $\mathcal{B}_{\varpi_1} \otimes \mathcal{B}_{\varpi_1}$. The crystal $\mathcal{B}_{\varpi_2}$ is isomorphic to the adjoint crystal depicted in Figure 5.7.*

Proof. The statements can be checked explicitly and are summarized in Table 5.7.   □

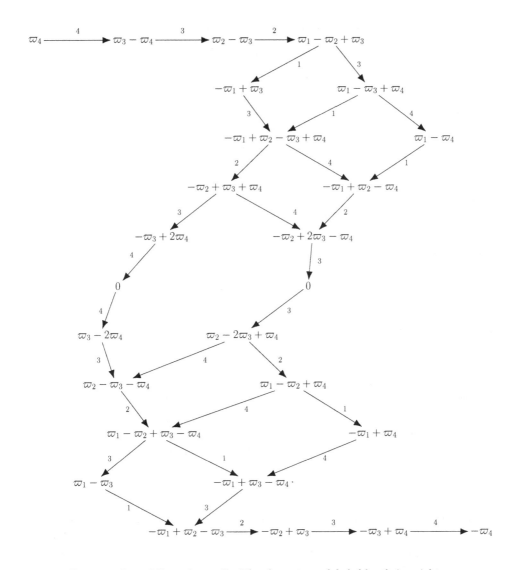

Fig. 5.8 Crystal $\mathcal{B}_{\varpi_4}$ of type F_4. The elements are labeled by their weights.

5.7 Normal crystals

As we have seen in Chapter 4 and Section 5.3, Stembridge crystals (for simply-laced types) and virtual crystals (for non-simply-laced types) have very similar properties. In our discussion of virtual crystals we discussed the semisimple, simply-connected weight lattice, but we now remove that assumption and call a crystal of Cartan type B_r, C_r, G_2 or F_4 with any weight lattice *virtualizable* if it is related (in the sense of Section 2.7) to a twist of a virtual crystal. We call $\mathcal{B}$ a *normal crystal* if it is either a Stembridge crystal or a virtualizable crystal.

Table 5.7   Fundamental crystals for type G_2.

fundamental crystal	construction	dimension
$\mathcal{B}_{\varpi_1}$	virtual	7
$\mathcal{B}_{\varpi_2}$	$\mathcal{B}_{\varpi_1} \otimes \mathcal{B}_{\varpi_1}$	14

In [Kashiwara (1995)], a normal crystal is defined to be one isomorphic to the crystal basis of a representation of a quantized enveloping algebra. This class of crystals has the properties asserted in Theorem 5.20 below, which characterizes the class of normal crystals, and so his class of normal crystals is equivalent to ours. (In [Kashiwara (1993)] the term normal is used to mean what in [Kashiwara (1995)] is called seminormal.)

We will later show that the characters of a connected normal crystals with given highest weight vector equals the character of the corresponding irreducible representation of a Lie group (see Corollary 13.9).

Theorem 5.20. *The class of normal crystals is closed under tensor products and twisting. Every connected crystal in the class has a unique highest weight element, and the highest weights give a bijection between dominant weights and connected crystals in the class.*

Proof. By Theorems 4.10 and 5.8 normal crystals are closed under tensor products. By the discussions of Section 2.7, normal crystals are closed under twisting. Theorems 4.12 and 5.9 show that each connected normal crystal has a unique highest weight element. By Theorems 4.13 and 5.10, connected components of normal crystals are uniquely characterized by their highest weight.

It remains to show that there exists a connected normal crystal for every dominant weight.

First we note that by the results of Sections 5.4 and 5.6 there is a "fundamental" crystal $\mathcal{B}_{\varpi_i}$ that is a connected normal crystal with highest weight ϖ_i for each fundamental weight ϖ_i with $i \in I$. Indeed, it is easy to check that the standard crystals $\mathcal{B}_{\varpi_1}$ of types A_r and D_r are Stembridge crystals. We checked in Section 5.2 that the standard crystals $\mathcal{B}_{\varpi_1}$ of types B_r and C_r can be realized as virtual crystals. By Proposition 5.13 the type D_r spin crystals are Stembridge crystals and by Proposition 5.14 the type B_r spin crystal can be realized as a virtual crystal. Hence for all nonexceptional types, the standard and spin crystals are normal. All other fundamental crystals of nonexceptional types can be obtained as connected components in tensor products of the standard by Proposition 5.11 and are hence all normal. We have also exhibited fundamental crystals for the exceptional types in Section 5.6 and showed that they are either Stembridge or virtualizable.

Now suppose that λ is an arbitrary dominant weight. Let λ' be the weight

$$\lambda' = \sum_{i \in I} k_i \varpi_i, \qquad k_i = \langle \lambda, \alpha_i^\vee \rangle.$$

Then there is a normal crystal with highest weight λ', namely if u_{ϖ_i} is the highest weight element of the fundamental crystal $\mathcal{B}_{\varpi_i}$, then $\bigotimes_{i\in I}\left(u_{\varpi_i}^{\otimes k_i}\right)$ is a highest weight crystal in a tensor product of normal crystals with weight λ'. Call this crystal $\mathcal{B}_{\lambda'}$.

Now $\theta = \lambda - \lambda'$ is orthogonal to all the coroots. We may therefore twist $\mathcal{B}_{\lambda'}$ by θ and obtain a normal crystal with highest weight λ. (See Section 2.7.) $\qquad\square$

Theorem 5.21 ([Kang, Kashiwara, Misra, Miwa, Nakashima and Nakayashiki (1992a, Proposition 2.4.4)]). *Let $\mathcal{C}$ be a finite crystal associated to the root system Φ. Assume that for every pair i, j of indices the crystal $\mathcal{C}$ is normal as a crystal for the rank two root system generated by α_i and α_j. Then $\mathcal{C}$ is a normal crystal.*

We will describe the hypothesis on $\mathcal{C}$ as *2-normality*, so this result may be expressed as saying that 2-normality implies normality.

Proof. We will make use of the obvious fact that a 2-normal crystal is seminormal.

Let u_λ be a highest weight element with weight λ. Let $\mathcal{C}'$ be the connected normal crystal with highest weight λ, and let u'_λ be the unique highest weight element of $\mathcal{C}'$ which exists by Theorem 5.20. We consider pairs (Ω, Ω') of subsets of $\mathcal{C}$ and $\mathcal{C}'$ respectively, such that $u_\lambda \in \Omega$ and $u'_\lambda \in \Omega'$ with a weight-preserving bijection $x \mapsto x'$ of Ω onto Ω' satisfying the following conditions. First, $u_\lambda \in \Omega$ and $x \mapsto x'$ takes u_λ to u'_λ. Second, if $x \in \Omega$ and $e_i x \neq 0$, then $e_i x \in \Omega$, and moreover $(e_i x)' = e_i x'$. Such pairs exist, since we could take $\Omega = \{u_\lambda\}$ and $\Omega' = \{u'_\lambda\}$. We assume that Ω is maximal with respect to this property. We will show that $\Omega' = \mathcal{C}'$. This shows that every highest weight element u_λ is contained in a connected subcrystal Ω of $\mathcal{C}$ that is isomorphic to a normal crystal, and since every connected component of $\mathcal{C}$ contains a highest weight element, this proves that $\mathcal{C}$ is normal.

If Ω' is a proper subset of $\mathcal{C}'$, let w' be a maximal element of its complement. We will define an element w of $\mathcal{C} \setminus \Omega$ such that we can enlarge Ω to $\Omega \cup \{w\}$ and Ω' to $\Omega' \cup \{w'\}$. This contradicts the assumed maximality of Ω.

Since $w' \notin \Omega'$, we have $w' \neq u'_\lambda$. Hence there is an index i such that $e_i w' \neq 0$. The assumed maximality of w' implies that $e_i w' \in \Omega'$. Thus $e_i w' = y'$ for some $y \in \Omega$. Now we claim that $\varphi_i(y) = \varphi_i(y')$. Indeed, since $\mathrm{wt}(y) = \mathrm{wt}(y')$, it is sufficient to show $\varepsilon_i(y) = \varepsilon_i(y')$. Now $y \in \Omega$ and $y' \in \Omega'$, so $e_i^k(y) \in \Omega$ and $e_i^k(y') \in \Omega'$ for all $k \geqslant 0$ as long as $e_i^k(y) \neq 0$, and in this case $e_i^k(y') = \left(e_i^k(y)\right)'$. Both crystals are seminormal so $\varepsilon_i(y)$, being the number of times e_i may be applied, equals $\varepsilon_i(y')$.

Now because $f_i(y') = w' \neq 0$, we see that $\varphi_i(y) = \varphi_i(y') > 0$ and therefore $y = e_i(w)$ for some $w \in \mathcal{C}$. We need to show that w does not depend on the choice of index i. So suppose that j is another index such that $e_j(w') \neq 0$. Arguing as with i shows that $e_j(w') = z'$ where $\varphi_j(z) = \varphi_j(z') > 0$. We need to show that $w = f_j(z)$. We restrict both crystals $\mathcal{C}$ and $\mathcal{C}'$ to the rank two root system $\Phi_{i,j}$ generated by α_i and α_j. Let $\mathcal{C}'_0$ be the connected component of $\mathcal{C}'$ containing w' when restricted to $\Phi_{i,j}$. Then $\mathcal{C}'_0$ contains both y' and z'. If v' is the highest weight element of $\mathcal{C}'_0$, then $v' \succcurlyeq y'$ and so $v' \in \Omega'$. Let v be the element of Ω that it corresponds to, and let $\mathcal{C}_0$ be the $\Phi_{i,j}$ connected component that contains v. Then $e_i v' = e_j v' = 0$ implies that $e_i v = e_j v = 0$. By hypothesis, $\mathcal{C}_0$ is a normal rank two crystal. Since $\mathcal{C}_0$ and $\mathcal{C}_0'$ are highest weight crystals whose highest weight elements v and v' have the same weight, and since both are normal, they are isomorphic. The isomorphism must agree with $x \mapsto x'$ on Ω. Because $f_j z' = f_i y' = w'$, it follows that $f_j z = f_i y = w$, as required.

Now we have found a way of extending the map $x \mapsto x'$ to a larger set $\Omega \cup \{w\}$, contradicting the assumed maximality of Ω. This proves that Ω is a subcrystal of $\mathcal{C}$ that is isomorphic to a normal crystal. Because it is seminormal, it is a connected component of $\mathcal{C}$. We have shown that every highest weight element of $\mathcal{C}$ is contained in a normal subcrystal, so $\mathcal{C}$ is normal. $\qquad\square$

5.8 Reducible Cartan types

For many purposes we may restrict ourselves to the study of irreducible (simple) root systems. Yet even if we are only interested in crystals for simple root systems, reducible ones come up naturally as Levi subsystems. For example, if we consider the GL(4) weight lattice and root system of type A_3, then removing the middle node in the Dynkin diagram gives the GL(2) $\times$ GL(2) root system, corresponding to the $A_1 \times A_1$ reducible Cartan type.

Suppose that Λ_1 and Λ_2 are weight lattices containing root systems Φ_1 and Φ_2, respectively. We may form the direct sum $\Lambda = \Lambda_1 \oplus \Lambda_2$, with $\Phi = \Phi_1 \cup \Phi_2$ a root system inside it. The Weyl group W of Φ is clearly the direct product of the Weyl groups W_1 and W_2 of Φ_1 and Φ_2. The index set I will be chosen to be the disjoint union of the index sets I_1 and I_2 of Φ_1 and Φ_2, respectively. We choose the Weyl group invariant inner product on Λ, so that the subspaces Λ_1 and Λ_2 are orthogonal. Other reducible root systems occur, but they will be isogenous to one of this type (Section 2.7).

Now suppose that $\mathcal{B}_1$ and $\mathcal{B}_2$ are crystals associated to the root systems Φ_1 and Φ_2. We may define a crystal $\mathcal{B}_1 \boxtimes \mathcal{B}_2$ associated to $\Phi = \Phi_1 \cup \Phi_2$. As a set, it is the Cartesian product of $\mathcal{B}_1$ and $\mathcal{B}_2$. If $x_i \in \mathcal{B}_i$ we denote the ordered pair (x_1, x_2) as $x_1 \boxtimes x_2$. We define $\mathrm{wt}(x_1 \boxtimes x_2) = \mathrm{wt}(x_1) + \mathrm{wt}(x_2)$, where we interpret this equation by the natural embeddings $\Lambda_1 \hookrightarrow \Lambda$ and $\Lambda_2 \hookrightarrow \Lambda$. For $i \in I$

$$\varphi_i(x_1 \boxtimes x_2) = \begin{cases} \varphi_i(x_1) & \text{if } i \in I_1, \\ \varphi_i(x_2) & \text{if } i \in I_2, \end{cases} \qquad f_i(x_1 \boxtimes x_2) = \begin{cases} f_i(x_1) \boxtimes x_2 & \text{if } i \in I_1, \\ x_1 \boxtimes f_i(x_2) & \text{if } i \in I_2, \end{cases}$$

and similarly for ε_i and e_i.

Suppose that Φ_1 and Φ_2 are simply-laced. We may normalize the inner product on Λ so that the roots of Φ_1 have the same length as the roots on Φ_2, and so we consider the root system Φ to be simply-laced. The results on Stembridge crystals are therefore applicable. More generally, the results on normal crystals apply. The virtual crystal construction is easily adapted to reducible Cartan types and Theorem 5.10 still holds.

Proposition 5.22. *With notation as above, any normal crystal $\mathcal{B}$ type Φ is of the form $\mathcal{B}_1 \boxtimes \mathcal{B}_2$, where $\mathcal{B}_1$ and $\mathcal{B}_2$ are normal crystals for Φ_1 and Φ_2.*

Proof. If λ is a highest weight of $\mathcal{B}$, then λ is dominant. Hence $\lambda = \lambda_1 + \lambda_2$ where λ_i is a dominant weight in Λ_i for $i = 1, 2$. We may find normal crystals $\mathcal{B}_{\lambda_i}$ with the given highest weights and then consider $\mathcal{B}_1 \boxtimes \mathcal{B}_2$. This has the same highest weight as the component of weight λ in $\mathcal{B}$, and so they are isomorphic by Theorem 5.20. $\quad\square$

5.9 Similarity of crystals

Let $\mathcal{B}_\lambda$ denote the normal crystal of highest weight λ for a fixed root system Φ with weight lattice Λ.

Proposition 5.23. *Let λ and μ be dominant weights in Λ. Then there exists an embedding*

$$\mathcal{B}_{\lambda+\mu} \hookrightarrow \mathcal{B}_\lambda \otimes \mathcal{B}_\mu.$$

Proof. Let u_λ and u_μ be highest weight elements of $\mathcal{B}_\lambda$ and $\mathcal{B}_\mu$, respectively. Then $u_\lambda \otimes u_\mu$ is a highest weight element in $\mathcal{B}_\lambda \otimes \mathcal{B}_\mu$, that is, $\varepsilon_i(u_\lambda \otimes u_\mu) = 0$ for all $i \in I$ by (2.10). The connected component of $\mathcal{B}_\lambda \otimes \mathcal{B}_\mu$ containing $u_\lambda \otimes u_\mu$ is a normal crystal with highest weight $\lambda + \mu$, and by Theorem 5.20 it is isomorphic to $\mathcal{B}_{\lambda+\mu}$. $\quad\square$

Now let us fix an integer n. Suppose that there exists a map $S \colon \mathcal{B}_\lambda \to \mathcal{B}_{n\lambda}$ with the following properties for $v \in \mathcal{B}_\lambda$ and all $i \in I$:

$$\mathrm{wt}(S(v)) = n\,\mathrm{wt}(v), \qquad \varphi_i(S(v)) = n\varphi_i(v), \qquad \varepsilon_i(S(v)) = n\varepsilon_i(v) \qquad (5.8)$$

and

$$S(e_i v) = e_i^n S(v), \qquad S(f_i v) = f_i^n S(v). \qquad (5.9)$$

We will call such a map a *similarity map of degree n*. It clearly has to take the highest weight element of $\mathcal{B}_\lambda$ to the highest weight element of $\mathcal{B}_{n\lambda}$, and from this it is easy to deduce that if a similarity map exists, it is unique.

[Kashiwara (1996)] proved the existence of similarity maps for all normal crystals and all Cartan types (including those for infinite-dimensional Kac–Moody Lie algebras). He applied it to proving that the crystals obtained by the Littelmann path method ([Littelmann (1995b)]) agree with those coming from crystal bases of representations.

Proposition 5.24. *Suppose that*

$$S_\lambda \colon \mathcal{B}_\lambda \to \mathcal{B}_{n\lambda}, \qquad S_\mu \colon \mathcal{B}_\mu \to \mathcal{B}_{n\mu}$$

are similarity maps. Then there exists a similarity map for $\mathcal{B}_{\lambda+\mu}$. Identifying $\mathcal{B}_{\lambda+\mu}$ with a subcrystal of $\mathcal{B}_\lambda \otimes \mathcal{B}_\mu$ as in Proposition 5.23, the similarity map on $\mathcal{B}_{\lambda+\mu}$ is defined by

$$S_{\lambda+\mu}(v \otimes w) = S_\lambda(v) \otimes S_\mu(w).$$

The proof is very similar to that of Proposition 5.8.

Proof. Abbreviating $S_{\lambda+\mu}$ as S, we have

$$\varphi_i(S(u \otimes v)) = \varphi_i(S(u) \otimes S(v)) = \max\{\varphi_i(S(u)), \varphi_i(S(u)) + \varphi_i(S(v)) - \varepsilon_i(S(u))\}$$
$$= \max\{n\varphi_i(u), n\varphi_i(u) + n\varphi_i(v) - n\varepsilon_i(u)\} = n\varphi_i(u \otimes v).$$

This proves the middle equation in (5.8); the last equation is similar, and the first is straightforward.

We must prove (5.9). Applying (2.10) n times to $S(v \otimes w)$ gives

$$f_i^n(S(v) \otimes S(w)) = \begin{cases} f_i^n(S(v)) \otimes S(w) & \text{if } \varphi_i(S(w)) \leqslant \varepsilon_i(S(v)), \\ v \otimes f_i^n(S(w)) & \text{if } \varphi_i(S(w)) \geqslant \varepsilon_i(S(v)) + n. \end{cases}$$

The intermediate cases

$$\varepsilon_i(S(v)) < \varphi_i(S(w)) < \varepsilon_i(S(v)) + n$$

cannot occur since $\varepsilon_i(S(v)) = n\varepsilon_i(v)$ and $\varphi_i(S(w)) = n\varphi_i(w)$ are multiples of n. Therefore

$$f_i^n(S(v \otimes w)) = \begin{cases} S(f_i v \otimes w) & \text{if } \varphi_i(w) \leqslant \varepsilon_i(v), \\ S(v \otimes f_i w) & \text{if } \varphi_i(w) \geqslant \varepsilon_i(v) + 1, \end{cases}$$

which means that $S(f_i(v \otimes w)) = f_i^n S(v \otimes w)$. This proves part of (5.9), and the other part is similar.   □

Proposition 5.25. *If the Cartan type is A_r, then $\mathcal{B}_\lambda$ has a similarity map.*

Proof. Let $\mathbb{B}$ be the standard $\mathrm{GL}(r+1)$ crystal. Then $\mathcal{B}_\lambda$ is related (as in Section 2.7) to a subcrystal of $\otimes^k \mathbb{B}$ for some k. Hence it is enough to show that $\mathbb{B}$ admits a similarity map to $\mathcal{B}_{n\varpi_1}$, which is the crystal of rows $\mathcal{B}_{(n)}$. In fact, it is easy to see that

$$\boxed{i} \mapsto \boxed{i}\,\cdots\,\boxed{i}$$

is a similarity map.   □

5.10   Levi branching of normal crystals

In this section we will prove that the class of normal crystals is preserved under Levi branching. For the simply-laced types, this is clear: the Stembridge axioms are clearly preserved under Levi branching. Therefore we must consider the case of virtual crystals. We must slightly expand the class of virtual crystals that we described in Section 5.2 to include examples where the simply-connected ambient crystal has a reducible Cartan type, and similarity of crystals is taken into account.

Example 5.26. *We need to expand the concept of virtual crystals to include the following situation. Let $\mathcal{B}$ be a crystal of simply-laced Cartan type X, and let $Y = X \times X$ be the corresponding reducible Cartan type. Then Y is also a simply-laced type. Let $I = I^X$ be the index set for X. The index set I^Y for Y is the disjoint union of two copies I_1 and I_2 of I. If $i \in I$, we will denote by i_1 and i_2 the corresponding elements of I_1 and I_2. We define a map $\sigma \colon I \longrightarrow (I_1 \cup I_2)/\operatorname{aut}$ by $\sigma(i) = \{i_1, i_2\}$, where aut is the automorphism of I^Y that interchanges I_1 and I_2. We will also take $\gamma_i = 1$ for all $i \in I$. Let $\widehat{V} = \mathcal{B} \boxtimes \mathcal{B}$, a crystal of type Y. We are now in a situation similar to Section 5.2, except that the Cartan type X is simply-laced. We see that $V = \{u \boxtimes u \mid u \in \mathcal{B}\}$ is a virtual crystal inside $\widehat{V}$.*

Proposition 5.27. *The class of normal crystals is preserved by Levi branching.*

Proof. If the Cartan type is simply-laced, then it is obvious that a Stembridge crystal, when branched to a Levi subsystem, remains a Stembridge crystal, since the axioms will obviously still be satisfied.

If the root system is not simply-laced, then we must consider a virtual crystal. Let V be a virtual crystal inside $\widehat{V}$. Let X and Y be the corresponding Cartan types, with I^X and I^Y the corresponding index sets. Then if J^X is a subset of I^X, let J^Y be the preimage of J^X under the map $I^Y \to I^X$ described in Section 5.1. Then corresponding to the subsets J^X and J^Y, we have Levi branchings to Levi Cartan types. It is easy to see that under branching, V becomes a virtual crystal inside $\widehat{V}$. Since the latter branched Stembridge crystal is Stembridge, it follows that the branched crystal of V is normal. $\qquad\square$

Let us do an example. A maximal Levi root system for type C_r is obtained by removing one node from the Dynkin diagram. The type will usually be reducible, of type $A_k \times C_{r-k-1}$. Thus, let us consider C_6, which, we recall, is embedded into A_{11}. The setup is shown in Figure 5.9.

Let us remove the third node from the Dynkin diagram and obtain a reducible Dynkin diagram of type $A_2 \times C_3$. Recalling that C_6 is embedded in A_{11}, the Levi type $A_2 \times C_3$ is embedded in the Levi subtype of A_{11} obtained by removing the two nodes in $\sigma(3) = \{3, 9\}$. This is a reducible Cartan type $A_2 \times A_5 \times A_2$. This setup is shown in Figure 5.10.

Branching the C_6 virtual crystal V inside the A_{11} virtual crystal $\widehat{V}$ to $A_2 \times C_3$ can be understood as follows. Branching $\widehat{V}$ to $A_2 \times A_5 \times A_2$ gives a Stembridge crystal that is stable under the automorphism aut corresponding to the graph automorphism of the A_{11} Dynkin diagram. Therefore, for each irreducible constituent $\widehat{V}_i$ of the $A_2 \times A_5 \times A_2$ branched crystal, we may find A_2 and A_5 crystals $\mathcal{B}_{A_2}$ and $\mathcal{B}_{A_5}$ so that $\widehat{V} \cong \mathcal{B}_{A_2} \boxtimes \mathcal{B}_{A_5} \boxtimes \mathcal{B}_{A_2}$. Now we have a virtual crystal $\mathcal{W}_{A_2}$ inside $\widehat{\mathcal{W}} = \mathcal{B}_{A_2} \boxtimes \mathcal{B}_{A_2}$, which is the setup of Example 5.26. Similarly we have a C_3 virtual crystal $\mathcal{U}$ inside $\widehat{\mathcal{U}} = \mathcal{B}_{A_5}$. Then inside $\widehat{\mathcal{W}} \boxtimes \widehat{\mathcal{U}}$ we have a virtual $A_2 \times C_3$ crystal $\mathcal{W} \boxtimes \mathcal{U}$, and the branched crystal is a union of such crystals. Hence the branched crystal is normal.

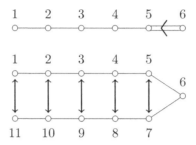

Fig. 5.9   The C_6 and A_{11} Cartan types.

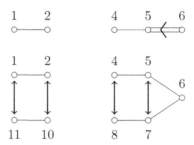

Fig. 5.10   The $A_2 \times C_3$ and $A_2 \times A_5 \times A_2$ Cartan types.

The problem of *non-Levi branching* of crystals seems to be an interesting question. For example, any Lie group G is embedded in $G \times G$ along the diagonal and the branching rule is just the tensor product of representations. So Kashiwara's tensor product rule (Section 2.3) is an example of a non-Levi branching rule. There is no general theory of non-Levi branching for crystals yet, but see [Naito and Sagaki (2005); Schumann and Torres (2016)] for another example.

Exercises

Exercise 5.1.

(i) Recall that the Weyl group W of a root system has a *long element* w_0 of order 2 such that $w_0(\Phi^+) = \Phi^-$ where Φ^+ and Φ^- are the positive and negative roots, respectively. Show that there exists a permutation $i \mapsto i'$ of the index set I such that

$$w_0(\alpha_i) = -\alpha_{i'} .$$

(ii) Show for type A_{r-1} (that is, $GL(r)$ or $SL(r)$) that $i' = r - i$.

(iii) Determine the map $i \mapsto i'$ for Cartan types B_r, C_r and D_r. Be careful in case D_r since the answer may depend on the parity of r.

Let $\mathcal{C}$ be a connected, finite, normal crystal. We say that $\mathcal{C}$ has a *crystal involution* if there is a map $S: \mathcal{C} \to \mathcal{C}$ such that

$$\mathrm{wt}(Sx) = w_0 \, \mathrm{wt}(x)$$

$$e_i(Sx) = f_{i'}(x), \qquad f_i(Sx) = e_{i'}(x),$$

$$\varepsilon_i(Sx) = \varphi_{i'}(x), \qquad \varphi_i(Sx) = \varepsilon_{i'}(x).$$

For crystals of tableaux in type A, an involution on the set of tableaux of given shape was constructed by [Schützenberger (1972)] using *jeu de taquin*, and it is therefore called the *Schützenberger involution*. See also [Edelman and Greene (1987); Kirillov and Berenstein (1995)]. For general Cartan types, the involution is sometimes called the *Lusztig involution*. See [Lusztig (1990b); Lenart (2007)].

Exercise 5.2. Let $\mathcal{C}$ be a connected normal crystal with highest weight λ. Let w_0 and $i \mapsto i'$ be as in Exercise 5.1. Define a new crystal structure on $\mathcal{C}$ as follows. The weight function $\mathrm{wt}': \mathcal{C} \to \Lambda$ is the map $\mathrm{wt}'(x) = w_0 \, \mathrm{wt}(x)$, and we also define

$$e_i'(x) = f_{i'}(x), \qquad f_i'(x) = e_{i'}(x),$$

$$\varepsilon_i'(x) = \varphi_{i'}(x), \qquad \varphi_i'(x) = \varepsilon_{i'}(x).$$

(i) Show that this makes $\mathcal{C}$ into a normal crystal. What is its highest weight?
(ii) Prove that a connected normal crystal has a unique crystal involution.
   Hint: Use Theorem 5.20.
(iii) Compute the involutions for the GL(3) crystals $\mathcal{B}_\lambda$ when $\lambda = (1,0,0)$, $\lambda = (2,0,0)$ and $\lambda = (2,1,0)$.

Exercise 5.3. Show that if $\lambda \in \Lambda^+$ is minuscule, then all weights in the W-orbit of λ are minuscule.

Exercise 5.4. Let $\mathcal{C}$ be a minuscule crystal associated to the simply-laced root system Φ and let α_i and α_j be two simple roots that are not orthogonal. Use the Stembridge axioms to show that there is no $x \in \mathcal{C}$ such that $\varepsilon_i(x) > 0$ and $\varepsilon_j(x) > 0$.

Exercise 5.5. In Proposition 5.25, we saw that type A normal crystals have similarity maps. Prove this for types B, C and D.

Chapter 6

Crystals of Tableaux II

In Chapter 3 we showed how to construct crystals of tableaux for $GL(n)$ crystals. In this chapter we generalize this to arbitrary classical types. We achieve this by first constructing crystals of tableaux for all fundamental crystals of Section 5.4 in terms of single columns and then combine them to arbitrary weights. This chapter may be skipped without serious loss of continuity.

The normal crystals for a particular Cartan type can be embedded in tensor powers of the standard crystal $\mathbb{B}$, except for the spin crystals in types B_r and D_r. We have already seen in type A_r that there may be more than one such embedding. (See Figure 3.3.) So one may approach the crystals of tableaux of a given Cartan type by embedding them into $\mathbb{B}^{\otimes k}$; with a modification, this approach will even account for the spin crystals.

For $GL(n)$ crystals, there can be many embeddings of $\mathcal{B}_\lambda$ into $\mathbb{B}^{\otimes k}$; these embeddings will be studied in Chapter 8. Two particular ways concern us here: the *row reading* which was the basis of the theory in Chapter 3, and the *column reading* that we will use in this chapter. These two readings coincide if the shape is a row or column. In either the row reading or the column reading, a row, consisting of a tableau

$$\boxed{j_1}\,\boxed{j_2}\,\cdots\,\boxed{j_k}\,,\qquad j_1 \leqslant \cdots \leqslant j_k$$

is embedded as

$$\boxed{j_1}\otimes\cdots\otimes\boxed{j_k}\,.$$

On the other hand, in either the row or column reading, a column

$$\begin{array}{c}\boxed{j_1}\\[-2pt]\boxed{\vdots}\\[-2pt]\boxed{j_k}\end{array}\,,\qquad j_1 < \cdots < j_k$$

is embedded as

$$\boxed{j_k}\otimes\cdots\otimes\boxed{j_1}\,.$$

This works, because the resulting subsets of $\mathbb{B}^{\otimes k}$ are closed under the crystal operations.

In the row reading, if $R_1, \ldots, R_\ell$ are the rows of a $\mathrm{GL}(n)$ tableau T ordered from top to bottom for a partition λ of k, we embed T in $\mathbb{B}^{\otimes k}$ as

$$R_\ell \otimes \cdots \otimes R_1,$$

where the rows are themselves embedded as already discussed. In the column reading, if $C_1, \ldots, C_m$ are the columns of T ordered from left to right, we embed T as

$$C_1 \otimes \cdots \otimes C_m.$$

Remarkably, with either the row or column reading a type $\mathrm{GL}(n)$ tableau embeds $\mathcal{B}_\lambda$ as a full connected subcrystal of $\mathbb{B}^{\otimes k}$, and the two embeddings give isomorphic subcrystals. We will prove this in Section 6.1.

As we already mentioned, applied to a row or column, the row and column readings give the same element of $\mathbb{B}^{\otimes k}$. Even in the case of the shape $\lambda = (2, 1)$, the row and column readings give the same embedding into $\mathbb{B}^{\otimes 3}$. (The second embedding in Figure 3.3 is neither the row nor column embeddding!)

The difference between the row and column readings appears with more complicated shapes, such as $(2, 2)$. For example, in the row reading, the tableau

$$\begin{array}{|c|c|} \hline 1 & 2 \\ \hline 3 & 4 \\ \hline \end{array}$$

is embedded as $\boxed{3} \otimes \boxed{4} \otimes \boxed{1} \otimes \boxed{2}$. In the column reading, it embeds as $\boxed{3} \otimes \boxed{1} \otimes \boxed{4} \otimes \boxed{2}$.

For Cartan types B_r, C_r and D_r, different theories of tableaux are possible. Such a theory associates a set of tableaux with every dominant weight, the number of tableaux being the dimension of the irreducible representation of the Lie group $\mathrm{SO}(2r + 1)$, $\mathrm{Sp}(2r)$ or $\mathrm{SO}(2r)$ in these respective cases. For symplectic tableaux, two essentially different classes of tableaux were defined by [De Concini (1979)] and [King (1976)]. Exhibiting a bijection between the two classes of tableaux is difficult, but was accomplished by [Sheats (1999)]. See also [Lecouvey (2002)].

[Kashiwara and Nakashima (1994)] described tableaux for all four classes A_r, B_r, C_r and D_r, with crystal graph structures for each. (For type C_r, the Kashiwara–Nakashima tableaux are related to the De Concini ones by a straightforward bijection.) In this chapter, we explain these *Kashiwara–Nakashima tableaux* and their crystal bases. See [Hong and Kang (2002)] for another account. Tableaux models for some exceptional types were studied by [Kang and Misra (1994); Jones and Schilling (2010)], the details of which we will not present here.

The Kashiwara–Nakashima tableaux may be embedded into tensor powers of the standard crystal by generalizing the column reading. Since in Chapter 3 we constructed the type A crystals by the row reading instead of the column reading, we first treat type A column reading in Section 6.1. We then proceed in Section 6.2 to columns of other classical nonexceptional types, and conclude in Section 6.3 with the general tableaux models.

6.1   Column reading in type A

In this section, we will consider $\mathrm{GL}(n)$ crystals again. Let $\mathbb{B}$ be the standard $\mathrm{GL}(n)$ crystal.

Theorem 6.1. *Let C be any connected Stembridge $\mathrm{GL}(n)$ crystal. Suppose that the highest weight vector of C has weight λ, where λ is partition. Then $C \cong \mathcal{B}_\lambda$.*

Proof. This follows from Theorem 4.13 since by Theorem 4.11, $\mathcal{B}_\lambda$ satisfies Stembridge's axioms.   □

Corollary 6.2. *Any connected component of $\mathbb{B}^{\otimes k}$ is isomorphic to $\mathcal{B}_\lambda$ for some $\lambda \vdash k$.*

Proof. Every full subcrystal of $\mathbb{B}^{\otimes k}$ satisfies Stembridge's axioms. It has a unique highest weight vector by Theorem 4.12. If λ is the corresponding weight, then λ is dominant by Proposition 2.16. Moreover, every element of $\mathbb{B}^{\otimes k}$ has weight of the form $(\lambda_1, \ldots, \lambda_n)$ with λ_i nonnegative integers whose sum is k. Since λ is dominant, it is a partition of k. The statement now follows from Theorem 6.1.   □

Dual to the row reading which we employed in Chapter 3, there is another map CR that serves to embed $\mathcal{B}_\lambda$ into $\mathbb{B}^{\otimes k}$. The main result of this section is that this *column reading* induces the same crystal structure on $\mathcal{B}_\lambda$ as the row reading RR.

If T is a row or column, then $\mathrm{CR}(T)$ agrees with $\mathrm{RR}(T)$ and is given by (3.2). For general tableaux, $\mathrm{CR}(T)$ is defined by reading each column of T in reverse order and taking the columns from left to right. Thus if

$$T = \begin{array}{|c|c|c|c|c|c|}\hline 1 & 1 & 2 & 2 & 2 & 4 \\ \hline \end{array}$$

(Tableau T with first row $1\,1\,2\,2\,2\,4$, second row $2\,3\,3$, third row 4.)

then

$$\mathrm{CR}(T) = \mathrm{CR}(C_1) \otimes \mathrm{CR}(C_2) \otimes \mathrm{CR}(C_3) \otimes \mathrm{CR}(C_4) \otimes \mathrm{CR}(C_5)$$

$$= \boxed{4} \otimes \boxed{2} \otimes \boxed{1} \otimes \boxed{3} \otimes \boxed{1} \otimes \boxed{3} \otimes \boxed{2} \otimes \boxed{2} \otimes \boxed{2} \otimes \boxed{4}.$$

Analogous to Theorem 3.2 we have:

Theorem 6.3. *Let λ be a partition of k of length $\leqslant n$. Then $\mathrm{CR}(\mathcal{B}_\lambda)$ is a connected component of $\mathbb{B}^{\otimes k}$. It has a unique highest weight element, namely $\mathrm{CR}(u_\lambda)$.*

Here the tableau u_λ was defined in Chapter 3: it is the tableau of shape λ in which every entry in the i-th row equals i.

Proof. Let $T \in \mathcal{B}_\lambda$. Assume that $f_i \, \mathrm{CR}(T) \neq 0$. We will show that $f_i \, \mathrm{CR}(T) = \mathrm{CR}(T')$ where T' is another tableau of shape λ.

Let $C_1, \ldots, C_m$ be the columns of T. Then

$$f_i \, \mathrm{CR}(T) = \mathrm{CR}(C_1) \otimes \cdots \otimes f_i \, \mathrm{CR}(C_j) \otimes \cdots \otimes \mathrm{CR}(C_m),$$

where j is the smallest integer such that

$$\varphi_i(C_1) + \varphi_i(C_2) + \cdots + \varphi_i(C_j) - \varepsilon_i(C_1) - \cdots - \varepsilon_i(C_{j-1})$$

takes its maximum. In particular, either $j = m$ or $\varphi_i(C_{j+1}) \leqslant \varepsilon_i(C_j)$. Now let T' be the tableau obtained by assembling the columns $C_1, \ldots, f_iC_j, \ldots, C_m$. We need to check that T' is semistandard. Since $f_iC_j \neq 0$, we know that C_j contains an entry i but no entry $i+1$. The tableau T' is obtained by changing the i in the j-th column C_j of T to $i+1$. The only way this could fail to be a semistandard tableau is if the corresponding entry in C_{j+1} is also equal to i. Thus $\varphi_i(C_{j+1}) = 1$. Since $\varphi_i(C_{j+1}) \leqslant \varepsilon_i(C_j)$ we see that $\varepsilon_i(C_j) = 1$, meaning that C_j contains an entry $i+1$. This is a contradiction, so T' is semistandard, and clearly $\mathrm{CR}(T') = f_i\,\mathrm{CR}(T)$.

Similarly, if $e_i\,\mathrm{CR}(T) \neq 0$, then $e_i\,\mathrm{CR}(T) = \mathrm{CR}(T'')$, where T'' is another semistandard tableau of shape λ. We leave the verification of this to the reader. We have shown that $\mathrm{CR}(\mathcal{B}_\lambda) \sqcup \{0\}$ is closed under the f_i. Similarly, the reader may check that it is closed under the e_i. Thus $\mathrm{CR}(\mathcal{B}_\lambda)$ is a full subcrystal of $\mathbb{B}^{\otimes k}$.

We leave it to the reader to check that u_λ is the unique highest weight vector in $\mathrm{CR}(\mathcal{B}_\lambda)$. As with Theorem 3.2, this implies that $\mathrm{CR}(\mathcal{B}_\lambda)$ is a connected crystal, and we are done. $\qquad\square$

Theorem 6.4. *Let $\lambda \vdash k$ be a partition of length $\leqslant n$. Then the crystal structures induced on $\mathcal{B}_\lambda$ by the row and column readings are the same.*

Proof. This follows from Theorems 6.1 and 6.3. $\qquad\square$

We have noted that for the crystals of rows and columns the embeddings RR and CR are the same. More generally, the two readings are the same for any shape whose Young diagram is of "hook" shape, that is, a partition of the form $(p, 1^q)$. Thus for the shape $\lambda = (2, 1)$, the row and column readings are actually the same map

$$\begin{array}{|c|c|}\hline a & b \\\hline c \\\cline{1-1}\end{array} \mapsto \boxed{c} \otimes \boxed{a} \otimes \boxed{b}\,.$$

This means that, referring to Figure 3.3, the second crystal $\mathcal{B}'_{(2,1)}$ does not correspond to either row or column readings. This shows that the row and column readings do not exhaust the possible embeddings of $\mathcal{B}_\lambda$ in $\mathbb{B}^{\otimes k}$. We will see later in Chapter 8, in particular in Theorem 8.7, that there is a nice way to understand all the different embeddings of the crystal $\mathcal{B}_\lambda$ into $\mathbb{B}^{\otimes k}$ via plactic relations.

6.2 Crystals of columns

In this section we give tableaux models for the fundamental crystals $\mathcal{B}_{\varpi_k}$ of Section 5.4 for types B_r, C_r, D_r. The spin crystals $\mathcal{B}_{\varpi_k}$ for $k = r$ in type B_r and $k = r-1, r$ in type D_r were already constructed explicitly in Section 5.4. Here we construct all others using the embedding into the k-fold tensor product of the standard crystal $\mathbb{B}^{\otimes k}$ using Proposition 5.11.

Recall the standard crystals for types B_r, C_r, and D_r from Examples 2.22, 2.23, and 2.24, respectively. Let

$$\mathcal{A}_{B_r} = \{1 \prec 2 \prec \cdots \prec r \prec 0 \prec \overline{r} \prec \cdots \prec \overline{2} \prec \overline{1}\}$$

$$\mathcal{A}_{C_r} = \{1 \prec 2 \prec \cdots \prec r \prec \overline{r} \prec \cdots \prec \overline{2} \prec \overline{1}\}$$

$$\mathcal{A}_{D_r} = \{1 \prec 2 \prec \cdots \prec r-1 \prec r, \overline{r} \prec \overline{r-1} \prec \cdots \prec \overline{2} \prec \overline{1}\}$$

be the corresponding (partially) ordered alphabets. Note that the alphabets for types B_r and C_r are in fact totally ordered, and only the letters r and $\overline{r}$ in type D_r are incomparable. Define a *column of height k* with $1 \leqslant k \leqslant r$ as a filling of the shape (1^k) with letters from the alphabet $\mathcal{A}_X$ in type $X \in \{B_r, C_r, D_r\}$ such that:

(1) The entries are strictly increasing from top to bottom with the exception that
   (a) the letter 0 in type B_r can be repeated, and
   (b) the letters r and $\overline{r}$ in type D_r can alternate.

(2) If both letters j and $\overline{j}$ appear in the column, and j is in the a-th box from the top whereas $\overline{j}$ is in the b-th box from the bottom, then $a + b \leqslant j$.

We call Condition 1 the type X-strictly increasing condition and Condition 2 the letter-j condition.

Example 6.5. *In type B_5, the following columns of height 4 satisfy Conditions 1 and 2*

$$\begin{array}{|c|}\hline 1 \\\hline 0 \\\hline 0 \\\hline \overline{2} \\\hline\end{array}, \quad \begin{array}{|c|}\hline 2 \\\hline 4 \\\hline 0 \\\hline \overline{2} \\\hline\end{array}, \quad \begin{array}{|c|}\hline 2 \\\hline 4 \\\hline \overline{4} \\\hline \overline{2} \\\hline\end{array}, \quad whereas \quad \begin{array}{|c|}\hline 2 \\\hline 4 \\\hline \overline{2} \\\hline \overline{1} \\\hline\end{array}$$

does not because it violates Condition 2 with $j = 2$ since $a = 1$, $b = 2$, but $a + b = 3 > j = 2$.

In type D_5, the following columns of height 3 satisfy Conditions 1 and 2

$$\begin{array}{|c|}\hline 5 \\\hline \overline{5} \\\hline 5 \\\hline\end{array}, \quad \begin{array}{|c|}\hline \overline{5} \\\hline 5 \\\hline \overline{5} \\\hline\end{array}, \quad whereas \quad \begin{array}{|c|}\hline 5 \\\hline \overline{5} \\\hline \overline{3} \\\hline\end{array},$$

does not since only multiple letters $r = 5$ can appear if they alternate with $\overline{r} = \overline{5}$.

Define $\mathcal{C}_k$ for $1 \leqslant k \leqslant r$ to be the set of columns of height k for type X. In addition, we define $\mathcal{C}_r^+$ (resp. $\mathcal{C}_r^-$) in type D_r to be the subset of $\mathcal{C}_r$ such that if r (or $\overline{r}$) appears in the j-th box from the top, then $r - j$ is even (or odd) (resp. odd or even).

Example 6.6. *For type D_4, the element*

$$\begin{array}{|c|}\hline 1 \\\hline 3 \\\hline \overline{4} \\\hline 4 \\\hline\end{array} \in \mathcal{C}_4^+, \quad whereas \quad \begin{array}{|c|}\hline 1 \\\hline \overline{4} \\\hline 4 \\\hline \overline{2} \\\hline\end{array} \in \mathcal{C}_4^-.$$

We embed $\mathcal{C}_k$ into $\mathbb{B}^{\otimes k}$ by the map

$$
\begin{array}{ccc}
\mathcal{C}_k & \longhookrightarrow & \mathbb{B}^{\otimes k} \\[6pt]
\boxed{\begin{array}{c} j_1 \\ \vdots \\ j_k \end{array}} & \longmapsto & \boxed{j_k} \otimes \cdots \otimes \boxed{j_1}.
\end{array}
\tag{6.1}
$$

We will prove momentarily that the image of this map is subcrystal, and so $\mathcal{C}_k$ inherits a crystal structure. Thus we call $\mathcal{C}_k$ for $1 \leqslant k \leqslant r$ the *column crystal* of height k for the Cartan type X.

The crystals $\mathcal{B}_\lambda$ in the following proposition are those constructed in Proposition 5.11.

Proposition 6.7. *The crystal $\mathcal{C}_k$ is connected for all $1 \leqslant k \leqslant r$, except for $k = r$ in type D_r. In addition, $\mathcal{C}_r^+$ and $\mathcal{C}_r^-$ are connected crystals in type D_r. Furthermore:*

(1) $\mathcal{C}_k \cong \mathcal{B}_{\varpi_k}$ for all $1 \leqslant k \leqslant r$ except for $k = r$ in type B_r and $k = r-1, r$ in type D_r.

(2) $\mathcal{C}_r \cong \mathcal{B}_{2\varpi_r}$ in type B_r.

(3) $\mathcal{C}_{r-1} \cong \mathcal{B}_{\varpi_{r-1}+\varpi_r}$ in type D_r.

(4) $\mathcal{C}_r^- \cong \mathcal{B}_{2\varpi_{r-1}}$ and $\mathcal{C}_r^+ \cong \mathcal{B}_{2\varpi_r}$ in type D_r.

Proof. We will prove this completely for Cartan types C_r and B_r, leaving the case D_r partly to the reader.

We begin by showing that $\mathcal{C}_k \sqcup \{0\}$ for $1 \leqslant k \leqslant r$ is closed under the crystal operators f_i and e_i for $i \in I = \{1, 2, \ldots, r\}$. We will only discuss the proof for f_i since the proof for e_i is similar.

When we apply f_i with $1 \leqslant i < r$ to a column $C \in \mathcal{C}_k$, if the result is nonzero, then either an i is changed to $i+1$ or $\overline{i+1}$ is changed to $\bar{i}$. In either case, we must verify that the result is still in $\mathcal{C}_k$. Let Σ be the intersection of the set $\{i, i+1, \overline{i+1}, \bar{i}\}$ with the entries in C. If C is identified with a tensor product via the column reading, the entries occur in descending order, and the application of f_i is determined by Lemma 2.33. The arguments we will give below are identical for type B_r, C_r with $1 \leqslant i < r$ and D_r with $1 \leqslant i \leqslant r-2$.

First let us suppose that f_i changes i to $i+1$. In this case, Lemma 2.33 implies that Σ is one of the following three possibilities:

$$
\{i\}, \qquad \{i, \overline{i+1}\}, \qquad \{i, \overline{i+1}, \bar{i}\}.
$$

If $\Sigma = \{i\}$ then changing i to $i+1$ to obtain $f_i C$ will not violate either Condition 1 nor 2. In the other two cases we must confirm that $f_i C$ does not violate Condition 2. Let a be the box from the top that contains i in C, and let b be the box from the bottom that contains $\overline{i+1}$. In $f_i C$, these boxes will contain $i+1$ and $\overline{i+1}$ so we need to argue that $a + b \leqslant i+1$. If $\Sigma = \{i, \overline{i+1}, \bar{i}\}$ then we may argue as follows. Since C satisfies Condition 2, and since the $\bar{i}$ is in the $(b-1)$-st box from the bottom, we have $a + b - 1 \leqslant i$ and, therefore, $a + b \leqslant i+1$. Thus we are left with the case

$\Sigma = \{i, \overline{i+1}\}$. If $a + b > i + 1$, then there are $a + b - 2 \geqslant i$ boxes that are above i in C or below $\overline{i+1}$. Since $\overline{i} \notin \Sigma$, the boxes above i or below $\overline{i+1}$ must contain entries x or $\overline{x}$ with $x \leqslant i - 1$, and by the pigeonhole principle, there must be both j and $\overline{j}$ for some $j \leqslant i - 1$. Let j be the largest such duplicate. Assume that j is in the c-th box from the top and that $\overline{j}$ is in the d-th box from the bottom. We will now count the number of boxes above the box containing i and below the box containing $\overline{i+1}$. We note that since C satisfies Condition 2 we must have $c + d \leqslant j$. There are $c + d - 2 \leqslant j - 2$ boxes above the one containing j or below the one containing $\overline{j}$. On the other hand, since j is the largest duplicate, for $j < x < i$ we can have a box containing x or $\overline{x}$ but not both. There are, therefore, at most $i - j - 1$ such entries. Finally there are the two boxes containing j and $\overline{j}$. We count at most $(j - 2) + (i - j - 1) + 2 = i - 1$ such boxes, which is a contradiction since as we have already noted there are $a + b - 2 \geqslant i$.

Next let us consider the case where f_i changes $\overline{i+1}$ to $\overline{i}$. In this case Lemma 2.33 implies that Σ equals one of

$$\{i, i+1, \overline{i+1}\}, \qquad \{i+1, \overline{i+1}\}, \qquad \{\overline{i+1}\}.$$

In the second two cases, it is straightforward that $f_i(C)$ satisfies both conditions. In the first case, we may argue as follows. Let i be in the a-th box from the top and $\overline{i+1}$ be in the b-th box from the bottom. We must show that $a + b \leqslant i$. Since C has $i+1$ in the $a+1$-st box from the top and $\overline{i+1}$ in the b-th box from the bottom, by Condition 2 we have $(a + 1) + b \leqslant i + 1$ and therefore $a + b \leqslant i$.

We must now consider the case where $i = r$ for type B_r, C_r and $i = r - 1, r$ for type D_r. In this case, it is necessary to consider the different Cartan types B_r, C_r and D_r separately.

The case of type C_r is straightforward: if $f_r C$ is nonzero, then C contains an r but not an $\overline{r}$, and $f_r C$ replaces the r by $\overline{r}$; both Conditions 1 and 2 are obvious.

In type $X = B_r$, if $f_r C \neq 0$, then we may deduce from Lemma 2.33 that C does not contain $\overline{r}$. If it contains an entry equal to r, then f_r changes that to 0; otherwise it may change a 0 to $\overline{r}$. But in no case can $f_r C$ contain both r and $\overline{r}$, so Condition 2 is not violated.

It remains to be shown that in type D_r, the set $\mathcal{C}_k \sqcup \{0\}$ is closed under f_{r-1} and f_r. This is left to the reader as an exercise, see Exercise 6.2.

Since $\mathcal{C}_k$ is closed under the crystal operations, it is a subcrystal of $\mathbb{B}^{\otimes k}$, and therefore it is a normal crystal. Thus its isomorphism class is determined by the weights of its highest weight vectors. These are easily found as follows. If (6.1) is a highest weight vector, then (2.15) gives a set of equalities that must be satisfied in order for ε_i applied to the column to vanish for all $i \in I$. Taking $j = 1$ in (2.15) gives $\varepsilon_i(j_1) = 0$ for all $i \in I$, and so $j_1 = 1$. Now assume by induction that $j_1 = 1, j_2 = 2, \ldots, j_\ell = \ell$ for some $1 \leqslant \ell < k$. Then by (2.15), we must have $j_{\ell+1} = \ell + 1$ or $j_{\ell+1} = \overline{\ell}$. However, note that $j_\ell = \overline{\ell}$ violates Condition 2. Thus there is exactly one highest weight vector

$$\boxed{\begin{matrix} j_1 \\ \vdots \\ j_k \end{matrix}} = \boxed{\begin{matrix} 1 \\ \vdots \\ k \end{matrix}} := u_{(1^k)}$$

with one exception: if $k = r$ in type D_r, the choice $\bar{r}$ does not violate Condition 2 and hence there exists a second vector with $\bar{r}$ replacing r. These two vectors give rise to the column crystals $\mathcal{C}_r^+$ and $\mathcal{C}_r^-$. The weight of the highest weight vector determines the structure of the crystal. $\qquad\square$

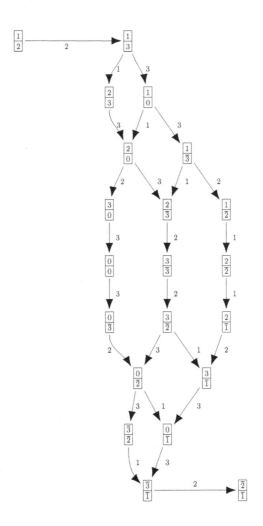

Fig. 6.1   The column crystal $\mathcal{C}_2$ of type B_3.

Example 6.8. *The column crystal $\mathcal{C}_2$ of type B_3 is depicted in Figure 6.1.*

6.3   Crystals of tableaux

As we have seen in Chapter 3, in type A_r the elements in the highest weight crystal $\mathcal{B}_\lambda$ can be represented by semistandard tableaux of shape λ. Recall that a semistandard tableau is weakly increasing along rows and strictly increasing along columns. Using the column reading of Section 6.1, a semistandard tableau can be viewed as the composition of its columns. In Section 6.2, we characterized the column tableaux in types B_r, C_r, and D_r. In this section, we will investigate how to characterize the connected components generated by the concatenation of highest weight columns. This will lead to the generalization of weak row increase beyond type A. The resulting tableaux are called *Kashiwara–Nakashima (KN) tableaux*.

6.3.1   *Crystal of tableaux: Type C_r*

Given two boxes in a tableau, we will define the *vertical distance* between the boxes to be $|p - q|$ where one box is in the p-th row and the other in the q-th row.

We start with type C_r. Let λ be a partition and let Tab_λ be the set of all tableaux of shape λ in the alphabet $\mathcal{A}_{C_r}$ such that each $T \in \text{Tab}_\lambda$ satisfies the following additional conditions:

C1. Each column of height k in T is in $\mathcal{C}_k$.
C2. Each row in T is weakly increasing.
C3. If T has two adjacent columns of either of the forms:

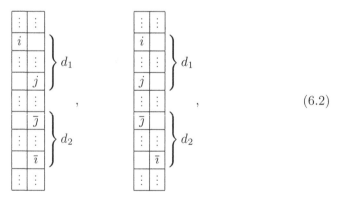

$$(6.2)$$

where $i \leqslant j$ and $d_1, d_2 \geqslant 0$ are the vertical distances between the boxes containing i, j, resp. $\bar{\imath}, \bar{\jmath}$, then we must have $d_1 + d_2 < j - i$.

Example 6.9. *Consider C_3 and the tableau*

$$
\begin{array}{|c|c|}
\hline
2 & 2 \\
\hline
3 & 3 \\
\hline
\bar{3} & \bar{2} \\
\hline
\end{array}
\, .
$$

This tableau is not in $\mathrm{Tab}_{(2,2,2)}$ *since with* $i = 2, j = 3$, *we have* $d_1 = 1, d_2 = 0$, *so that* $d_1 + d_2 = 1 \not< j - i = 1$.

On the other hand

$$
\begin{array}{|c|c|}
\hline
1 & \bar{1} \\
\hline
3 \\
\cline{1-1}
\bar{3} \\
\cline{1-1}
\end{array}
\quad \in \mathrm{Tab}_{(2,1,1)}
$$

since the first column is in $\mathcal{C}_3$, *the second column is in* $\mathcal{C}_1$, *the first row is weakly increasing, and none of the configurations* (6.2) *occur (note that for the second configuration the letter* $\bar{\imath}$ *needs to be weakly below* $\bar{\jmath}$ *which is not the case).*

Define a crystal structure on Tab_λ via the embedding into the tensor product of column crystals

$$
\begin{aligned}
\mathrm{Tab}_\lambda &\hookrightarrow \mathcal{C}_{k_1} \otimes \cdots \otimes \mathcal{C}_{k_\ell} \\
T &\mapsto C_1 \otimes \cdots \otimes C_\ell
\end{aligned}
\tag{6.3}
$$

where λ corresponds to the weight $\varpi_{k_1} + \cdots + \varpi_{k_\ell}$ with $r \geqslant k_1 \geqslant \cdots \geqslant k_\ell \geqslant 1$ and $T = C_1 \cdots C_\ell$ is the concatenation of the columns.

Theorem 6.10. *Let* $\lambda \in \Lambda$ *be a dominant weight in the weight lattice of type* C_r. *Then* $\mathrm{Tab}_\lambda \cong \mathcal{B}_\lambda$ *is isomorphic to the highest weight crystal of type* C_r *with highest weight* λ.

Proof. First note that if $T \in \mathrm{Tab}_\lambda$ contains two adjacent columns of either of the forms

then the condition $d_1 + d_2 < \max(j, k) - \min(i, \ell)$ must be satisfied. See Exercise 6.4.

To prove the assertion of the theorem, we need to show that Tab_λ is closed under the crystal operators e_i and f_i for $i \in I = \{1, 2, \ldots, r\}$. In addition, we need to show that if $T \in \mathrm{Tab}_\lambda$ is highest weight, that is, $e_i T = 0$ for all $i \in I$, then T is the concatenation of the columns $u_{(1^{h_1})}, \ldots, u_{(1^{h_{\lambda_1}})}$, where $h_1 \geqslant \cdots \geqslant h_{\lambda_1} > 0$ are the column heights of λ.

First we show that Tab_λ is closed under f_i. The proof for e_i is similar, see Exercise 6.5. By Proposition 6.7, each column of height k in $f_i T$ is in $\mathcal{C}_k$. Note that $f_i T$ differs from T either by changing a letter i to $i+1$ or $\overline{i+1}$ to $\overline{i}$. Since the crystal structure on T is defined by the column reading of T and the tensor product rule, the rightmost unbracketed such letter is changed. From this it is not hard to show that $f_i T$ is also weakly increasing in rows. Finally, consider condition (6.2). If f_i changes i to $i+1$, then potentially $f_i T$ could contain two adjacent columns of the form

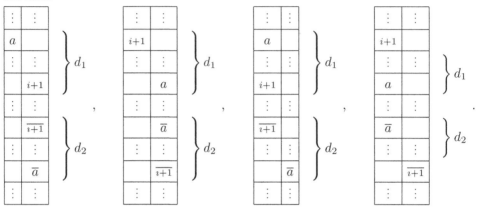

In this case, T must have been of the form

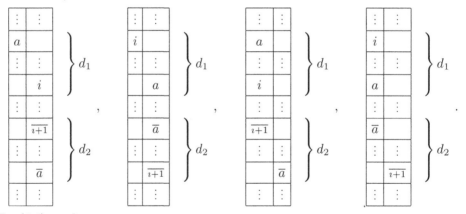

By (6.4), we have

$$d_1 + d_2 < i + 1 - a, \qquad (\text{resp. } d_1 + d_2 < a - i)$$

in the first and third (resp. second and fourth) cases. This implies that $f_i T$ satisfies (6.2).

The case when f_i changes a letter $\overline{i}$ to $\overline{i+1}$ is left as Exercise 6.5.

We conclude with the proof that if T is a highest weight vector in Tab_λ, then it is the concatenation of highest weight columns of the form $u_{(1^k)}$. By (2.15), the last column of T has to be highest weight, and since it is an element in $\mathcal{C}_k$ it must be of the form $u_{(1^k)}$ by Proposition 6.7. Now assume by induction that the last ℓ columns of T are of the form $u_{(1^k)}$ for some k. If the ℓ-th column from the right in T has height k, then by the weak row increase, the top k entries in the $(\ell+1)$-th column from the right have to be $12 \ldots k$. Again by (2.15) the next letter in the $(\ell + 1)$-th column is either $k + 1$ or $\bar{h}$, where h is the column height of one of the ℓ rightmost columns in T. If the letter is $\bar{h}$, then the letter-h condition of Section 6.2 is violated. Hence the letter has to be $k + 1$. Repeating the argument shows that the $(\ell + 1)$-th column from the right is also of the form $u_{(1^{k'})}$ for some k'. The statement follows. □

6.3.2 *Crystal of tableaux: Type B_r*

Let λ be a dominant weight in the weight lattice of type B_r. We can view λ as a partition with possibly one column of width $1/2$ of height r (this happens when λ, written as an integer sum of the fundamental weights ϖ_i, has an odd coefficient for ϖ_r). The column crystals C_k for $1 \leqslant k \leqslant r$ were defined in Section 6.2. The spin crystal $C_r^{\text{spin}} \cong B_{\varpi_r}$ was constructed in Proposition 5.14. Here we want to represent the elements in C_r^{spin} by a column of height r of width $1/2$ filled increasingly from top to bottom such that each letter $1, 2, \ldots, r$ appears precisely once, either barred or unbarred. Compared to the spin representation of Section 5.4, the letter i is barred if the entry in position i is $-$ and unbarred if the entry in position i is $+$.

Now Tab_λ is the set of all tableaux of shape λ in the alphabet $\mathcal{A}_{B_r}$ such that each $T \in \text{Tab}_\lambda$ satisfies the following additional conditions:

B1. Each column of full width of height $1 \leqslant k \leqslant r$ in T is in C_k. If there is a column of width $1/2$ and height r, it is in C_r^{spin}.

B2. Each row in T is weakly increasing, but the letter 0 cannot be repeated.

B3. Condition C3 holds for $1 \leqslant i \leqslant j < r$.

B4. If T has two adjacent columns of the form

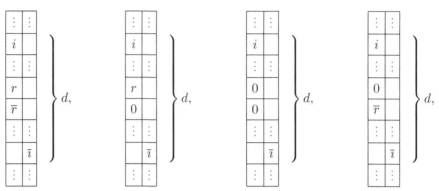

where $i \leqslant r$ and $d > 0$ is the vertical distance between the boxes containing i and $\bar{\imath}$, then we must have $d - 1 < r - i$.

Note that by Conditions B1, B2, and B4 with $i = r$ it follows that T cannot have two adjacent columns of the form

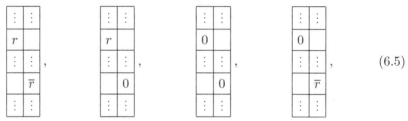
(6.5)

where the two entries are in two different rows.

Example 6.11. *In type B_3, the tableau*

3	3
0	$\bar{3}$

is not in $\mathrm{Tab}_{(2,2)}$ *since it violates* (6.5)*. In type B_5, we have*

2	2
3	0
4	0
5	2

$\notin \mathrm{Tab}_{(2,2,2,2)}$, *but*

1	2
3	0
4	0
5	$\bar{1}$

$\in \mathrm{Tab}_{(2,2,2,2)}$

since the left tableau violates B4 with $i = 2$ since $d = 4$ and hence $d - 1 = 3 \not< r - i = 5 - 2 = 3$. On the other hand, the right tableau does not violate B4: $d - 1 = 3 < r - 1 = 4$.

The conditions also apply to spin columns: in type B_5

2	2
3	5
4	$\bar{5}$
5	$\bar{2}$
$\bar{1}$	

$\notin \mathrm{Tab}_{(\frac{3}{2},\frac{3}{2},\frac{3}{2},\frac{3}{2},\frac{1}{2})}$, *where the shaded first column is a spin column of width $\frac{1}{2}$, since condition B4 is violated. Namely, for $i = 2$ we have $d = 4$, but $d - 1 = 3 \nleqslant r - i = 5 - 2 = 3$.*

Theorem 6.12. *Let $\lambda \in \Lambda$ be a dominant weight in the weight lattice of type B_r. Then $\mathrm{Tab}_\lambda \cong \mathcal{B}_\lambda$ is isomorphic to the highest weight crystal of type B_r with highest weight λ.*

Proof. See Exercise 6.6. $\qquad\qquad\qquad\qquad\qquad\qquad\qquad\qquad\qquad\qquad$ □

6.3.3 Crystal of tableaux: Type D_r

Let $\lambda = a_1 \varpi_1 + \cdots + a_r \varpi_r$ be a dominant weight in the weight lattice of type D_r. We can view λ as a partition with possibly one column of width $\frac{1}{2}$ of height r as follows: for each $1 \leqslant i < r - 1$, the partition has a_i columns of height i; in addition it has $\bar{a}_{r-1} := \min(a_{r-1}, a_r)$ columns of height $r - 1$,

$$\bar{a}_r = \lfloor \tfrac{1}{2}\big(\max(a_{r-1}, a_r) - \min(a_{r-1}, a_r)\big) \rfloor$$

columns of height r, and possibly one column of height r and width $\frac{1}{2}$ if $\max(a_{r-1}, a_r) - \min(a_{r-1}, a_r)$ is odd. Here $\lfloor \cdot \rfloor$ is the floor function. The column crystals $\mathcal{C}_k$ for $1 \leqslant k < r$ and $\mathcal{C}_r^-$ and $\mathcal{C}_r^+$ were defined in Section 6.2. The columns of height r are associated to $\mathcal{C}_r^-$ (resp. $\mathcal{C}_r^+$) crystals if $a_{r-1} > a_r$ (resp. $a_r > a_{r-1}$). In addition we have two spin crystals $\mathcal{C}_r^{\mathrm{spin}-} \cong \mathcal{B}_{\varpi_{r-1}}$ and $\mathcal{C}_r^{\mathrm{spin}+} \cong \mathcal{B}_{\varpi_r}$ as constructed in Section 5.4. Here we want to represent the elements in $\mathcal{C}_r^{\mathrm{spin}\pm}$ by a column of height r of width $\frac{1}{2}$ filled increasingly from top to bottom such that each letter $1, 2, \ldots, r$ appears precisely once, either barred or unbarred. In addition, for $\mathcal{C}_r^{\mathrm{spin}+}$ the letter r (resp. $\bar{r}$) appears at height h where $r - h$ is even (resp. odd), whereas for $\mathcal{C}_r^{\mathrm{spin}-}$ the letter r (resp. $\bar{r}$) appears at height h where $r - h$ is odd (resp. even). Compared to the spin representation of Section 5.4, the letter i is barred if the entry in position i is $-$ and unbarred if the entry in position i is $+$.

Now Tab_λ is the set of all tableaux of shape λ in the alphabet $\mathcal{A}_{D_r}$ such that each $T \in \mathrm{Tab}_\lambda$ satisfies the following additional conditions:

D1. Each column of height $1 \leqslant k < r$ in T is in $\mathcal{C}_k$. If $a_{r-1} > a_r$ (resp. $a_{r-1} < a_r$), the columns of height r are in $\mathcal{C}_r^-$ (resp. $\mathcal{C}_r^+$). Depending on whether $|a_{r-1} - a_r|$ is odd, the single spin column is in $\mathcal{C}_r^{\mathrm{spin}-}$ (resp. $\mathcal{C}_r^{\mathrm{spin}+}$).

D2. Each row in T is weakly increasing (and hence r and $\bar{r}$ cannot appear simultaneously since they are incomparable in $\mathcal{A}_{D_r}$).

D3. Condition C3 holds for $1 \leqslant i \leqslant j < r$.

D4. If T has two adjacent columns of the form

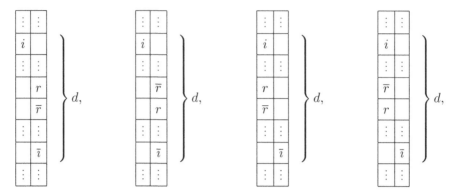

where $i \le r$ and $d > 0$ is the vertical distance between the boxes containing i and $\bar{\imath}$, then we must have $d - 1 < r - i$.

D5. T cannot have two adjacent columns of the form

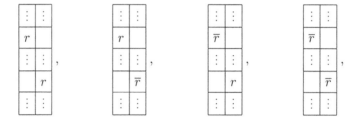

where the two entries are in two different rows.

D6. If T has two adjacent columns of the form

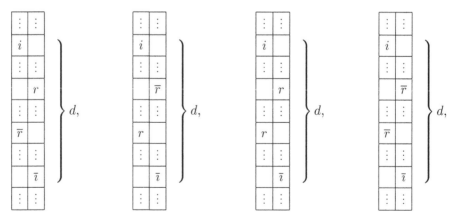

where $i < r$, $d > 0$ is the vertical distance between the boxes containing i and $\bar{\imath}$, and the vertical distance between r and $\bar{r}$ in the first two configurations is odd and the vertical distance between r and r (or $\bar{r}$ and $\bar{r}$) in the third (or fourth) configuration is even, then we must have $d < r - i$.

Example 6.13. *For type D_6, we have*

$$
\begin{array}{|c|c|}
\hline
1 & 2 \\
\hline
3 & 6 \\
\hline
\bar{6} & \bar{5} \\
\hline
\bar{3} & \bar{1} \\
\hline
\end{array}
\in \mathrm{Tab}_{(2,2,2,2)} \,, \qquad but \qquad
\begin{array}{|c|c|}
\hline
2 & 2 \\
\hline
3 & 6 \\
\hline
\bar{6} & \bar{5} \\
\hline
\bar{3} & \bar{2} \\
\hline
\end{array}
\notin \mathrm{Tab}_{(2,2,2,2)} \,,
$$

since in the second case D6 is violated with $i = 2$ because $d = 4 \nless r - i = 6 - 2 = 4$.

Theorem 6.14. *Let $\lambda \in \Lambda$ be a dominant weight in the weight lattice of type D_r. Then $\mathrm{Tab}_\lambda \cong \mathcal{B}_\lambda$ is isomorphic to the highest weight crystal of type D_r with highest weight λ.*

Proof. See Exercise 6.7. □

Exercises

Exercise 6.1. Take the column

$$
\begin{array}{|c|}
\hline
2 \\
\hline
4 \\
\hline
\bar{2} \\
\hline
\bar{1} \\
\hline
\end{array}
$$

for type B_5 of Example 6.5 and explicitly construct a sequence of crystal raising operators e_i that take it to a highest weight element. Which highest weight vector do you find? Try to do the exercise in SAGE as well.

Exercise 6.2. Let $\mathcal{C}_k$ for $1 \leqslant k \leqslant r$ be a column crystal of type $X \in \{B_r, C_r, D_r\}$ as defined in Section 6.2. Complete some of the arguments presented in the proof of Proposition 6.7:

(a) For $X = D_r$, prove that if $C \in \mathcal{C}_k$, then $f_{r-1}C, f_r C \in \mathcal{C}_k \sqcup \{0\}$ in analogy to the arguments presented in the proof of Proposition 6.7 for $f_i C$ when $1 \leqslant i \leqslant r - 2$.
(b) Prove that if $C \in \mathcal{C}_k$ and $\varepsilon_i(C) > 0$, then $e_i C \in \mathcal{C}_k$ for all $1 \leqslant i \leqslant r$.

Exercise 6.3. Prove Proposition 6.7 for the column crystals $\mathcal{C}_r^-$ and $\mathcal{C}_r^+$ of type D_r.

Exercise 6.4. Prove that if $T \in \mathrm{Tab}_\lambda$ contains adjacent columns of the form (6.4), then $d_1 + d_2 < \max(j, k) - \min(i, \ell)$ must hold.

Exercise 6.5. Show that Tab_λ is closed under the crystal operators e_i for all $i \in I$. In addition complete the proof of Theorem 6.10 for the case when f_i changes a letter $\bar{i}$ to $\overline{i+1}$.

Exercise 6.6. Proof Theorem 6.12.

Exercise 6.7. Proof Theorem 6.14.

Chapter 7

Insertion Algorithms

We now turn to combinatorial analogs of many of the results in the previous chapters. These are based on remarkable properties of the Robinson–Schensted–Knuth (RSK) algorithm. The main reference is [Knuth (1970)], but see also [Fulton (1997), Chapter 1], [Knuth (1998), Section 5.1.4], and [Stanley (1999), Section 7.11]. We consider group-theoretic correspondences which are discussed in the appendices. The RSK algorithm gives combinatorial analogs of these, which may be understood in terms of crystals of type A.

The RSK algorithm is introduced in Section 7.1. Its counterpart, called the dual RSK algorithm, is subject of Section 7.2. The Edelman–Greene correspondence [Edelman and Greene (1987)] is a variant of RSK adapted to the setting of reduced words for elements of the symmetric group. This is the topic of Section 7.3.

7.1 The RSK algorithm

We begin by mentioning several constructions in representation theory to motivate the combinatorial algorithms of this section.

Example 7.1. *If G is a finite group, then the group algebra $\mathbb{C}[G]$ is a $(G \times G)$-bimodule, with one copy of G acting on the left, the other on the right. We know that*

$$\mathbb{C}[G] \cong \bigoplus_{\pi} \pi \otimes \hat{\pi},$$

where the sum is over all irreducible representations of G, and $\hat{\pi}$ is the contragredient. If $G = S_k$ is the symmetric group, then $\hat{\pi} \cong \pi$, and $\pi = \pi_\lambda$ is indexed by partitions λ of k. Thus

$$\mathbb{C}[S_k] \cong \bigoplus_{\lambda \vdash k} \pi_\lambda^{S_k} \otimes \pi_\lambda^{S_k}.$$

Example 7.2 (Schur–Weyl duality). *Let $\mathrm{GL}(n, \mathbb{C})$ act on $V = \mathbb{C}^n$ by the standard representation. We have commuting actions of S_k and $\mathrm{GL}(n, \mathbb{C})$ on $V^{\otimes k}$. According* *to Schur–Weyl duality (Theorem A.10)*

$$V^{\otimes k} = \bigoplus_{\substack{\lambda \vdash k \\ \ell(\lambda) \leqslant n}} \pi_\lambda^{S_k} \otimes \pi_\lambda^{\mathrm{GL}(n,\mathbb{C})}. \tag{7.1}$$

Example 7.3 (Cauchy identity). *Consider the action of* $\mathrm{GL}(r,\mathbb{C}) \otimes \mathrm{GL}(n,\mathbb{C})$ *on* $\mathbb{C}^r \otimes \mathbb{C}^n$ *by the tensor product of the standard representations. This is equivalent to the action on* $\mathrm{Mat}_{r \times n}(\mathbb{C})$ *via:*

$$(g,h): X \mapsto g \cdot X \cdot {}^t h, \qquad g \in \mathrm{GL}(r), \ h \in \mathrm{GL}(n).$$

(Here ${}^t h$ *is the matrix transpose.) By Theorem B.3, the symmetric algebra has the following decomposition:*

$$\bigvee (\mathrm{Mat}_{r \times n}(\mathbb{C})) \cong \bigoplus_{\ell(\lambda) \leqslant \min(r,n)} \pi_\lambda^{\mathrm{GL}(r)} \otimes \pi_\lambda^{\mathrm{GL}(n)}.$$

The sum is over all partitions of length $\leqslant$ *both* r, n. *If* α_i *are the eigenvalues of* g *and* β_j *are the eigenvalues of* h, *then taking the trace of* (g,h) *gives the identity*

$$\prod_{i,j}(1 - \alpha_i \beta_j)^{-1} = \sum_\lambda s_\lambda(\alpha) \, s_\lambda(\beta)$$

where s_λ *is the Schur function.*

Example 7.4 (Dual Cauchy identity). *Similarly (Theorem B.17), the exterior algebra*

$$\bigwedge (\mathrm{Mat}_{r \times n}(\mathbb{C})) \cong \bigoplus_{\substack{\ell(\lambda) \leqslant r \\ \ell(\lambda') \leqslant n}} \pi_\lambda^{\mathrm{GL}(r)} \otimes \pi_{\lambda'}^{\mathrm{GL}(n)}.$$

The sum is over partitions λ *whose Young diagram fits in an* $r \times n$ *box. Here* λ' *is the conjugate partition. Taking traces,*

$$\prod_{i,j}(1 + \alpha_i \beta_j) = \sum_\lambda s_\lambda(\alpha) \, s_{\lambda'}(\beta).$$

All of these facts have combinatorial analogs, which are due to [Knuth (1970)]. We will review these. They are based on *Schensted insertion*, which we explain next.

By a *location* in a tableau T of shape λ, we mean a box in the Young diagram $\mathrm{YD}(\lambda)$. We may think of the tableau as a map from the Young diagram to the alphabet. If the box is in the i-th row and the j-th column, we will write $T(i,j)$ for the corresponding entry of T.

Let λ and μ be partitions. Assume that $\mathrm{YD}(\mu) \supseteq \mathrm{YD}(\lambda)$. Then we say that the pair (μ, λ) is a *skew shape* and denote it μ/λ. The diagram $\mathrm{YD}(\mu/\lambda)$ is defined to be the set-theoretic difference $\mathrm{YD}(\mu) \setminus \mathrm{YD}(\lambda)$. The skew shape μ/λ is called a *horizontal strip* of length $|\mu| - |\lambda|$ if $\mathrm{YD}(\mu/\lambda)$ has no two boxes in the same column, and a *vertical strip* if $\mathrm{YD}(\mu/\lambda)$ has no two boxes in the same row.

If T is a semistandard tableau of shape λ and i is given, we define a new semi-standard tableau $T \leftarrow i$. The shape of $T \leftarrow i$ is the same as T, but with one new box in its Young diagram. The tableau $T \leftarrow i$ has the same entries as T and one new one, equal to i. The new entry i is somewhere in the first row. Possibly it is added at the end of the row, or possibly it displaces or bumps an entry somewhere in the middle, in which case the entries of T will have to be moved around to make room for it.

The algorithm by which $T \leftarrow i$ is produced is called *Schensted insertion*, and we will explain it now.

If $i \geqslant T(1, \lambda_1)$ (which is the last entry in the first row), then we place i at the end, and the algorithm terminates.

Otherwise, let j be the smallest entry in the first row such that $j > i$. We replace the first occurrence of j in the first row by i. We say that j is *bumped* by i. Then we insert j into the second row by repeating the process. That is, if the bumped value $j \geqslant T(2, \lambda_2)$ (which is the last entry in the second row), it is placed at the end, and the algorithm terminates.

Otherwise, let k be the smallest entry in the second row such that $k > j$. We replace the first occurrence of k in the second row by j. As before, k is bumped by i. So we insert k into the third row by repeating the process. The algorithm continues this way until a bumped entry can be placed at the end of a row. Since there are only finitely many rows in T, the algorithm will eventually terminate.

Example 7.5. *Let us compute* $T \leftarrow 1$, *where* T *is*

1	1	1	2	2	4	4
2	2	3	3	4		
3	3					
4						

.

The 1 cannot be placed at the end of the first row, so it bumps the first occurrence of 2 in the first row. Since 2 was bumped, we insert it into the second row. It cannot go at the end, so it bumps the first 3, which is then inserted in the third row. It can go at the end, and so the algorithm terminates. We see that $T \leftarrow 1$ *is the following tableau, where we have shaded the changed entries:*

1	1	1	1	2	4	4
2	2	2	3	4		
3	3	3				
4						

.   (7.2)

Proposition 7.6. *If* T *is semistandard, then so is* $T \leftarrow i$.

Proof. Indeed, it is clear that the insertion into a weakly increasing row produces another weakly increasing row. To see that the resulting tableau has strictly increasing columns, we must consider what happens when the entry j that is bumped from some row is directly above an entry equal to $j + 1$. Since the j that is bumped is its first occurrence in its row, the $j + 1$ is the first occurrence of $j + 1$ in its row, and it will be bumped by the j, so column strictness is not disturbed by this algorithm. $\qquad\square$

If μ is the shape of $T \leftarrow i$, then μ/λ is a skew shape consisting of a single box. This box is called the *new location* for the insertion of i into T. The sequence of locations where entries are bumped, together with the new location, is called the *bumping path*. In Example 7.5, the bumping path consists of the shaded locations in (7.2). We will call the entry in the bumping path in the first (or top) row the *landing location*.

Using this insertion algorithm, we will exhibit several combinatorial bijections that were first defined in [Knuth (1970)], though Example 7.7 was known earlier to Robinson and Schensted. Collectively these bijections are known as the RSK algorithm. We will discuss these, and explain how they give combinatorial analogs to Examples 7.1 through 7.4.

Let $\lambda \vdash k$. By a *standard tableau* of shape λ we mean a tableau in the alphabet $[k]$ with strictly increasing rows and columns, in which every entry $1, 2, 3, \ldots, k$ occurs exactly once.

Example 7.7. *There is a bijection between elements of S_k and pairs (P, Q) of standard tableaux of the same shape $\lambda \vdash k$, with the alphabet $[k]$.*

Example 7.8. *There is a bijection between words $w = (i_1, \ldots, i_k)$ with entries in $[n]$ and pairs (P, Q) of semistandard tableaux of the same shape $\lambda \vdash k$, where P is semistandard in the alphabet $[n]$, and Q is standard in the alphabet $[k]$.*

Example 7.9. *There is a bijection between $r \times n$ matrices with entries in $\mathbb{N} = \{0, 1, 2, \ldots\}$ and pairs (P, Q) of tableaux with the same shape λ, such that P is in the alphabet $[n]$ and Q is in the alphabet $[r]$.*

Example 7.10. *There is a bijection between $r \times n$ matrices with entries in $\{0, 1\}$ and pairs (P, Q) of tableaux such that P is in the alphabet $[n]$ and Q is in the alphabet $[r]$, and P and Q are of conjugate shape.*

We begin by discussing Example 7.8. Note that Example 7.7 is a special case of Example 7.8, when the word $(i_1, \ldots, i_k)$ is in fact an element in S_k. To construct the bijection, we let P be the *insertion tableau* $\varnothing \leftarrow i_1 \leftarrow i_2 \leftarrow \cdots \leftarrow i_k$ obtained by successively inserting the letters of the word w starting with the empty tableau $\varnothing$. The tableau Q, called the *recording tableau*, is numbered $1, 2, \ldots, k$, where each entry r is the location of the new box after the last insertion in

$$\varnothing \leftarrow i_1 \leftarrow i_2 \leftarrow \cdots \leftarrow i_r.$$

For example, let us consider the word $(3, 1, 2, 1, 4)$. Inserting $3, 1, 2, 1, 4$ in sequence gives the following sequence of tableaux:

$$
\boxed{3}
\qquad
\begin{array}{c}
\boxed{1}\\
\boxed{3}
\end{array}
\qquad
\begin{array}{cc}
\boxed{1}&\boxed{2}\\
\boxed{3}
\end{array}
\qquad
\begin{array}{cc}
\boxed{1}&\boxed{1}\\
\boxed{2}\\
\boxed{3}
\end{array}
\qquad
\begin{array}{ccc}
\boxed{1}&\boxed{1}&\boxed{4}\\
\boxed{2}\\
\boxed{3}
\end{array}
$$

The last tableau here is P. The recording tableau Q shows where new boxes are added in the succession of insertions.

$$
Q =
\begin{array}{ccc}
\boxed{1}&\boxed{3}&\boxed{5}\\
\boxed{2}\\
\boxed{4}
\end{array}
\ .
$$

The map $w \mapsto (P,Q)$ just described is called the *RSK map*.

Proposition 7.11. *Let n and k be positive integers. The RSK map from words w of length k in the alphabet $[n]$ to pairs (P,Q) of tableaux of the same shape $\lambda \vdash k$ in alphabets $[n]$ and $[k]$ with P semistandard and Q standard is a bijection.*

We defer the proof of this since it is a special case of Theorem 7.14. But let us consider how this statement is analogous to Example 7.2. Observe that the cardinality of the set of words w is n^k, which is the dimension of $V^{\otimes k}$ where, as in Example 7.2, $V = \mathbb{C}^n$. On the other hand, if $\lambda \vdash k$ has length $\leqslant n$, then the number of semistandard P with shape λ in the alphabet $[n]$ equals the degree of $\pi_\lambda^{GL(n,\mathbb{C})}$, and the number of standard Q with shape λ equals the degree of $\pi_\lambda^{S_k}$. So the cardinalities of the two sets in this bijection are the same as the dimensions of the vector spaces in the Schur–Weyl duality isomorphism (7.1).

We observe that Proposition 7.11 implies the correspondence in Example 7.7. Assuming that $n = k$, the word w is a permutation of $[k]$ if and only if P is standard. This is obvious since the entries in P are exactly the inserted entries.

Before returning to the proof of Proposition 7.11 we turn to Example 7.9, which is a combinatorial analog of Example 7.3. First we define a mapping from the set $\mathrm{Mat}_{r \times n}(\mathbb{N})$ of $r \times n$ matrices with nonnegative integer entries to pairs (P,Q) of semistandard tableaux having the same shape λ, where P and Q are in the alphabets $[n]$ and $[r]$, respectively. Given a matrix X we label the columns $1, 2, \ldots, n$ and the rows $1, 2, \ldots, r$. We make a *two-rowed array* A whose columns are pairs $\binom{i}{j}$ with $1 \leqslant i \leqslant r$ and $1 \leqslant j \leqslant n$, and where the column $\binom{i}{j}$ appears x_{ij} times. These are ordered lexicographically. For example, if

$$
X = \begin{pmatrix} 1 & 3 & 0 \\ 2 & 0 & 2 \end{pmatrix}
$$

then the two-rowed array is

$$
A = \begin{pmatrix} 1 & 1 & 1 & 1 & 2 & 2 & 2 & 2 \\ 1 & 2 & 2 & 2 & 1 & 1 & 3 & 3 \end{pmatrix}.
$$

Now we insert the entries in the bottom row in order, starting with the empty tableau. Thus in the example, the insertion tableau P is

$$\varnothing \leftarrow 1 \leftarrow 2 \leftarrow 2 \leftarrow 2 \leftarrow 1 \leftarrow 1 \leftarrow 3 \leftarrow 3 \quad = \quad \begin{array}{|c|c|c|c|c|c|} \hline 1 & 1 & 1 & 2 & 3 & 3 \\ \hline 2 & 2 \\ \cline{1-2} \end{array}.$$

The entries in the top row of the two-rowed array A go into the recording tableau in order. In this example after inserting $\varnothing \leftarrow 1 \leftarrow 2 \leftarrow 2 \leftarrow 2$, which are the entries of the bottom row corresponding to 1's in the top row, we obtain:

$$\begin{array}{|c|c|c|c|} \hline 1 & 2 & 2 & 2 \\ \hline \end{array}.$$

Thus the 1's from the top row go into the recording tableau at these locations. The 2's from the top row of A go into the remaining locations, and the recording tableau Q is:

$$\begin{array}{|c|c|c|c|c|c|} \hline 1 & 1 & 1 & 1 & 2 & 2 \\ \hline 2 & 2 \\ \cline{1-2} \end{array}.$$

Lemma 7.12. *The recording tableau is semistandard.*

Proof. It is clear that the rows and columns are weakly increasing, but we must check that the columns are *strictly* increasing. It is enough to show that no two i's are in the same column. Thus we must consider the new locations where the entries j of the bottom row of the two-rowed array that are below i in the top row fall. With i fixed, the entries below i in the two-rowed array are in weakly increasing order. It is easy to see that this implies that each bumping path for one of these insertions lies strictly to the right of the previous one, and so the succession of new locations are in successively increasing columns. Hence, no two new locations for these insertions will be in the same column. □

Theorem 7.13 (Knuth). *This map from* $\mathrm{Mat}_{r \times n}(\mathbb{N})$ *to pairs* (P, Q) *of semistandard tableaux with the same shape* λ, *where* P *is in the alphabet* $[n]$ *and* Q *is in the alphabet* $[r]$, *is a bijection.*

Note that this implies Proposition 7.11, whose proof we postponed.

Proof. To construct an inverse correspondence, we need to explain how, given P and Q we can reconstruct the matrix X algorithmically. Clearly it is enough to show how we reconstruct the two-rowed array A. This depends on the following point. Let $\binom{i}{j}$ be the last column of A, and let A' be the two-rowed array obtained by deleting this last column. We claim that, if we know P and Q, then we can determine the last column of A, and the tableaux P', Q' corresponding to A'. If we know how to do this, we may then reconstruct the two-rowed array by repeating this analysis until A is completely known.

First we note that we may determine the new location for the insertion. This is determined by Q alone. Because Q is semistandard, if u is the largest entry in

Q, then the locations where u appears in Q form a horizontal strip, that is, a skew tableau with no two entries in the same column. It is not hard to see that the new (insertion) location is the location farthest to the right in this horizontal strip.

So now we will show that since we know P, Q and the new location for the insertion $P' \leftarrow j$, then we can recover P', Q', i and j. Of course i is just the entry of Q at the new location, and Q' can be recovered from Q by removing that box. So we have only to determine P' and j. It may be useful to have an example in mind as we explain this. Consider the case where $j = 1$ and

$$P' = \begin{array}{|c|c|c|c|c|c|} \hline 1 & 1 & 2 & 2 & 3 & 4 \\ \hline 2 & 3 & 4 & 4 \\ \cline{1-4} 3 & 4 \\ \cline{1-2} \end{array} \quad , \quad P = \begin{array}{|c|c|c|c|c|c|} \hline 1 & 1 & 1 & 2 & 3 & 4 \\ \hline 2 & 2 & 4 & 4 \\ \cline{1-4} 3 & 3 \\ \cline{1-2} 4 \\ \cline{1-1} \end{array} \quad .$$

The new location is the dark shaded box in the fourth row.

In general, if the insertion location is in the k-th row, it is at the end of the row, and the bumping path consists of a sequence $(1, c_1), (2, c_2), (3, c_3), \ldots, (k, c_k)$ where $c_1 \geqslant c_2 \geqslant \cdots \geqslant c_k$.

First we show that knowing P and the new location determines the bumping path, and then we show how to recover P' and j. The new location is the last location (k, c_k) in the bumping path, and from that we can recover the entire path as follows. Note that c_{k-1} is the largest integer such that $P(k-1, c_{k-1}) < P(k, c_k)$. Thus in the example, since we know that $P(k, c_k) = P(4, 1) = 4$, we infer that $(k - 1, c_{k-1}) = (3, 2)$ since $c_{k-1} = 2$ is the largest integer such that $P(3, c_3) < P(4, c_4) = 4$. Then c_{k-2} is the largest integer such that $P(k - 2, c_{k-2}) < P(k - 1, c_{k-1})$ and so forth. Once the bumping path has been determined we may recover P' since all entries of P' are the same as those in P, except those on the bumping path. And for those on the bumping path (excluding the bumping location, which is not in P') we have $P'(i, c_i) = P(i + 1, c_{i+1})$. It is now clear that if we know the bumping location, then we know P'. Moreover, since we know the bumping path we know $j = P(1, c_1)$. $\qquad \square$

Theorem 7.14 (Knuth). *If $X \in \mathrm{Mat}_{r \times n}(\mathbb{N})$ corresponds to (P, Q) then the transpose matrix ${}^t X \in \mathrm{Mat}_{n \times r}(\mathbb{N})$ corresponds to (Q, P).*

The argument that we give follows [Knuth (1970)]. A more "geometric" argument was given by [Viennot (1977)]. (See also [Sagan (2001)].) Although Viennot's paper only considers RSK for permutations, it extends to the full RSK algorithm. Fomin's theory of growth diagrams ([Fomin (1995); Sagan (2001)]) generalizes Viennot's construction, as does the "matrix ball" construction ([Fulton (1997)]).

Proof. This is a symmetry of the RSK algorithm that is not apparent from its description. Following Knuth, we give an alternative description that makes this symmetry apparent.

We start with the two-rowed array

$$A = \begin{pmatrix} x_1 & x_2 & x_3 & \cdots & x_l \\ y_1 & y_2 & y_3 & \cdots & y_l \end{pmatrix}.$$

We define a partial order on the columns of A. Let $Z = \{z_i\}$ be the set of columns. If $z \in Z$, we denote by $x(z)$ and $y(z)$ the two elements of the column z. In other words, $x(z) = x_i$ and $y(z) = y_i$ where i is such that $z = z_i$.

Even if $x_i = x_j$ and $y_i = y_j$ we regard z_i and z_j as being distinct if $i \neq j$. Thus the cardinality of Z equals the number of columns of A even if some columns are repeated. We want a partial order on Z such that $z_i \leqslant z_j$ if and only if $x_i \leqslant x_j$ and $y_i \leqslant y_j$ with the following exception. If $i \neq j$ but $x_i = x_j$ and $y_i = y_j$, then we want exactly one of $z_i < z_j$ and $z_j < z_i$ to be true. Thus if (x, y) are fixed, we totally order the *level set* consisting of those z_i with $x_i = x$ and $y_i = y$. We refer to the resulting partial order of Z as an *admissible* partial order.

One particular admissible partial order will be required in the future. This is the partial order in which $z_i \leqslant z_j$ if and only if $x_i \leqslant x_j$ and $y_i \leqslant y_j$, and, in the case where $x_i = x_j$ and $y_i = y_j$, we also require that the column z_i preceed the column z_j in the matrix A. We refer to this as the *admissible order determined by A*.

An admissible partial order determines a digraph structure on Z. We draw an arrow $z_i \to z_j$ if $z_i < z_j$ and if there is no third element z_k such that $z_i < z_k < z_j$. We observe that the digraph structure is essentially independent of the choice of totally ordering of the level sets in the following sense. If we choose another admissible order, then there is a permutation $z \to z'$ of Z transporting the first digraph structure to the second, such that $x(z) = x(z')$ and $y(z) = y(z')$. We simply have to permute the level sets. So in this sense the digraph structure on Z is essentially unique, and independent of the choice of admissible order. This is important since when we consider the point of the theorem, that transposition of Z corresponds to the interchange of P and Q, we will be interchanging the rows of the two-rowed array, then permuting the columns. We need to know that this permutation of the columns does not change the digraph.

First we show how to compute the top rows of P and Q from the digraph Z. We describe the algorithm, give an example, then prove that it is correct. If $z \in Z$ we define the *rank* of z to be the length r of the longest chain

$$z^{(1)} \to z^{(2)} \to \cdots \to z^{(r)} = z$$

with $z^{(i)} \in Z$, terminating in z. Let Z^r be the set of elements of rank r. Let

$$p_r = \min\{y(z) \mid z \in Z^r\}, \qquad q_r = \min\{x(z) \mid z \in Z^r\}.$$

We will show that $\boxed{p_1}\boxed{p_2}\boxed{p_3}\boxed{\cdots}$ and $\boxed{q_1}\boxed{q_2}\boxed{q_3}\boxed{\cdots}$ are the top rows of P and Q, respectively.

For example, suppose that

$$A = \begin{pmatrix} 1 & 1 & 2 & 2 & 3 & 3 \\ 2 & 3 & 2 & 3 & 1 & 2 \end{pmatrix}.$$

The reader may check that

$$
P = \begin{array}{|c|c|c|}\hline 1 & 2 & 2 \\\hline 2 & 3 \\\cline{1-2} 3 \\\cline{1-1}\end{array} \quad , \qquad Q = \begin{array}{|c|c|c|}\hline 1 & 1 & 2 \\\hline 2 & 3 \\\cline{1-2} 3 \\\cline{1-1}\end{array} \quad .
$$

Here is the digraph:

$$
\begin{pmatrix}1\\2\end{pmatrix} \longrightarrow \begin{pmatrix}2\\2\end{pmatrix} \qquad \begin{pmatrix}3\\1\end{pmatrix}
$$
$$
\downarrow \qquad\qquad \downarrow \quad \searrow \quad \downarrow
$$
$$
\begin{pmatrix}1\\3\end{pmatrix} \longrightarrow \begin{pmatrix}2\\3\end{pmatrix} \qquad \begin{pmatrix}3\\2\end{pmatrix}
$$

We have

$$
Z^1 = \left\{\begin{pmatrix}1\\2\end{pmatrix},\begin{pmatrix}3\\1\end{pmatrix}\right\}, \qquad p_1 = \min(2,1) = 1, \qquad q_1 = \min(1,3) = 1,
$$

$$
Z^2 = \left\{\begin{pmatrix}1\\3\end{pmatrix},\begin{pmatrix}2\\2\end{pmatrix}\right\}, \qquad p_2 = \min(3,2) = 2, \qquad q_2 = \min(1,2) = 1,
$$

$$
Z^3 = \left\{\begin{pmatrix}2\\3\end{pmatrix},\begin{pmatrix}3\\2\end{pmatrix}\right\}, \qquad p_3 = \min(3,2) = 2, \qquad q_3 = \min(2,3) = 2.
$$

Thus we may read off the top rows of P and Q.

To prove the correctness of this description, since it is independent of admissible ordering, we choose the admissible ordering of Z determined by A. Any two elements of Z^r are incomparable in the partial order, so we may arrange Z^r in an order to be described as follows. We write $Z^r = \{z_1^r, \dots, z_{k_r}^r\}$, and since no two elements of Z^r are comparable, we may order these so that

$$
x_1^r \leqslant \cdots \leqslant x_{k_r}^r, \qquad y_1^r \geqslant \cdots \geqslant y_{k_r}^r.
$$

What we must prove is that the top row of P is $(y_{k_1}^1, y_{k_2}^2, y_{k_3}^3, \dots)$ and the top row of Q is $(x_1^1, x_1^2, x_1^3, \dots)$.

Let $P_1, P_2, \dots, P_N = P$ be the sequence of tableaux obtained by successive insertions of the bottom row entries of A. Let $z \in Z$. There is a first tableau in this sequence where a given bottom row element $y = y(z)$ appears. In this tableau, it appears in the first row, though later it may be bumped by another element. Let r be the location in the first row where y first appears. We call this the *landing location* of z. It is not hard to see that Z^r is the set of elements of Z whose landing location is r. The first insertion to land at the location r is z_1^r, so the r-th entry in Q is x_1^r. Each time we insert a subsequent y_j^r ($j > 1$), we bump y_{j-1}^r, and the resulting tableau has y_j^r in the r-th position. The final tableau P will have $y_{k_r}^r$ in its r-th position. This proves the correctness of the description of the two top rows of P and Q.

Let P' and Q' be the tableaux obtained from P and Q by removing their top rows. We will also describe the two-rowed array A' that corresponds to P' and Q'

by the RSK algorithm. The corresponding digraph D has one fewer entry than Z in each rank, so that $|D^r| = k_r - 1$. The elements of D^r are $w_1^r, \ldots, w_{k_r-1}^r$ and satisfy

$$x(w_i^r) = x(z_{i+1}^r), \qquad y(w_i^r) = y(z_i^r).$$

To see that this is correct, observe that P' can be built up from the empty tableau by successively inserting each entry which is bumped from the top row of P. Considering only those entries whose landing location is the r-th position of the top row, when we insert y_{i+1}^r we bump y_i^r. So the two-rowed array with columns

$$\begin{pmatrix} x_{i+1}^r \\ y_i^r \end{pmatrix}$$

arranged into lexicographical order will have exactly the desired effect.

Now consider the two-rowed array corresponding to the transposed matrix $\,^tX$. This is obtained by interchanging the two rows of A and reordering its columns. The corresponding digraph will be the same, except that the functions x and y will be switched. The symmetry of the above description of the RSK algorithm makes it apparent that the corresponding tableaux P and Q are interchanged. □

7.2 The dual RSK algorithm

Next we turn next to the dual RSK algorithm of Example 7.10.

We have pointed out that the RSK algorithm gives a combinatorial analog of the following representation-theoretic fact. Let $\mathrm{GL}(r, \mathbb{C}) \times \mathrm{GL}(n, \mathbb{C})$ act on $\mathbb{C}^r \otimes \mathbb{C}^n$ by the tensor product of the standard representation, or equivalently on $\mathrm{Mat}_{r \times n}(\mathbb{C})$ by:

$$(g, h) \colon X \mapsto g \cdot X \cdot \,^th.$$

Then the tensor algebra of this module decomposes as a sum of $\pi_\lambda^{\mathrm{GL}(r)} \otimes \pi_\lambda^{\mathrm{GL}(n)}$, where λ runs over partitions of length $\leqslant \min(r, n)$.

Similarly the exterior algebra decomposes as a direct sum of $\pi_\lambda^{\mathrm{GL}(r)} \otimes \pi_{\lambda'}^{\mathrm{GL}(n)}$, where now λ is a partition whose Young diagram fits in a $r \times n$ rectangle and λ' is the conjugate partition. The combinatorial analog of this statement is the *dual RSK algorithm*, also due to Knuth.

Let us say that a tableau is *dual semistandard* if the rows are strictly increasing and the columns are weakly increasing. Thus T is dual semistandard if and only if its transpose $\,^tT$ is semistandard.

Dual Schensted insertion takes a dual semistandard tableau T and inserts a new entry j. To distinguish it from ordinary Schensted insertion, we denote this by $T \Leftarrow j$. In this operation, it is assumed that T is row strict, and it will remain row strict after insertion. If j is strictly greater than the last entry in the first row of T, it is placed at the end. Otherwise, it bumps the largest entry that is $\geqslant j$.

A $(0, 1)$-*matrix* is a matrix with each entry equal to either 0 or 1. We may now describe a bijection between $(0, 1)$-matrices in $\mathrm{Mat}_{r \times n}(\mathbb{Z})$ and pairs of tableaux P

and Q such that P is dual semistandard in the alphabet $[n] = \{1, 2, \ldots, n\}$ and Q is semistandard in the alphabet $[r]$.

The bijection is exactly the same as standard RSK except that we use dual Schensted insertion. Let $X \in \mathrm{Mat}_{r \times n}(\mathbb{Z})$ be a $(0, 1)$-matrix. We make the two-rowed array as before, successively insert the entries in the bottom row, starting with the empty tableau, and put the top row entries in the recording tableau.

For example, suppose

$$X = \begin{pmatrix} 0 & 1 & 1 \\ 0 & 0 & 1 \\ 0 & 1 & 0 \\ 1 & 0 & 1 \end{pmatrix}.$$

The two-rowed array is

$$A = \begin{pmatrix} 1 & 1 & 2 & 3 & 4 & 4 \\ 2 & 3 & 3 & 2 & 1 & 3 \end{pmatrix}$$

After inserting the first two entries into the empty tableau, we have to insert 3 into:

$$\boxed{2\ 3}$$

In ordinary RSK we would place the 3 at the end, but in dual RSK, inserting 3 bumps the 3 at the end. Continuing, the sequence of tableaux obtained by successive insertions is

The recording tableau Q fills the sequence of new locations from the top row of the two-rowed array, so

Note that P is dual semistandard and Q is semistandard, see Exercise 7.1. Since the transpose ${}^t P$ is semistandard, the bijection may also be stated as with pairs of semistandard tableaux of conjugate shape.

Knuth points out that, unlike the usual RSK algorithm, taking the transpose of X does not interchange the insertion and recording tableaux. Compare the results for the transpose in the above example:

$$^t X = \begin{pmatrix} 0 & 0 & 0 & 1 \\ 1 & 0 & 1 & 0 \\ 1 & 1 & 0 & 1 \end{pmatrix} \qquad A' = \begin{pmatrix} 1 & 2 & 2 & 3 & 3 & 3 \\ 4 & 1 & 3 & 1 & 2 & 4 \end{pmatrix}$$

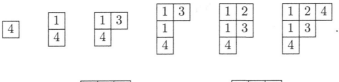

Thus

$$P = \begin{array}{|c|c|c|} \hline 1 & 2 & 4 \\ \hline 1 & 3 \\ \cline{1-2} 4 \\ \cline{1-1} \end{array} \qquad \text{and} \qquad Q = \begin{array}{|c|c|c|} \hline 1 & 2 & 3 \\ \hline 2 & 3 \\ \cline{1-2} 3 \\ \cline{1-1} \end{array} .$$

7.3 Edelman–Greene insertion

Edelman–Greene (EG) insertion is a variant of RSK that is useful for studying reduced words for elements of the symmetric group.

Let n be some integer. Let $s_1, \ldots, s_{n-1}$ be the simple reflections in the Coxeter group W. A sequence $(i_1, i_2, \ldots, i_r)$ of integers $1 \leqslant i_k \leqslant n$ is called a *word*. It is a *reduced word for* $w \in W$ if $w = s_{i_1} \cdots s_{i_r}$ and this expression is as short as possible. Then r is called the *length* of w. We will apply this when $W = S_n$ is the symmetric group.

A semistandard tableau T is called *strict* if the rows, as well as the columns are strictly increasing. If the tableau is a row, let

$$\text{word}(T) = (i_1, \ldots, i_s), \qquad T = \boxed{i_1 \, i_2 \cdots i_s}. \tag{7.3}$$

More generally, if $T_1, \ldots, T_r$ are the rows of T, let $\text{word}(T)$ be obtained by concatenating $\text{word}(T_r), \ldots, \text{word}(T_1)$. Thus $\text{word}(T)$ is the *reading word* of T, reading row by row, starting with the bottom row and reading each row from left to right. Then T is called *reduced* if $\text{word}(T)$ is a reduced word for w where $w = s_{i_1} \cdots s_{i_r}$. In this case we will also write $w(T) = w$. Clearly a reduced tableau is strict.

We may now explain EG insertion. Let T be a semistandard tableau and k an integer. We assume that T is reduced. Let $\text{word}(T) = (i_1, \ldots, i_s)$ and assume that $\text{word}(T)$ is a reduced word for $w = s_{i_1} \cdots s_{i_s}$. Let k be a right ascent for w, so that the length $\ell(ws_k) = s + 1$. We will describe a new semistandard tableau $T \twoheadleftarrow k$ obtained by the following insertion algorithm.

First suppose that T is the row (7.3). If $k > i_s$ then we define $T \twoheadleftarrow k$ to be

$$\boxed{i_1 \, i_2 \cdots i_s \, k}.$$

On the other hand if $k \leqslant i_s$ then our assumption that k is an ascent for w implies that $k < i_s$. Let u be the smallest positive integer such that $k < i_u$. Let $k' = i_u$. There are two cases. If $u = 1$ or if $u > 1$ and $i_{u-1} < k$, then $T \twoheadleftarrow k$ is

$$\begin{array}{|c|c|c|c|c|c|c|} \hline i_1 & \cdots & i_{u-1} & k & i_{u+1} & \cdots & i_s \\ \hline k' \\ \cline{1-1} \end{array} \tag{7.4}$$

(This coincides with RSK insertion.) On the other hand if $u > 1$ and $i_{u-1} = k$ then our assumption that k is an ascent for w implies that $i_u = k + 1$, since otherwise

s_k commutes with $s_{i_u} \cdots s_{i_s}$ and therefore $w s_k = s_{i_1} \cdots s_{i_{u-2}} s_{i_u} \cdots s_{i_s}$ has length $< \ell(w)$. In this case $T \leftsquigarrow k$ is

i_1	$\cdots$	i_{u-1}	i_u	i_{u+1}	$\cdots$	i_s
k'						

$$i_{u-1} = k, i_u = k' = k+1. \qquad (7.5)$$

This differs from RSK insertion in that the top row is unchanged.

In situations (7.4) and (7.5), we also use the notation $k' \leftsquigarrow T'$ for the tableau $T \leftsquigarrow k$. In these last two cases, we say that k' is *bumped* on insertion. We denote by $\beta(T \leftsquigarrow k) = u$ the location of the bumped entry.

Now that we have defined EG insertion in the case the tableau is a row, let us consider the general case. Let $T_1, \ldots, T_r$ be the rows of the reduced tableau T. We replace several of these, $T_1, \ldots, T_q$ by new rows $T'_1, \ldots, T'_q$ and $T \leftsquigarrow k$ is the tableau with rows $T'_1, \ldots, T'_q, T_{q+1}, \ldots, T_r$.

If $T_1 \leftsquigarrow k$ is a row T'_1, the algorithm terminates. That is, $q = 1$ and $T'_1 = T_1 \leftsquigarrow k$. We have simply added an entry k at the end of the first row. Otherwise, suppose that k_1 is bumped, so $(T_1 \leftsquigarrow k) = (k_1 \leftsquigarrow T'_1)$ for a row T'_1. Either one entry i_u in T_1 has been replaced by a smaller entry k, or $T'_1 = T_1$. Now let us insert k_1 into T_2. If $T_2 \leftsquigarrow k_1$ is a tableau T'_2, then $q = 2$ and the algorithm terminates. We have added k_1 at the end of the second row. On the other hand, if k_2 is bumped, so that $(T_2 \leftsquigarrow k_1) = (k_2 \leftsquigarrow T'_2)$ then we continue inserting k_2 into T_3, etc.

Lemma 7.15. $T'_1, \ldots, T'_q, T_{q+1}, \ldots, T_r$ *are the rows of a semistandard tableau $T \leftsquigarrow k$.*

Proof. Let $\beta_i = \beta(T_i, k_{i-1})$ be the sequence of bump locations. (Here we denote $k_0 = k$.) Each insertion of type (7.4) in a row T_i with $i > 1$ decreases the β_i-th entry of T_i to k_{i-1}, and what must be checked is that the value above it in T'_{i-1} is less than k_{i-1}. Let $u_1, u_2, \ldots$ be the entries in T_{i-1} and let $v_1, v_2, \ldots$ be the entries in T_i. It follows from the definition of $T_i \leftsquigarrow k_{i-1}$ that $\beta_i \leqslant \beta_{i-1}$ since $k_{i-1} = u_{\beta_i}$ cannot be inserted to the right of $v_{\beta_{i-1}}$ which is $> u_{\beta_{i-1}}$. There are two cases to consider. If $\beta_i < \beta_{i-1}$, then $u_{\beta_i} < u_{\beta_{i-1}} = k_{i-1}$ since T is strict. Therefore, suppose that $\beta_i = \beta_{i-1}$. If the insertion in T_{i-1} is of type (7.4), then the β_{i-1} entry has already been decreased from k_{i-1} in T'_{i-1}. Thus let us assume that the insertion in T_{i-1} is of type (7.5). Then $u_{\beta_i} = k_{i-1}$, $u_{\beta_i - 1} = k_{i-1} - 1$ and since T is column strict $v_{\beta_i - 1} \geqslant k_{i-1}$. We also have $v_{\beta_i - 1} \leqslant k_{i-1}$, since otherwise β_i would be $< \beta_{i-1}$. So $v_{\beta_i - 1} = k_{i-1}$. Because we are assuming that the insertion in row T_i is of type (7.4), it follows that T'_i has two consecutive entries $v_{\beta_i - 1}$ and v_{β_i} that are both equal to $k_i - 1$. This is a contradiction, since T'_i is strict. $\qquad\square$

For example, suppose that

$$T = \begin{array}{|c|c|} \hline 1 & 2 \\ \hline 3 \\ \cline{1-1} \end{array}.$$

Then $w(T) = s_3 s_1 s_2$ and $k = 1$. The tableau $T \leftsquigarrow 1$ is

$$\begin{array}{|c|c|}\hline 1 & 2 \\\hline 2 \\\cline{1-1} 3 \\\cline{1-1}\end{array}\ .$$

Lemma 7.16. *Suppose that T is a reduced semistandard tableau and that k is a right ascent of $w(T)$. Then $w(T \leftsquigarrow k) = w(T)\, s_k$.*

Proof. We reduce easily to the case where T is a row. There are three cases to consider, where $T \leftsquigarrow k$ is a row, and where k' is bumped, as in (7.4) and (7.5). In the second two cases, we must show that if T' is the top row of $T \leftsquigarrow k$ then $s_{k'} w(T') = w(T) s_k$. In case (7.4), $s_{i_u} = s_{k'}$ commutes with $s_{i_1} \cdots s_{i_{u-1}}$ while s_k commutes with $s_{i_{u+1}} \cdots s_{i_s}$, and so

$$w(T)s_k = s_{i_1} \cdots s_{i_{u-1}} s_{k'} s_{i_{u+1}} \cdots s_{i_s} s_k = s_{k'} s_{i_1} \cdots s_{i_{u-1}} s_k s_{i_{u+1}} \cdots s_{i_s} = s_{k'} w(T).$$

In case (3), we have

$$w(T)s_k = s_{i_1} \cdots s_{i_{u-2}} s_k s_{k+1} s_{i_{u+1}} \cdots s_{i_s} s_k = s_{i_1} \cdots s_{i_{u-2}} s_k s_{k+1} s_k s_{i_{u+1}} \cdots s_{i_s}.$$

Now using the braid relation $s_k s_{k+1} s_k = s_{k'} s_k s_{k+1}$ with $k' = k+1$

$$s_{i_1} \cdots s_{i_{u-2}} s_{k'} s_k s_{k+1} s_{i_{u+1}} \cdots s_{i_s} = s_{k'} s_{i_1} \cdots s_{i_{u-2}} s_k s_{k+1} s_{i_{u+1}} \cdots s_{i_s} = s_{k'} w(T).$$

$\square$

Now let $\mathbf{i} = (i_1, \ldots, i_m)$ be a reduced word for $w \in S_n$. We define two tableaux $P(\mathbf{i})$ and $Q(\mathbf{i})$ of the same shape λ, a partition of m. The tableau $P(\mathbf{i})$ will be semistandard and reduced, and the tableau $Q(\mathbf{i})$ will be standard. We build up $P(\mathbf{i})$ by successively inserting $i_1, \ldots, i_m$ into the empty tableau $\varnothing$; that is,

$$P(\mathbf{i}) = \varnothing \leftsquigarrow i_1 \leftsquigarrow i_2 \leftsquigarrow \cdots \leftsquigarrow i_m.$$

By Lemma 7.16 we have $w = w(P(\mathbf{i}))$. The tableaux $Q(\mathbf{i})$ is the *recording tableau*, as in the usual RSK algorithm: the entries are the locations of the successive insertions.

For example, suppose that $\mathbf{i} = (2, 3, 2, 1, 2)$. Then

$$P(\mathbf{i}) = \begin{array}{|c|c|}\hline 1 & 2 \\\hline 2 & 3 \\\hline 3 \\\cline{1-1}\end{array}\ , \qquad Q(\mathbf{i}) = \begin{array}{|c|c|}\hline 1 & 2 \\\hline 3 & 5 \\\hline 4 \\\cline{1-1}\end{array}\ .$$

Proposition 7.17 ([Edelman and Greene (1987)]). *Let w be a permutation. Then the map $\mathbf{i} \mapsto (P(\mathbf{i}), Q(\mathbf{i}))$ is a bijection between reduced words for w and pairs (P, Q) of tableaux of the same shape such that P is reduced and semistandard and Q standard with $w = w(P)$.*

Note that $w(P(\mathbf{i}))$ may be different from $\mathbf{i}$, though they are both reduced words for w.

Proof. To see that this map is a bijection, we observe that there is an inverse construction. Let P and Q be given. It is assumed that they are tableau with the same shape λ, a partition of m, with P reduced and Q standard; and $w = w(P)$. The location of m in Q is an extremal location for the shape λ. Let λ_{m-1} be the shape obtained by removing this location from the shape λ, and let Q_{m-1} be the shape obtained by removing this location from Q.

By running the EG construction in reverse, we may find a tableau P_{m-1} of shape λ_{m-1} and a value i_m such that $P = P_{m-1} \leftsquigarrow i_m$. We repeat the process with P and Q replaced by P_{m-1} and Q_{m-1} to reconstruct the unique sequence $\mathbf{i} = (i_1, \ldots, i_m)$ such

$$(P(\mathbf{i}), Q(\mathbf{i})) = (P, Q). \qquad \square$$

To illustrate this, in the above example, $m = 5$ and λ_4 is the partition $(2, 1, 1)$. The insertions are to terminate in the second row at the location of 5 in Q. Thus the second row of P_4 should be $\boxed{2}$ and the top row should be $\boxed{1\,3}$ since $\boxed{1\,3} \leftsquigarrow 2$ equals $3 \leftsquigarrow \boxed{1\,2}$. We find that

$$P_4 = \begin{array}{c} \boxed{1\,3} \\ \boxed{2} \\ \boxed{3} \end{array} \qquad \text{and} \qquad Q_4 = \begin{array}{c} \boxed{1\,2} \\ \boxed{3} \\ \boxed{4} \end{array}.$$

Theorem 7.18 ([Stanley (1984)]). *Let $n \geqslant 2$. The number of reduced words of the long element $w_0 \in S_n$ equals the number of standard tableaux of shape $\rho = (n - 1, n - 2, \ldots, 3, 2, 1)$.*

Proof. In the bijection of Proposition 7.17, there is only one possible choice for P, since there is only one strict tableau with shape ρ in the alphabet $\{1, 2, 3, \ldots, n-1\}$. Hence the number of reduced words is the number of standard tableaux Q with this shape. $\qquad \square$

Exercises

Exercise 7.1. Show that the insertion tableau P for dual RSK is dual semistandard and the recording tableau Q for dual RSK is semistandard. See Section 7.2 for the definitions.

Exercise 7.2. Prove the identity

$$\sum_{\lambda \vdash k} f_\lambda s_\lambda(x_1, \ldots, x_n) = (x_1 + \cdots + x_n)^k,$$

where $s_\lambda(x_1, \ldots, x_n)$ is the Schur polynomial in n variables and f_λ is the number of standard Young tableaux of shape λ.

Exercise 7.3. For the six permutations in S_3, compute the insertion tableau P and the recording tableau Q. Which permutations have the same P-tableau (resp. Q-tableau)?

Exercise 7.4. Compute the recording tableaux Q for all 16 reduced words of the long element $w_0 \in S_4$, verifying Theorem 7.18 in this case.

Chapter 8

The Plactic Monoid

The *plactic monoid* was first introduced by [Lascoux and Schützenberger (1981)] as a multiplicative structure on the set of semistandard Young tableaux. See also Chapter 5 (by Lascoux, Leclerc and Thibon) in [Lothaire (2002)] and [Fulton (1997)]. As we have seen in Chapter 3, semistandard Young tableaux can be used to describe the elements in highest weight $GL(n)$ crystals. In this chapter, we introduce the plactic monoid using $GL(n)$ crystals.

In Section 8.1 we define the plactic monoid in terms of crystals and state the main results. In Section 8.2 we prove the relation between the plactic monoid and Knuth equivalence, which is an equivalence relation that can be described by local operations on words. In Section 8.3 we explain how the Schensted insertion defined in Chapter 7 is related to crystals. We conclude in Section 8.4 with a treatment of skew crystals and show how they give rise to the Littlewood–Richardson coefficients.

8.1 The definition of the plactic monoid

In this chapter, we denote by $\mathbb{B}$ the standard $GL(n)$ crystal, and if λ is a partition of k of length $\leqslant n$, we will embed the crystal $\mathcal{B}_\lambda$ of tableaux by the row reading RR that was defined in Chapter 3.

The multiplicative structure on tableaux is essentially just the tensor product of crystals. Thus let T and U be semistandard tableaux. Then $T \in \mathcal{B}_\lambda$ and $U \in \mathcal{B}_\mu$ for some λ and μ. Their tensor product $T \otimes U$ is in some connected component of $\mathcal{B}_\lambda \otimes \mathcal{B}_\mu$. This subcrystal is isomorphic to $\mathcal{B}_\nu$ for some ν, and if we define $T \cdot U$ to be the image in $\mathcal{B}_\nu$ of $T \otimes U$ under this isomorphism, then $T \cdot U$ is a well-defined tableau. We have defined a monoid structure on the set of all semistandard tableaux. This is one realization of the plactic monoid.

Here is another way of explaining the plactic monoid. Let us make the following definition.

Definition 8.1. Let $\mathcal{C}_1$ and $\mathcal{C}_2$ be normal crystals of the same Cartan type, and let $x_i \in \mathcal{C}_i$. Let $\mathcal{C}_i'$ be the connected component of $\mathcal{C}_i$ that contains x_i. If $\mathcal{C}_1'$ is isomorphic to $\mathcal{C}_2'$, and if the unique isomorphism $\mathcal{C}_1' \to \mathcal{C}_2'$ takes x_1 to x_2, then we say that x_1 and x_2 are *plactically equivalent*, and write $x_1 \equiv x_2$.

If $x_1 \equiv x_2$ and $y_1 \equiv y_2$ then clearly $x_1 \otimes y_1 \equiv x_2 \otimes y_2$, so plactic equivalence gives a monoid structure on any class of crystal elements that is closed under tensor product.

In particular, specializing this definition to the case of $GL(n)$ crystals gives the plactic monoid. Indeed, every $\mathcal{B}_\lambda$ where $\lambda \vdash k$ is a subcrystal of $\mathbb{B}^{\otimes k}$, although there are different embeddings of $\mathcal{B}_\lambda$ into $\mathbb{B}^{\otimes k}$, as we saw in Example 3.2. But by Corollary 6.2, every element of $\mathbb{B}^{\otimes k}$ is plactically equivalent to a unique element of a unique $\mathcal{B}_\lambda$. This means that the disjoint union of all $\mathbb{B}^{\otimes k}$ modulo the relation of plactic equivalence becomes a monoid. It is obvious that this second construction of the plactic monoid is equivalent to the first, since both are derived from the tensor product.

If
$$x = \boxed{u_1} \otimes \cdots \otimes \boxed{u_k} \in \mathbb{B}^{\otimes k}, \qquad (8.1)$$
define the *reading word* $\text{word}(x) = u_1 \ldots u_k$. We will define another equivalence relation, called *Knuth equivalence*, on words, that is, on sequences in the alphabet $[n] = \{1, 2, \ldots, n\}$. Knuth equivalence is generated by certain elementary *Knuth operations* which we now describe. Under certain circumstances we are allowed to switch two adjacent letters in a word. First we consider words with three letters:

(1) If $a < b \leqslant c$, then bac is Knuth equivalent to bca and vice versa.
(2) If $a \leqslant b < c$, then acb is Knuth equivalent to cab and vice versa.

More generally, these replacements are allowed at any place in the middle of a longer word provided a, b, c are consecutive entries. Such a replacement is called an *elementary Knuth operation*. We say that x is Knuth equivalent to y, and write $x \equiv_K y$, if there is a sequence of such elementary operations connecting x to y.

In this chapter, we will prove that $x, y \in \mathbb{B}^{\otimes k}$ are plactically equivalent if and only if their reading words are Knuth equivalent. Our main tool will be RSK in the form of Example 7.8.

We will also see that the subcrystals of $\mathbb{B}^{\otimes k}$ that are isomorphic to $\mathcal{B}_\lambda$ are parametrized by standard tableaux of shape λ. To explain how this works, let $u_1 \ldots u_k$ be the reading word of $x \in \mathbb{B}^{\otimes k}$. We do RSK insertion of u_i starting with the empty tableau and define
$$P(x) = (\varnothing \leftarrow u_1 \leftarrow u_2 \leftarrow \cdots \leftarrow u_k).$$
As we explained in Chapter 7, this procedure produces, in addition to the insertion tableau $P(x)$, the recording tableau $Q(x)$, which is a standard tableau in the alphabet $[k]$. We will prove that x is plactically equivalent to $P(x)$, and moreover, that if y is another element of $\mathbb{B}^{\otimes k}$, then x and y lie in the same connected subcrystal of $\mathbb{B}^{\otimes k}$ if and only if $Q(x) = Q(y)$.

8.2 The plactic monoid and Knuth equivalence

By Theorem 6.1 and Corollary 6.2, the crystal $\mathbb{B}^{\otimes k}$ decomposes into crystals each isomorphic to $\mathcal{B}_\lambda$, and each of these has a unique highest weight vector. So knowing the highest weight vectors in $\mathcal{B}_\lambda$ will be useful in understanding the plactic equivalence. Let us say that the word $u_1 \ldots u_k$ is a *Yamanouchi word* if for every i each final segment $\{u_j, \ldots, u_k\}$ contains at least as many i's as it does $i + 1$'s. (Yamanouchi words are also called *reverse lattice words* in the literature.)

Proposition 8.2. *The element* (8.1) *is a highest weight element if and only if its reading word is a Yamanouchi word.*

Proof. To test whether $e_i x = 0$, we apply (2.15) with $x_j = u_{k+1-j}$, so that

$$\varepsilon_i(x) = \max_{j=1}^{k} \left(\sum_{h=j}^{k} \varepsilon_i(u_h) - \sum_{h=j+1}^{k} \varphi_i(u_h) \right).$$

If the maximum is not zero, then the maximum is attained where $u_j = i + 1$, so we may as well include $\varphi_i(u_j)$ in the second summation on the right. The right-hand side then equals the number of $i + 1$'s in a final segment, and so the condition is exactly that word(x) is a Yamanouchi word. $\square$

In the following discussion we *identify* $\mathcal{B}_\lambda$ with $\mathrm{RR}(\mathcal{B}_\lambda)$. As we have seen in Section 3.2, there are elements of $\mathbb{B}^{\otimes k}$ that are not in the union of these $\mathcal{B}_\lambda$ embedded via the row reading map RR. For example $\boxed{1} \otimes \boxed{3} \otimes \boxed{2}$ is not in $\mathrm{RR}(\mathcal{B}_\lambda)$ for any partition λ of 3. Still it does lie in a subcrystal isomorphic to $\mathcal{B}_{(2,1)}$. In Section 3.2 we assumed that $n = 3$, but we now show that decomposition of $\mathbb{B}^{\otimes 3}$ is really the same if $n \geqslant 3$.

Proposition 8.3. *Let $\mathbb{B}$ be the standard $\mathrm{GL}(n)$ crystal. Assume that $n \geqslant 3$. Then $\mathbb{B}^{\otimes 3}$ decomposes into four irreducible crystals. One each is isomorphic to $\mathcal{B}_{(3)}$ and $\mathcal{B}_{(1,1,1)}$, which we recall are identified with subcrystals of $\mathbb{B}^{\otimes 3}$. The other two are isomorphic to $\mathcal{B}_{(2,1)}$.*

Proof. By Theorem 6.1, the tensor product $\mathbb{B}^{\otimes k}$ decomposes into a direct sum of crystals each isomorphic to $\mathcal{B}_\lambda$ for some λ. Since each of these contains a unique highest weight element, that is, an element annihilated by e_i for all $i \in I = \{1, 2, \ldots, n\}$, we may find these subcrystals by looking for highest weight vectors. We know in advance that we will find $\mathcal{B}_{(3)}$, $\mathcal{B}_{(1,1,1)}$ and $\mathcal{B}_{(2,1)}$, which we have defined as subcrystals of $\mathbb{B}^{\otimes 3}$.

Thus we must determine when the element $x = \boxed{p} \otimes \boxed{q} \otimes \boxed{r}$ such that pqr is a Yamanouchi word. Clearly from the definition we must have $r = 1$, q must be 2 or 1 and r must be 3, 2 or 1. We find four Yamanouchi words and four highest weight vectors:

$$\boxed{1} \otimes \boxed{1} \otimes \boxed{1}, \qquad \boxed{3} \otimes \boxed{2} \otimes \boxed{1},$$
$$\boxed{2} \otimes \boxed{1} \otimes \boxed{1}, \qquad \boxed{1} \otimes \boxed{2} \otimes \boxed{1}.$$

The first three are the highest weight vectors for $\mathcal{B}_{(3)}$, $\mathcal{B}_{(1,1,1)}$ and $\mathcal{B}_{(2,1)}$. The fourth vector has weight $(2, 1, 0)$ so it is the highest weight vector of another crystal isomorphic to $\mathcal{B}_{(2,1)}$. $\square$

As we noted before, we identify $\mathcal{B}_{(2,1)}$ with its image in $\mathbb{B}^{\otimes 3}$ via the embedding RR. We will denote the other subcrystal of $\mathbb{B}^{\otimes 3}$ that is isomorphic to $\mathcal{B}_{(2,1)}$ by $\mathcal{B}'_{(2,1)}$.

Theorem 8.4. *Let $x, y \in \mathbb{B}^{\otimes k}$. If $w(x) \equiv_K w(y)$, then x and y are plactically equivalent.*

Proof. We will prove this first when $k = 3$ and then deduce the general case. By Proposition 8.3, the tensor product $\mathbb{B}^{\otimes 3}$ is the disjoint union of four crystals $\mathcal{B}_{(3)}$, $\mathcal{B}_{(1,1,1)}$, $\mathcal{B}_{(2,1)}$ and $\mathcal{B}'_{(2,1)}$, where $\mathcal{B}'_{(2,1)}$ may be characterized as the set theoretic complement of the other three crystals. Thus $\boxed{a} \otimes \boxed{b} \otimes \boxed{c}$ lies in $\mathcal{B}_{(3)}$ if $a \leqslant b \leqslant c$; it lies in $\mathcal{B}_{(1,1,1)}$ if $a > b > c$; it lies in $\mathcal{B}_{(2,1)}$ if $a > b$ and $b \leqslant c$. In these three respective cases it represents the tableau

$$\boxed{a\,b\,c}, \qquad \begin{array}{c}\boxed{c}\\\boxed{b}\\\boxed{a}\end{array}, \qquad \begin{array}{c}\boxed{b\,c}\\\boxed{a}\end{array}.$$

If none of these possibilities pertain, then the element is in $\mathcal{B}'_{(2,1)}$.

Define $\theta\colon \mathcal{B}_{(2,1)} \longrightarrow \mathcal{B}'_{(2,1)}$ to be the unique crystal isomorphism. We show that if $a \leqslant b < c$, then $x = \boxed{c} \otimes \boxed{a} \otimes \boxed{b}$ is plactically equivalent to $y = \boxed{a} \otimes \boxed{c} \otimes \boxed{b}$. (The reading words are Knuth equivalent.) In this case, $x \in \mathcal{B}_{(2,1)}$ and $y \in \mathcal{B}'_{(2,1)}$. We will prove that $y = \theta(x)$. We establish this by induction on b with a being fixed. If $b = a$, then x is the unique element of $\mathcal{B}_{(2,1)}$ with weight $\mathrm{wt}(x)$ and since $y \in \mathcal{B}'_{(2,1)}$ has the same weight, y is the unique element of $\mathcal{B}'_{(2,1)}$ with that weight, and so $\theta(x) = y$ in this case. Now arguing by induction, we may assume that $a < b < c$ and that $\theta(x_1) = y_1$ where

$$x_1 = \boxed{c} \otimes \boxed{a} \otimes \boxed{b-1}, \qquad y_1 = \boxed{a} \otimes \boxed{c} \otimes \boxed{b-1}.$$

We now apply f_{b-1} to this identity. We have $\varepsilon_{b-1}(c) = 0$ since $b < c$. Since

$$\varphi_{b-1}\!\left(\boxed{a} \otimes \boxed{b-1}\right) = \varphi_{b-1}\!\left(\boxed{a\,\,b-1}\right) \geqslant 1$$

we have

$$f_{b-1}x_1 = \boxed{c} \otimes f_{b-1}\boxed{a\,\,b-1} = \boxed{c} \otimes \boxed{a\,\,b} = x.$$

On the other hand, $\varepsilon_{b-1}(a) = 0$ since $a < b$ and $\varepsilon_{b-1}(c) = 0$ since $b < c$, so

$$f_{b-1}y_1 = \boxed{a} \otimes \boxed{c} \otimes f_{b-1}\boxed{b-1} = y.$$

Therefore $\theta(x) = \theta(f_{b-1}x_1) = f_{b-1}\theta(x_1) = f_{b-1}y_1 = y$, completing the induction. We see that x and y are plactically equivalent.

Finally, we must show that if $a < b \leqslant c$, then $x = \boxed{b} \otimes \boxed{a} \otimes \boxed{c}$ is plactically equivalent to $y = \boxed{b} \otimes \boxed{c} \otimes \boxed{a}$. Now $x \in \mathcal{B}_{(2,1)}$ and $y \in \mathcal{B}'_{(2,1)}$. We will prove that $\theta(x) = y$. The case where $c = b$ is clear since then x and y are the unique elements of the crystal of weight $\mathrm{wt}(x)$. Thus by induction we may assume that $a < b < c$ and that

$$\theta\!\left(\boxed{b} \otimes \boxed{a} \otimes \boxed{c-1}\right) = \boxed{b} \otimes \boxed{c-1} \otimes \boxed{a}.$$

We leave it to the reader to check that

$$f_{c-1}\left(\boxed{b}\otimes\boxed{a}\otimes\boxed{c-1}\right) = x, \qquad f_{c-1}\left(\boxed{b}\otimes\boxed{c-1}\otimes\boxed{a}\right) = y.$$

The result is now proved when $k = 3$. We will deduce from this the general case. If the reading words word(x) and word(y) are connected by an elementary Knuth operation, then we may write $x = u \otimes x_1 \otimes v$ and $y = u \otimes y_1 \otimes v$, where $x_1, y_1 \in \mathbb{B}^{\otimes 3}$ are such that word(x_1) $\equiv_K$ word(y_1) and u, v are in $\mathbb{B}^{\otimes \ell}$, $\mathbb{B}^{\otimes m}$ with $\ell + m + 3 = k$. Thus x_1 and y_1 are plactically equivalent, so if $\mathcal{C}$ and $\mathcal{D}$ are the full subcrystals of $\mathbb{B}^{\otimes 3}$ containing x_1 and y_1 there is an isomorphism $\mathcal{C} \to \mathcal{D}$ taking x_1 to y_1. This induces an isomorphism

$$\left(\mathbb{B}^{\otimes \ell}\right) \otimes \mathcal{C} \otimes \left(\mathbb{B}^{\otimes m}\right) \to \left(\mathbb{B}^{\otimes \ell}\right) \otimes \mathcal{D} \otimes \left(\mathbb{B}^{\otimes m}\right)$$

that takes x to y, so x and y are plactically equivalent. $\qquad \square$

The relationship between the plactic monoid and crystals for other root systems (beyond type A) were studied in [Lecouvey (2002, 2003); Kim and Shin (2004); Cain, Gray and Malheiro (2014)].

8.3   Crystals and Schensted insertion

Let us explain a relationship between Schensted insertion described in Chapter 7 and crystals. If $\mathcal{C}$ and $\mathcal{D}$ are crystals, we will denote their direct sum as crystals as $\mathcal{C} \oplus \mathcal{D}$; as a set, this is the disjoint union. In the following proposition, $\mathcal{B}_\lambda$ denotes the GL(n) crystals of highest weight λ and $\mathbb{B}$ is the standard GL(n) crystal.

Proposition 8.5. *We have an isomorphism $\mathcal{B}_{(k)} \otimes \mathbb{B} \cong \mathcal{B}_{(k+1)} \oplus \mathcal{B}_{(k,1)}$. In this isomorphism, if $T \in \mathcal{B}_{(k)}$ is a row tableau, then $T \otimes \boxed{j}$ corresponds to the tableau $T \leftarrow j$. Thus $T \otimes \boxed{j} \equiv T \leftarrow j$.*

Proof. First let us show that $\mathcal{B}_{(k)} \otimes \mathbb{B} \cong \mathcal{B}_{(k+1)} \oplus \mathcal{B}_{(k,1)}$. By Theorem 6.1, we know that $\mathcal{B}_{(k)} \otimes \mathbb{B}$ is a disjoint union of crystals that are isomorphic to crystals of tableaux. Since each crystal of tableaux has a unique highest weight vector, we can determine which ones appear by finding all highest weight vectors in $\mathcal{B}_{(k)} \otimes \mathbb{B}$. These are elements $T \otimes \boxed{j}$ such that $\varepsilon_i\left(T \otimes \boxed{j}\right) = 0$ for all $i \in I = \{1, 2, \ldots, n\}$. Since

$$\varepsilon_i(T \otimes \boxed{j}) = \max\left(\varepsilon_i(\boxed{j}), \varepsilon_i(T) + \varepsilon_i(\boxed{j}) - \varphi_i(\boxed{j})\right),$$

we must have $\varepsilon_i(\boxed{j}) = 0$ for all $i \in I$. Therefore $j = 1$. We have $\varphi_i(\boxed{1}) = 0$ unless $i = 1$, and so $\varepsilon_i(T) = 0$ for all T if $i \neq 1$, while $\varepsilon_1(T) = 0$ or 1. This means that T can be either $\boxed{1}\cdots\boxed{1}$ or $\boxed{1}\cdots\boxed{1}\boxed{2}$, and there are no other possibilities. Hence $\mathcal{B}_{(k)} \otimes \mathbb{B}$ contains two highest weight vectors:

$$\boxed{1}\cdots\boxed{1}\otimes\boxed{1}, \qquad \boxed{1}\cdots\boxed{1}\boxed{2}\otimes\boxed{1}. \qquad\qquad (8.2)$$

The weights of the two elements in (8.2) are $(k+1, 0, \ldots, 0)$ and $(k, 1, 0, \ldots, 0)$, so the two crystals of tableaux that appear in the decomposition are $\mathcal{B}_{(k+1)}$ and $\mathcal{B}_{(k,1)}$. This proves that $\mathcal{B}_{(k)} \otimes \mathbb{B} \cong \mathcal{B}_{(k+1)} \oplus \mathcal{B}_{(k,1)}$.

We must show that in this isomorphism $T \otimes \boxed{j}$ corresponds to the tableau $T \leftarrow j$. Let $T = \boxed{i_1 \cdots i_k}$. First suppose that $j \geqslant i_k$. Then

$$\boxed{i_1 \cdots i_k} \otimes \boxed{j} = \boxed{i_1} \otimes \cdots \otimes \boxed{i_k} \otimes \boxed{j} = \boxed{i_1 \cdots i_k \, j} = T \leftarrow j \,,$$

so the elements of $\mathcal{B}_{(k)} \otimes \mathbb{B}$ with $j \geqslant i_k$ correspond to the elements of $\mathcal{B}_{(k+1)}$.

It remains to consider the elements of $\mathcal{B}_{(k)} \otimes \mathbb{B}$ with $j < i_k$. We know that these correspond to elements of $\mathcal{B}_{(k,1)}$ since they do not correspond to elements of $\mathcal{B}_{(k+1)}$, which are already accounted for. By Theorem 8.4 it is enough to show that the reading words of $T \otimes \boxed{j}$ and $T \leftarrow j$ are Knuth equivalent, since this will mean that $T \leftarrow j$ is the unique element of $\mathcal{B}_{(k,1)}$ that is plactically equivalent to $T \otimes \boxed{j}$.

Let $s \leqslant k$ be the smallest integer such that $j < i_s$. Then i_s is bumped under Schensted insertion. Thus the reading word of $T \otimes \boxed{j}$ is $i_1 i_2 \ldots i_k j$, while the reading word of $T \leftarrow j$ is

$$i_s i_1 \ldots i_{s-1} \, j \, i_{s+1} \ldots i_k.$$

Rather than give a formal proof that these words are Knuth equivalent, we do a typical example. Suppose that

$$T = \boxed{1\,1\,2\,3\,3\,3\,4}, \qquad j = 2.$$

We need to show the Knuth equivalence of the words 11233342 and 31122334. First we use (three times) the equivalence $bca \equiv_K bac$ if $a < b \leqslant c$, obtaining the equivalences

$$11233342 \equiv_K 11233324 \equiv_K 11233234 \equiv_K 11232334.$$

After this we use the equivalence $cab \equiv_K acb$ if $a \leqslant b < c$ until we are finished:

$$11232334 \equiv_K 11322334 \equiv_K 13122334 \equiv_K 31122334.$$

For a proof in the general case, see Exercise 8.2. □

Both the P and Q tableau in the RSK algorithm have a crystal interpretation. Let $x = \boxed{i_1} \otimes \cdots \otimes \boxed{i_k}$. Inserting $i_1, i_2, \ldots$ in order starting with the empty tableau, let $P(x)$ and $Q(x)$ be the insertion and reading tableau. We will prove next that $P(x) = P(y)$ if and only if x and y are plactically equivalent. On the other hand, we will prove in Theorem 8.7 below that $Q(x) = Q(y)$ if and only if x and y lie in the same connected subcrystal.

Theorem 8.6. *The crystal $\mathbb{B}^{\otimes k}$ decomposes into a disjoint union of crystals, each isomorphic to $\mathcal{B}_\lambda$, where λ is a partition of k of length $\leqslant n$. Let $x \in \mathbb{B}^{\otimes k}$.*

(i) We have $x \equiv P(x)$.
(ii) If λ is the shape of $P(x)$ and $Q(x)$, then x lies in a subcrystal isomorphic to $\mathcal{B}_\lambda$.

(iii) If $P(x) = P(y)$, then x and y are plactically equivalent.

Proof. Parts (ii) and (iii) follow from (i) and Theorem 8.4. We prove (i). By induction, it is sufficient to show that $T \otimes \boxed{j} \equiv T \leftarrow j$ for any tableau T. Indeed, if $x = \boxed{i_1} \otimes \cdots \otimes \boxed{i_k}$, then arguing inductively $x \equiv T \otimes \boxed{j}$ where $j = i_k$ and T is

$$\varnothing \leftarrow i_1 \leftarrow i_2 \leftarrow \cdots \leftarrow i_{k-1}.$$

If we knew that $T \otimes \boxed{j} \equiv T \leftarrow j$, then indeed $x \equiv P(x)$ since

$$P(x) = \varnothing \leftarrow i_1 \leftarrow i_2 \leftarrow \cdots \leftarrow i_k.$$

Let us now show that $T \otimes \boxed{j} \equiv T \leftarrow j$. If T consists of a single row, then this is Proposition 8.5. To prove the general case, suppose that T has rows $T_1, \ldots, T_r$. By Proposition 8.5, $T_1 \otimes \boxed{j} \equiv T_1 \leftarrow j$. Let k_1 be the length of T_1. If $T_1 \leftarrow j$ is a single row of length $k_1 + 1$, then we may just tensor with $T_r \otimes \cdots \otimes T_1 \otimes \boxed{j}$ to obtain $T \otimes \boxed{j} \equiv T \leftarrow j$. On the other hand, if $T_1 \leftarrow j$ has shape $\mathcal{B}_{(k_1,1)}$, then we know that Schensted insertion of j bumps some j', and $T_1 \leftarrow j$ is $\boxed{j'} \otimes T_1'$, an element of the crystal of tableaux $\mathcal{B}_{(k_1,1)}$, where j' is the bumped entry of T_1, and T_1' is T_1 with j' replaced by j. But now

$$T \otimes \boxed{j} = T_r \otimes \cdots \otimes T_1 \otimes \boxed{j} \equiv T_r \otimes \cdots \otimes T_2 \otimes \boxed{j'} \otimes T_1',$$

and we may repeat this process (which mirrors the RSK algorithm) until we are done. $\square$

Let P be built up from the empty tableau by a series of insertions:

$$P = (\varnothing \leftarrow i_1 \leftarrow \cdots \leftarrow i_k).$$

Let $P_1, P_2, \ldots$ be the sequence of intermediate tableaux. For the next proof we remind the reader of some terminology from Chapter 7. When we insert i_t, the location in the top row where i_t is inserted is the *landing location*. If inserting i_t bumps some value, which in turn may bump another value, the final location where the last bumped value is inserted at the end of a row is called the *new location*. So the new location is the unique location in $\mathrm{YD}(P_t) \setminus \mathrm{YD}(P_{t-1})$. The landing location (in the top row) and the new location are the first and last locations in the bumping path, which is the sequence of locations where P_t differs from P_{t-1}.

Theorem 8.7. *Suppose that $x, y \in \mathbb{B}^{\otimes k}$. Then $Q(x) = Q(y)$ if and only if x and y lie in the same connected subcrystal.*

Proof. First suppose that x and y lie in the same connected subcrystal $\mathcal{C}$. Then we show that $Q(x) = Q(y)$. Since every element of $\mathcal{C}$ is obtained from its highest weight element by applying f_j in some sequence, it is sufficient to show that if $y = f_j x$, then $Q(y) = Q(x)$.

We will make use of the bracketing procedure introduced in Section 2.4 to explain how to determine which j in

$$x = \boxed{i_1} \otimes \cdots \otimes \boxed{i_k}$$

is promoted to $j + 1$ by the application of f_j. We saw there that certain elements of the sequence $\{i_1, \ldots, i_k\}$ which are equal to j are paired or "bracketed" with certain elements of the sequence that are equal to $j + 1$. (See Example 2.34 for an example.) Every $j + 1$ that is bracketed lies to the left of the j that it is paired with. Every unbracketed $j + 1$ lies to the left of every unbracketed j, and the j that is changed to $j + 1$ in y is the last unbracketed j.

Let us assume first that every value of i_t is either j or $j + 1$. Afterwards we will see what modifications are needed when other values occur. We know that every $j + 1$ that occurs in the initial segment $\{i_1, i_2, \ldots, i_m\}$ is bracketed with a later j. These are all bumped, and so the new position of $i_m = j$ is the same as its landing location on the first row. Replacing $i_m = j$ by $j + 1$ will clearly not change this. So up to m, the recording tableau $Q(y)$ will be the same as $Q(x)$. Both recording tableaux will have m in the same location on the first row.

But we also have to consider insertions that occur after i_m. Changing i_m from j to $j + 1$ may cause it to be bumped by a later $i_t = j$ that is inserted with $t > m$. These insertions and bumpings will clearly be unaffected by changing i_m from j to $j + 1$. Therefore $Q(y) = Q(x)$.

We have discussed the case where all the i_t are equal to either j or $j + 1$. Now let us consider what happens when we add numbers to the sequence that are either $< j$ or $> j + 1$. Let $\mathbf{i} = (i_1, \ldots, i_k)$ and let $\mathbf{i}_{j,j+1}$ be the subsequence of $\mathbf{i}$ in which the entries not equal to j or $j + 1$ are discarded. A difference from the simpler case just considered is that some of the elements of $\mathbf{i}_{j,j+1}$ may be bumped by insertions of entries $< j$. These do not distinguish between j and $j + 1$. Hence although the unbracketed j that is inserted at i_m is changed to $j + 1$ in y, if that j is bumped when f_i is applied to x, then the corresponding $j + 1$ is bumped when f_i is applied to y. So it is clear that the first rows of $Q(y)$ and $Q(x)$ are the same.

Let $P'(x)$ and $Q'(x)$ be the tableaux obtained from $P(x)$ and $Q(x)$ by removing their top rows. We still have to show that $Q'(x) = Q'(y)$. Let $\mathbf{i}'(x)$ be the sequence of elements that are bumped from the first row. Thus $P'(x)$ is built up from the empty tableau by inserting these elements. If the first unbracketed $j = i_m$ is not bumped, then there is nothing to prove, since $\mathbf{i}'(x) = \mathbf{i}'(y)$. The sequence $\mathbf{i}'(y)$ differs from $\mathbf{i}'(x)$ in that one entry is changed from j to $j + 1$. We will argue that this is the last unbracketed j in the sequence $\mathbf{i}'(x)$. Then the statement will follow by induction.

Let i_t be the last j that is bumped from the first row. Because we assume that the j that is inserted at i_m is bumped, we have $t \geqslant m$. Now before this i_t can be bumped from the first row, every j or $j + 1$ before it must also be bumped. Hence $\mathbf{i}'_{j,j+1}$ consists of an initial segment of $\mathbf{i}_{j,j+1}$, together possibly with some later values all of which are equal to $j + 1$. Since we thus see that $\mathbf{i}'_{j,j+1}$ is obtained from $\mathbf{i}_{j,j+1}$ by taking an initial segment and then appending some $j + 1$'s at the end, it is clear that the last unbracketed j in $\mathbf{i}'$ is the one that is changed to $j + 1$ in $\mathbf{i}'(y)$.

We have shown that if x and y are in the same connected subcrystal then $Q(x) = Q(y)$. To show the converse, suppose that $Q(x) = Q(y)$. To show that x and y lie in the same connected crystal, since we have shown that Q is constant on these, we may assume that x and y are highest weight vectors. If λ is the shape of Q, then P must be the highest weight vector in $\mathcal{B}_\lambda$, regarded as a tableau. But then $P(x) = P(y)$ and since x is determined by $P(x)$ and $Q(x)$, it follows that $x = y$. □

8.4 Crystals of skew tableaux

Recall from Chapter 7 that a *skew shape* is an ordered pair of partitions (λ, μ) such that $\mathrm{YD}(\lambda) \supseteq \mathrm{YD}(\mu)$, denoted λ/μ. The Young diagram $\mathrm{YD}(\lambda/\mu)$ is just the set theoretic difference $\mathrm{YD}(\lambda) \setminus \mathrm{YD}(\mu)$. By a *skew tableau* with shape λ/μ we mean an assignment of values from the alphabet $[n]$ to the boxes in $\mathrm{YD}(\lambda/\mu)$. We will say that the tableau is *semistandard* if the rows are weakly increasing and the columns strictly increasing. Note that usual semistandard tableaux are the special case when $\mu = \varnothing$.

The *reading word* is the set of entries read row by row, taking the rows from bottom to the top, and reading each row from left to right. For a tableau, this is the same as the row reading that we used in Chapter 3.

Thus if $\lambda/\mu = (5,4,4,3)/(2,1)$, consider the tableau

$$T = \begin{array}{ccccc} & & \boxed{1}\boxed{2}\boxed{2} \\ & \boxed{1}\boxed{2}\boxed{4} \\ \boxed{1}\boxed{2}\boxed{3}\boxed{5} \\ \boxed{2}\boxed{4}\boxed{4} \end{array} \ .$$

The reading word $\mathrm{word}(T) = 2441235124122$. Then we define the *row reading* $\mathrm{RR}(T)$ to be the element of $\mathbb{B}^{\otimes k}$ (where $k = |\lambda| - |\mu|$) obtained by tensoring the elements of the reading word in order. So in our example

$$\mathrm{RR}(T) = \boxed{2} \otimes \boxed{4} \otimes \boxed{4} \otimes \boxed{1} \otimes \boxed{2} \otimes \boxed{3} \otimes \boxed{5} \otimes \boxed{1} \otimes \boxed{2} \otimes \boxed{4} \otimes \boxed{1} \otimes \boxed{2} \otimes \boxed{2} \ .$$

With n fixed, let $\mathcal{B}_{\lambda/\mu}$ be the set of semistandard skew tableaux of shape λ/μ.

Theorem 8.8. *The subset $\mathrm{RR}(\mathcal{B}_{\lambda/\mu})$ is a subcrystal of $\mathbb{B}^{\otimes k}$.*

Proof. The proof is identical to that of Theorem 3.2. We leave the details to the reader. □

We give $\mathcal{B}_{\lambda/\mu}$ the crystal structure that it inherits from $\mathbb{B}^{\otimes k}$ via the embedding RR. Thus we have defined crystals of skew tableaux. By Corollary 6.2, $\mathcal{B}_{\lambda/\mu}$ can be decomposed into a direct sum of crystals isomorphic to $\mathcal{B}_\nu$, where ν is a partition of k. Our next task is to describe this decomposition more precisely.

Definition 8.9. Let λ/μ be a skew partition, and let ν be another partition. The *Littlewood–Richardson coefficient* $c_{\mu\nu}^\lambda$ is defined to be the number of skew tableaux of shape λ/μ with weight ν whose reading words are Yamanouchi words.

It is clear that there can only be a skew tableau T of shape λ/μ and weight ν only if $|\lambda| = |\mu| + |\nu|$, so $c^\lambda_{\mu\nu} = 0$ if this is not the case. The alphabet for these tableaux can be $[n]$ for any sufficiently large n, because if the reading word of T is Yamanouchi, it can only involve entries from $[k]$, where k is the number of rows in λ/μ.

Proposition 8.10. *For n sufficiently large, we have*

$$\mathcal{B}_{\lambda/\mu} \cong \bigoplus_\nu \mathcal{B}_\nu^{\oplus c^\lambda_{\mu\nu}}. \tag{8.3}$$

Proof. The number of connected components of $\mathcal{B}_{\lambda/\mu}$ that are isomorphic to $\mathcal{B}_\nu$ equals the number of highest weight elements of this weight. By Proposition 8.2, these are the tableaux whose reading words are Yamanouchi words. $\square$

Generalizing (3.3), we may interpret the character $s_{\lambda/\mu}$ of $\mathcal{B}_{\lambda/\mu}$ as a symmetric polynomial. These are the *skew Schur functions*. We see that

$$s_{\lambda/\mu} = \sum_\nu c^\lambda_{\mu\nu} s_\nu .$$

Example 8.11. *Consider $\lambda/\mu = (2,2)/(1)$. In this case there is a unique Yamanouchi word $(1,2,1)$ that may be realized as the reading word of a tableau of this shape, namely the tableaux*

$$\begin{array}{cc} & \boxed{1} \\ \boxed{1}\boxed{2} & \end{array}.$$

Thus $\mathcal{B}_{(2,2)/(1)} \cong \mathcal{B}_{(2,1)}$ and therefore

$$s_{(2,2)/(1)} = s_{(2,1)} .$$

Embedded via the row reading into $\mathbb{B}^{\otimes 3}$, this gives us the second crystal in Figure 3.3 (when $n = 3$) or in the proof of Proposition 8.3 (for general n).

Example 8.12. *Consider $\lambda/\mu = (3,2)/(1)$. We find that there are two tableaux*

$$\begin{array}{cc} \boxed{1}\boxed{1} \\ \boxed{1}\boxed{2} \end{array} \quad and \quad \begin{array}{cc} \boxed{1}\boxed{1} \\ \boxed{2}\boxed{2} \end{array}$$

with this shape whose reading words 1211 and 2211 are Yamanouchi words, and hence

$$s_{(3,2)/(1)} = s_{(3,1)} + s_{(2,2)}.$$

Crystals of skew tableaux can be used to describe the Levi branching of $\mathcal{B}_\lambda$ from $\mathrm{GL}(n)$ to $\mathrm{GL}(r) \times \mathrm{GL}(n-r)$. (See Section 2.8.)

As usual, we are identifying the $\mathrm{GL}(n)$ weight lattice $\Lambda = \Lambda_{\mathrm{GL}(n)}$ with $\mathbb{Z}^n$. The weight lattice $\Lambda_{\mathrm{GL}(r)} \times \Lambda_{\mathrm{GL}(n-r)}$ of $\mathrm{GL}(r) \times \mathrm{GL}(n-r)$ is also $\mathbb{Z}^n$, decomposed into a direct product $\Lambda \cong \mathbb{Z}^n = \mathbb{Z}^r \otimes \mathbb{Z}^{n-r}$. Thus if $\mu = (\mu_1, \ldots, \mu_n) \in \Lambda$, we may write $\mu = (\mu', \mu'')$ where $\mu' = (\mu_1, \ldots, \mu_r)$ and $\mu'' = (\mu_{r+1}, \ldots, \mu_n)$. Let $I = \{1, 2, \ldots, n-1\}$ be the $\mathrm{GL}(n)$ index set. The $\mathrm{GL}(r) \times \mathrm{GL}(n-r)$ index set is obtained by deleting the index r. The $\mathrm{GL}(n)$ simple roots $\alpha_1, \ldots, \alpha_{r-1}$ are interpreted as the simple roots for $\mathrm{GL}(r)$, and the simple roots $\alpha_{r+1}, \ldots, \alpha_{n-1}$ are interpreted as simple roots for $\mathrm{GL}(n-r)$.

Now the connected Stembridge $\mathrm{GL}(r) \times \mathrm{GL}(n-r)$ crystals may be described. Let $\mathcal{C}$ and $\mathcal{D}$ be connected Stembridge crystals for $\mathrm{GL}(r)$ and $\mathrm{GL}(n-r)$, respectively. We shift the indices for $\mathrm{GL}(n-r)$ from their usual $1, \ldots, n-r-1$ to $r+1, \ldots, n-1$. Let

$$\mathcal{C} \boxtimes \mathcal{D} = \{x \boxtimes y \mid x \in \mathcal{C}, y \in \mathcal{D}\}$$

be the Cartesian product of the two sets $\mathcal{C}$ and $\mathcal{D}$. As in Section 5.8, if $i \in I \setminus \{r\}$, $x \in \mathcal{C}$ and $y \in \mathcal{D}$ define

$$f_i(x \boxtimes y) = \begin{cases} f_i(x) \boxtimes y & \text{if } i < r, \\ x \boxtimes f_i(y) & \text{if } i > r, \end{cases}$$

with similar definitions for e_i, φ_i and ε_i. Thus $\mathcal{C} \boxtimes \mathcal{D}$ becomes a $\mathrm{GL}(r) \times \mathrm{GL}(n-r)$ crystal.

Lemma 8.13. *Every connected Stembridge* $\mathrm{GL}(r) \times \mathrm{GL}(n-r)$ *crystal is of the form* $\mathcal{C} \boxtimes \mathcal{D}$, *where* $\mathcal{C}$ *and* $\mathcal{D}$ *are Stembridge* $\mathrm{GL}(r)$ *and* $\mathrm{GL}(n-r)$ *crystals, respectively.*

Proof. Let μ be the highest weight of a given $\mathrm{GL}(r) \times \mathrm{GL}(n-r)$ crystal. Thus μ is a dominant weight. If we write $\mu = (\mu', \mu'')$, then μ' and μ'' are dominant weights for $\mathrm{GL}(r)$ and $\mathrm{GL}(n-r)$, respectively, so we may find connected Stembridge crystals $\mathcal{C}$ and $\mathcal{D}$ with the given weights. It follows from Theorem 4.13 that the given crystal is isomorphic to $\mathcal{C} \boxtimes \mathcal{D}$. $\square$

Now let λ be a partition of k of length $\leqslant n$, and let $\mathcal{B}_\lambda$ be the usual $\mathrm{GL}(n)$ crystal of tableaux. By identifying $\Lambda_{\mathrm{GL}(n)}$ with $\Lambda_{\mathrm{GL}(r)} \otimes \Lambda_{\mathrm{GL}(n-r)}$ as above, we may branch $\mathcal{B}_\lambda$ to obtain a $\mathrm{GL}(r) \times \mathrm{GL}(n-r)$ crystal.

Theorem 8.14. *Let* λ *be a partition of* k *of length* $\leqslant n$. *Branching* $\mathcal{B}_\lambda$ *to* $\mathrm{GL}(r) \times$ $\mathrm{GL}(n-r)$ *gives*

$$\bigoplus_{\substack{\mu \\ \mathrm{YD}(\mu) \subseteq \mathrm{YD}(\lambda)}} \left(\mathcal{B}_\mu \boxtimes \mathcal{B}_{\lambda/\mu}\right) \cong \bigoplus_{\substack{\mu, \nu \\ |\mu| + |\nu| = k \\ \ell(\mu) \leqslant n, \ell(\nu) \leqslant n-r}} \left(\mathcal{B}_\mu \boxtimes \mathcal{B}_\nu\right)^{\oplus c_{\mu\nu}^\lambda}.$$

Proof. Given a tableau $T \in \mathcal{B}_\lambda$, let $T(i, j)$ denote the value of T in the box in the i-th row and j-th column of T, when $(i, j) \in \mathrm{YD}(\lambda)$. Now, let μ be the partition determined by the condition that $(i, j) \in \mathrm{YD}(\mu)$ if and only if $T(i, j) \leqslant r$, while $(i, j) \in \mathrm{YD}(\lambda/\mu)$ if and only if $T(i, j) \geqslant r + 1$. Then we may reorganize T into the union of the tableau of shape μ and the skew tableau of shape λ/μ. Thus we have obtained a bijection

$$\mathcal{B}_\lambda \longrightarrow \bigoplus_{\mathrm{YD}(\mu) \subseteq \mathrm{YD}(\lambda)} \left(\mathcal{B}_\mu \boxtimes \mathcal{B}_{\lambda/\mu}\right),$$

and we leave it to the reader to check that this is a morphism of $\mathrm{GL}(r) \times \mathrm{GL}(n-r)$ crystals. The second decomposition follows from (8.3). $\square$

The character of the branched crystal can be conveniently described by partitioning the variables $t_1, \ldots, t_n$ into two parts. Thus denote $t_i = u_i$ if $1 \leqslant i \leqslant r$ and $t_i = v_{i-r}$ if $r + 1 \leqslant i \leqslant n$. We may then interpret

$$s_\lambda(t) = s_\lambda(u_1, \ldots, u_r, v_1, \ldots, v_{n-r}) = s_\lambda(u, v)$$

as the character of the branched crystal. Then Theorem 8.14 implies

$$s_\lambda(u, v) = \sum_{\mu, \nu} c^\lambda_{\mu\nu} s_\mu(u) \, s_\nu(v). \tag{8.4}$$

Since the Schur polynomial s_λ is symmetric, it follows that

$$c^\lambda_{\mu\nu} = c^\lambda_{\nu\mu}.$$

Exercises

Exercise 8.1. Prove Theorem 8.8.

Exercise 8.2. Give a formal proof of the following statement that appears in the proof of Proposition 8.5.
Suppose $1 \leqslant i_1 \leqslant i_2 \leqslant \cdots \leqslant i_k \leqslant n$ and $1 \leqslant j < i_s$ for some $1 \leqslant s \leqslant k$. Then the following two words are Knuth equivalent:

$$i_1 i_2 \ldots i_k \, j \equiv_K i_s i_1 \ldots i_{s-1} \, j \, i_{s+1} \ldots i_k.$$

Exercise 8.3.

(i) Let $n \geqslant 3$. Find two skew tableaux of shape $(2, 2, 1)/(1)$ whose reading words are Yamanouchi words, and deduce that

$$\mathcal{B}_{(2,2,1)/(1)} \cong \mathcal{B}_{(2,2)} \oplus \mathcal{B}_{(2,1,1)}.$$

(ii) Suppose that $n = 3$. Draw the crystal graphs of $\mathcal{B}_{(2,2,1)/(1)}$, $\mathcal{B}_{(2,2)}$ and $\mathcal{B}_{(2,1,1)}$.

(ii) Still with $n = 3$, describe how each of the three crystals in (ii) can be embedded in $\mathbb{B}^{\otimes 4}$. For $\mathcal{B}_{(2,2)}$ you have a choice of the row reading or the column reading. Describe them both.

Recall from Chapter 7 that the skew shape λ/μ is called a *horizontal strip* of length $|\lambda| - |\mu|$ if $\mathrm{YD}(\lambda/\mu)$ has no two boxes in the same column, and a *vertical strip* if $\mathrm{YD}(\lambda/\mu)$ has no two boxes in the same row.

Exercise 8.4. Let λ/μ be a skew partition and let $k = |\lambda| - |\mu|$. Show that

$$c^\lambda_{\mu\,(k)} = \begin{cases} 1 & \text{if } \lambda/\mu \text{ is a horizontal strip,} \\ 0 & \text{otherwise,} \end{cases}$$

and

$$c^\lambda_{\mu\,(1^k)} = \begin{cases} 1 & \text{if } \lambda/\mu \text{ is a vertical strip,} \\ 0 & \text{otherwise.} \end{cases}$$

Exercise 8.5. Let $\mu \vdash k$ and $\lambda \vdash (k+r)$ such that $\mathrm{YD}(\lambda) \supset \mathrm{YD}(\mu)$. Assume that the length of λ is $\leqslant n$. Prove that there exists a tableau T in $\mathcal{B}_\mu$ such that $T' = T \leftarrow r \leftarrow (r-1) \leftarrow \cdots \leftarrow 3 \leftarrow 2 \leftarrow 1$ is the highest weight element T' of $\mathcal{B}_\lambda$ if and only if λ/μ is a vertical strip. If this is so, there is a unique such T.

For example, suppose that $\mu = (2,2,2,2)$ and $r = 3$. Then

$$
\begin{array}{|c|c|}\hline 1&1\\\hline 2&2\\\hline 3&3\\\hline 4&4\\\hline\end{array}
\leftarrow 3 \leftarrow 2 \leftarrow 1 \quad = \quad
\begin{array}{|c|c|c|}\hline 1&1&1\\\hline 2&2&2\\\hline 3&3&3\\\hline 4&4\\\cline{1-2}\end{array}\,,
$$

$$
\begin{array}{|c|c|}\hline 1&1\\\hline 2&2\\\hline 3&4\\\hline 4&5\\\hline\end{array}
\leftarrow 3 \leftarrow 2 \leftarrow 1 \quad = \quad
\begin{array}{|c|c|c|}\hline 1&1&1\\\hline 2&2&2\\\hline 3&3\\\cline{1-2}4&4\\\cline{1-2}5\\\cline{1-1}\end{array}\,,
$$

$$
\begin{array}{|c|c|}\hline 1&1\\\hline 2&4\\\hline 3&5\\\hline 4&6\\\hline\end{array}
\leftarrow 3 \leftarrow 2 \leftarrow 1 \quad = \quad
\begin{array}{|c|c|c|}\hline 1&1&1\\\hline 2&2\\\cline{1-2}3&3\\\cline{1-2}4&4\\\cline{1-2}5\\\cline{1-1}6\\\cline{1-1}\end{array}\,,
$$

$$
\begin{array}{|c|c|}\hline 1&4\\\hline 2&5\\\hline 3&6\\\hline 4&7\\\hline\end{array}
\leftarrow 3 \leftarrow 2 \leftarrow 1 \quad = \quad
\begin{array}{|c|c|}\hline 1&1\\\hline 2&2\\\hline 3&3\\\hline 4&4\\\hline 5\\\cline{1-1}6\\\cline{1-1}7\\\cline{1-1}\end{array}\,.
$$

The shaded band is the insertion path for the final inserted 1.

Exercise 8.6. Let $\mu \vdash k$ and let λ be another partition such that $n \geqslant \ell(\lambda) + r$. Use Exercise 8.5 to show that

$$
\mathcal{B}_\mu \otimes \mathcal{B}_{(1^r)} \cong \bigoplus_{\substack{\lambda \vdash (k+r) \\ \lambda/\mu \text{ a vertical strip}}} \mathcal{B}_\lambda.
$$

Exercise 8.7. Let $\mu \vdash k$ and let λ be another partition such that $n \geqslant \ell(\lambda) + r$. Show that there exists a tableau T in $\mathcal{B}_\mu$ such that

$$
T \leftarrow \overbrace{1 \leftarrow \cdots \leftarrow 1}^{r \text{ times}} = T',
$$

where T' is the highest weight element of $\mathcal{B}_\lambda$ if and only if λ/μ is a horizontal strip. Deduce that

$$
\mathcal{B}_\mu \otimes \mathcal{B}_{(r)} \cong \bigoplus_{\substack{\lambda \vdash (k+r) \\ \lambda/\mu \text{ a horizontal strip}}} \mathcal{B}_\lambda.
$$

Chapter 9

Bicrystals and the Littlewood–Richardson Rule

We took the first steps towards interpreting the Littlewood–Richardson rule in terms of crystals in Chapter 8. In this chapter, we will take up this topic again, with a new tool. This is the bicrystal structure, which we will explain on $\mathrm{Mat}_{r \times n}(\mathbb{N})$, the set of $r \times n$ matrices whose entries are nonnegative integers. We will describe a "crystal see-saw" and as an application, we will prove the Littlewood–Richardson rule. This chapter may be skipped without loss of continuity.

Let λ be a partition of length $\leqslant n$. In this chapter we will encounter crystals of tableaux for $\mathrm{GL}(n)$ with different n. Thus, we will denote by $\mathcal{B}_\lambda^{(n)}$ the $\mathrm{GL}(n)$ crystal of tableaux with highest weight λ. Similarly, we will denote by $\pi_\lambda^{\mathrm{GL}(n)}$ the irreducible representation of $\mathrm{GL}(n, \mathbb{C})$ with highest weight λ.

In Chapter 8 we defined the Littlewood–Richardson coefficient $c_{\mu\nu}^\lambda$ to be the number of skew tableaux of shape λ/μ of weight ν whose reading words were Yamanouchi. In Theorem 8.14 we saw that $c_{\mu\nu}^\lambda$ is the multiplicity of the $\mathrm{GL}(r) \times \mathrm{GL}(s)$ crystal $\mathcal{B}^{(r)} \boxtimes \mathcal{B}^{(s)}$ in the Levi branching of the $\mathrm{GL}(r+s)$ crystal $\mathcal{B}^{(r+s)}$. Taking the character, this gives the identity of Schur functions (8.4).

On the other hand, the coefficients $c_{\mu\nu}^\lambda$ also appear in the identity

$$s_\mu s_\nu = \sum_\lambda c_{\mu\nu}^\lambda s_\lambda. \tag{9.1}$$

This is the *Littlewood–Richardson rule*. We will prove this rule in this chapter by deducing it from (8.4). The relationship between the two formulas comes from a "see-saw" that has a representation-theoretic form and a crystal form. The representation-theoretic form will be discussed in greater generality in Appendix B, culminating in Theorem B.5. We will review the situation here.

Let $\mathrm{GL}(n, \mathbb{C}) \times \mathrm{GL}(r, \mathbb{C})$ act on $\mathrm{Mat}_{r \times n}(\mathbb{C})$ by right and left multiplication. Consider the symmetric algebra on $\mathrm{Mat}_{r \times n}(\mathbb{C})$. In Appendix B we will consider its decomposition into irreducible $(\mathrm{GL}(n, \mathbb{C}) \times \mathrm{GL}(r, \mathbb{C}))$-modules. We will prove (see Theorem B.3) that $\bigvee (\mathrm{Mat}_{r \times n}(\mathbb{C}))$ decomposes as

$$\bigoplus_\lambda \pi_\lambda^{\mathrm{GL}(n)} \otimes \pi_\lambda^{\mathrm{GL}(r)},$$

where the direct sum is over all partitions whose length does not exceed r or n.

A basis of the symmetric algebra on $\mathrm{Mat}_{r \times n}(\mathbb{C})$ is in bijection with the additive monoid $\mathrm{Mat}_{r \times n}(\mathbb{N})$ of matrices with entries in the nonnegative integers $\mathbb{N}$. In this bijection, $X = (x_{i,j}) \in \mathrm{Mat}_{r \times n}(\mathbb{N})$ consists of the monomial

$$\prod_{i,j} t_{ij}^{x_{ij}},$$

where t_{ij} form the standard basis of $\mathrm{Mat}_{r \times n}(\mathbb{C})$ as a vector space. In view of this, it is natural to ask for a crystal structure on $\mathrm{Mat}_{r \times n}(\mathbb{N})$. And since $\bigvee(\mathrm{Mat}_{r \times n}(\mathbb{C}))$ is a $(\mathrm{GL}(n, \mathbb{C}) \times \mathrm{GL}(r, \mathbb{C}))$-module, $\mathrm{Mat}_{r \times n}(\mathbb{N})$ should be a $\mathrm{GL}(n) \times \mathrm{GL}(r)$ crystal.

The RSK algorithm gives a solution to this problem. Indeed by Theorem 7.13, RSK gives a bijection between $\mathrm{Mat}_{r \times n}(\mathbb{N})$ and

$$\bigcup_\lambda \mathcal{B}_\lambda^{(n)} \boxtimes \mathcal{B}_\lambda^{(r)},$$

and so we give $\mathrm{Mat}_{r \times n}(\mathbb{N})$ the unique crystal structure which makes the RSK bijection into a crystal morphism. It turns out that the crystal operations are pleasantly simple for this crystal. This *bicrystal* structure was studied by [van Leeuwen (2006)] and by [Danilov and Koshevoy (2004)].

Returning to the representation-theoretic point of view, let us consider the reason for the equivalence of the two interpretations of $c_{\mu\nu}^\lambda$, as the multiplicity in the tensor product rule for representations and as the multiplicity in the Levi branching rule. This proof, which is explained in greater detail in Section B.2 below, may be understood using the see-saw:

$$\begin{array}{ccc}
\mathrm{GL}(n) \times \mathrm{GL}(n) & & \mathrm{GL}(r + s) \\
\uparrow & \times & \uparrow \\
\mathrm{GL}(n) & & \mathrm{GL}(r) \times \mathrm{GL}(s)
\end{array} \qquad (9.2)$$

Here the vertical arrows are inclusions, for $\mathrm{GL}(n)$ is included diagonally in $\mathrm{GL}(n) \times \mathrm{GL}(n)$ and $\mathrm{GL}(r) \times \mathrm{GL}(s)$ is included as a Levi subgroup in $\mathrm{GL}(r + s)$.

All four groups in (9.2) act on $\mathrm{Mat}_{(r+s) \times n}(\mathbb{C})$ and hence on its symmetric algebra $\bigvee(\mathrm{Mat}_{(r+s) \times n})$. The diagonal lines indicate that in these actions, $\mathrm{GL}(n) \times \mathrm{GL}(n)$ is the centralizer of $\mathrm{GL}(r) \times \mathrm{GL}(s)$, and $\mathrm{GL}(n)$ is the centralizer of $\mathrm{GL}(r + s)$. We explain how the groups act on $\mathrm{Mat}_{(r+s) \times n}(\mathbb{C})$. First, we have commuting actions of $\mathrm{GL}(n)$ and $\mathrm{GL}(r + s)$ by right and left translations, namely if $g \in \mathrm{GL}(n)$ and $h \in \mathrm{GL}(r + s)$, then

$$(g, h) \colon U \mapsto h \cdot U \cdot {}^t g \qquad U \in \mathrm{Mat}_{(r+s) \times n}(\mathbb{C})$$

where ${}^t g$ is the transpose matrix. We have also commuting actions of $(g', g'') \in \mathrm{GL}(n) \times \mathrm{GL}(n)$ and $(h', h'') \in \mathrm{GL}(r) \times \mathrm{GL}(s)$ in which, splitting U into two blocks $U' \in \mathrm{Mat}_{r \times n}(\mathbb{C})$ and $U'' \in \mathrm{Mat}_{s \times n}(\mathbb{C})$:

$$\begin{pmatrix} U' \\ U'' \end{pmatrix} \mapsto \begin{pmatrix} h' \cdot U' \cdot {}^t g' \\ h'' \cdot U'' \cdot {}^t g'' \end{pmatrix}.$$

Computing the decomposition of $\bigvee(\mathrm{Mat}_{(r+s)\times n}(\mathbb{C}))$ restricted to $\mathrm{GL}(n,\mathbb{C}) \times \mathrm{GL}(r,\mathbb{C}) \times \mathrm{GL}(s,\mathbb{C})$ in two different ways leads to the identity of the two definitions of the Littlewood–Richardson coefficients. See Theorem B.4. We will find a crystal analog of this see-saw in Section 9.2.

9.1 The $\mathrm{GL}(n) \times \mathrm{GL}(r)$ bicrystal

Since by Theorem 7.13 the set $\mathrm{Mat}_{r\times n}(\mathbb{N})$ of $r \times n$ integer matrices with nonnegative entries is in bijection with the pairs of tableaux (P, Q) of the same shape in the alphabets $[n]$ and $[r]$, by transportation of structure, this set has the structure of a $\mathrm{GL}(n) \times \mathrm{GL}(r)$ crystal, with the $\mathrm{GL}(n)$ Kashiwara operators acting on P and leaving Q fixed, and conversely for the $\mathrm{GL}(r)$ Kashiwara operators.

Let us directly implement the $\mathrm{GL}(n)$ crystal structure on $\mathrm{Mat}_{r\times n}(\mathbb{N})$. We will define a weight map and Kashiwara operators, then show that it agrees with the operators that are already defined on P. The $\mathrm{GL}(r)$ statement is similar, interchanging row and column operations, by Theorem 7.14. We identify the $\mathrm{GL}(n)$ weight lattice Λ with $\mathbb{Z}^n$. If $X \in \mathrm{Mat}_{r\times n}(\mathbb{N})$, let $\mathrm{wt}(X) \in \mathbb{Z}^n$ be the vector of column sums. Also, if $1 \leqslant i \leqslant n-1$, define vectors $\psi_i(X)$ and $\delta_i(X)$ in $\mathbb{Z}^r$ as follows. We denote the entries of the matrix X by $x_{i,j}$.

$$\psi_i(X) = (\psi_{i,1}(X), \ldots, \psi_{i,r}(X)), \qquad \psi_{i,j}(X) = \sum_{k\leqslant j} x_{k,i} - \sum_{k<j} x_{k,i+1},$$

$$\delta_i(X) = (\delta_{i,1}(X), \ldots, \delta_{i,r}(X))), \qquad \delta_{i,j}(X) = \sum_{k\geqslant j} x_{k,i+1} - \sum_{k>j} x_{k,i}.$$

Let

$$\varphi_i(X) = \max_{j=1}^{r}\{\psi_{i,j}(X)\}, \qquad \varepsilon_i(X) = \max_{j=1}^{r}\{\delta_{i,j}(X)\}.$$

Let $\alpha_i = (0, 0, \ldots, 1, -1, 0, \ldots) \in \mathbb{Z}^n$ be the i-th simple root. If $\varphi_i(X) = 0$, then we define f_iX to be null, and if $\varepsilon_i(X) = 0$, then we define e_iX to be null. Assuming $\varphi_i(X) > 0$, define f_iX to be the element of $\mathrm{Mat}_{r\times n}(\mathbb{N})$ obtained by subtracting α_i from the j-th row of X, where j is the first value where $\psi_{ij}(X)$ attains its maximum. Similarly if $\varepsilon_i(X) > 0$, then let e_iX be obtained from X by adding α_i to the j-th row, where j is the last value where $\delta_{ij}(X)$ attains its maximum. We will prove in Theorem 9.1 that these crystal operators make $\mathrm{Mat}_{r\times n}(\mathbb{N})$ into a $\mathrm{GL}(n)$ crystal.

The $\mathrm{GL}(r)$ operators are obtained through matrix transpose. Let $\Lambda' = \mathbb{Z}^r$ be the $\mathrm{GL}(r)$ weight lattice. We define $\mathrm{wt}' \colon \mathrm{Mat}_{r\times n}(\mathbb{N}) \longrightarrow \Lambda'$ by taking $\mathrm{wt}'(X)$ to be the vector of row sums. We denote the remaining crystal maps as e_i', f_i', ε_i' and φ_i' ($i = 1, \ldots, r-1$). Thus define

$$\psi_i'(X) = (\psi_{i,1}'(X), \ldots, \psi_{i,n}'(X)), \qquad \psi_{i,j}'(X) = \sum_{k\leqslant j} x_{i,k} - \sum_{k<j} x_{i+1,k},$$

$$\delta_i'(X) = (\delta_{i,1}'(X), \ldots, \delta_{i,n}'(X)), \qquad \delta_{ij}'(X) = \sum_{k\geqslant j} x_{i+1,k} - \sum_{k>j} x_{i,k}.$$

Let

$$\varphi_i'(X) = \max_{j=1}^{r} \{\psi_{i,j}'(X)\}, \qquad \varepsilon_i'(X) = \max_{j=1}^{r} \{\delta_{i,j}'(X)\}.$$

Let $\alpha_i' = (0, 0, \ldots, 1, -1, 0, \ldots) \in \mathbb{Z}^r$ be the i-th simple root. If $\varphi_i'(X) = 0$, then we define $f_i'X = 0$, and if $\varepsilon_i'(X) = 0$, then we define $e_i'X = 0$. Assuming $\varphi_i'(X) > 0$, define $f_i'X$ to be the element of $\mathrm{Mat}_{r \times n}(\mathbb{N})$ obtained by subtracting α_i' from the j-th column of X, where j is the first value where $\psi_{i,j}'(X)$ attains its maximum, and similarly for e_i'.

We denote by $P(X)$ and $Q(X)$ the tableaux obtained from X by RSK, as in Theorem 7.14.

Theorem 9.1. *The above definitions make* $\mathrm{Mat}_{r \times n}(\mathbb{N})$ *into a* $\mathrm{GL}(n) \times \mathrm{GL}(r)$ *crystal. We have*

$$P(f_i X) = f_i P(X), \qquad P(e_i X) = e_i P(X),$$

$$Q(f_i X) = Q(X), \qquad Q(e_i X) = Q(X),$$

$$Q(f_i' X) = f_i Q(X), \qquad Q(e_i' X) = e_i Q(X),$$

$$P(f_i' X) = P(X), \qquad P(e_i' X) = P(X).$$

Proof. In view of Theorem 7.14, we have only to prove the first two identities. The second two will then follow by applying the first two identities to the matrix transpose of X.

We have a crystal structure on $\mathrm{Mat}_{r \times n}(\mathbb{N})$ by virtue of its bijection with pairs of tableaux, in which we let the weight function and Kashiwara operators act by transportation of structure on $P(X)$, and trivially on $Q(X)$. However, we have to show that wt, $\varphi_i, \varepsilon_i, e_i$ and f_i obtained this way agree with the ones we have defined above.

Let R_j be the row tableau containing x_{jk} copies of $\boxed{k}$. For example, if

$$X = \begin{pmatrix} 2 & 1 & 0 \\ 1 & 0 & 1 \\ 0 & 1 & 1 \\ 1 & 0 & 2 \end{pmatrix}$$

then

$$R_1 = \boxed{1\,1\,2} = \boxed{1} \otimes \boxed{1} \otimes \boxed{2},$$
$$R_2 = \boxed{1\,3} = \boxed{1} \otimes \boxed{3},$$
$$R_3 = \boxed{2\,3} = \boxed{2} \otimes \boxed{3},$$
$$R_4 = \boxed{1\,3\,3} = \boxed{1} \otimes \boxed{3} \otimes \boxed{3}.$$

By Theorem 8.6, we have $R_1 \otimes R_2 \otimes \cdots \otimes R_r \equiv P(X)$. Moreover $\varphi_i(R_j)$ is the number of entries equal to i in R_j, while $\varepsilon_i(R_j)$ is the number of $i + 1$. In other words

$$\varphi_i(R_j) = x_{j,i}, \qquad \varepsilon_i(R_j) = x_{j,i+1}.$$

By Lemma 2.33,

$$\varphi_i(R_1 \otimes R_2 \otimes \cdots \otimes R_r) = \max_{j=1}^{r} \left(\sum_{k \leqslant j} x_{k,i} - \sum_{k < j} x_{k,i+1} \right) = \max_{j=1}^{r} \{\psi_{i,j}(X)\}.$$

Since $R_1 \otimes R_2 \otimes \cdots \otimes R_r \equiv P(X)$, the same is true for $P(X)$; and similarly for ε_i. This shows that φ_i and ε_i obtained by transportation of structure agree with those we have defined. As for the f_i and e_i, these also are determined by Lemma 2.33 as follows. The lemma shows that

$$f_i(R_1 \otimes R_2 \otimes \cdots \otimes R_r \cdots) = R_1 \otimes \cdots \otimes f_i(R_j) \otimes \cdots \otimes R_r,$$

where j is the first value where $\psi_{i,j}(X)$ attains its maximum. Since $R_1 \otimes R_2 \cdots \otimes R_r \equiv P(X)$, this is plactically equivalent to $f_i P(X)$. The case of e_i is similar.

Finally, we need to show that the f_i and e_i thus described have no effect on $Q(X)$. We will deduce this from Theorem 8.7. That theorem describes the output of the RSK algorithm when applied to the sequence $\mathbf{R}$ obtained by concatenating the rows $R_1, R_2, \ldots, R_r$. The bottom row of the two-rowed array for this calculation is $\mathbf{R}$, and the top row is $1, 2, 3, \ldots$. By contrast, the two-rowed array for applying RSK to X has the same bottom row, but the top row has $|R_1|$ entries equal to 1, followed by $|R_2|$ entries equal to 2, and so forth. Hence, applying RSK to the sequence $\mathbf{R}$ produces the same $P(X)$ as applying it to the matrix X, while the recording tableaux are different. Applying RSK to $\mathbf{R}$ produces a standard recording tableau $\widetilde{Q}$. Clearly, $Q(X)$ can be obtained from $\widetilde{Q}$ by replacing the first $|R_1|$ integers $1, 2, \ldots$ by 1, then replacing the next $|R_2|$ integers by 2 and so forth. Now Theorem 8.7 implies that $\widetilde{Q}$ is unchanged by crystal operation applied to $\mathbf{R}$, and since $Q(X)$ is obtained from $\widetilde{Q}$ by the procedure described above, the same is true for $Q(X)$. $\square$

We may now give a bicrystal interpretation of the Cauchy identity.

Corollary 9.2. *As* $\mathrm{GL}(n) \times \mathrm{GL}(r)$ *crystals, we have*

$$\mathrm{Mat}_{r \times n}(\mathbb{N}) \cong \bigoplus_{\substack{\lambda \\ \ell(\lambda) \leqslant \max\{r,n\}}} \left(\mathcal{B}_\lambda^{(n)} \boxtimes \mathcal{B}_\lambda^{(r)} \right).$$

Proof. The formulas in Theorem 9.1 imply that $X \mapsto P(X) \boxtimes Q(X)$ is a morphism of $\mathrm{GL}(n) \times \mathrm{GL}(r)$ crystals from $\mathrm{Mat}_{r \times n}(\mathbb{N})$ to $\bigoplus_\lambda \left(\mathcal{B}_\lambda^{(n)} \boxtimes \mathcal{B}_\lambda^{(r)} \right)$. Since it is a bijection by Theorem 7.13, it is an isomorphism. $\square$

9.2   The crystal see-saw and the Littlewood–Richardson rule

Consider the see-saw (9.2). We gave the symmetric algebra $\bigvee \left(\mathrm{Mat}_{(r+s)n}(\mathbb{C}) \right)$ the structure of a $\left(\mathrm{GL}(n, \mathbb{C}) \right) \times \mathrm{GL}(n, \mathbb{C})$-module, and of a $\mathrm{GL}(r + s)(\mathbb{C})$-module. But these actions did not commute, so it is not true that this symmetric algebra is a module for $\mathrm{GL}(n, \mathbb{C}) \times \mathrm{GL}(n, \mathbb{C}) \times \mathrm{GL}(r+s, \mathbb{C})$. Instead, if we branch the $\mathrm{GL}(n, \mathbb{C}) \times \mathrm{GL}(n, \mathbb{C})$-structure down to $\mathrm{GL}(n, \mathbb{C})$ (which is embedded diagonally in $\mathrm{GL}(n, \mathbb{C}) \times \mathrm{GL}(n, \mathbb{C})$), we obtain an action of $\mathrm{GL}(n, \mathbb{C})$ that commutes with the action of $\mathrm{GL}(r+s, \mathbb{C})$. Similarly, if we branch the $\mathrm{GL}(r + s, \mathbb{C})$-structure to $\mathrm{GL}(r, \mathbb{C}) \times \mathrm{GL}(s, \mathbb{C})$, we get an action that commutes with the $\mathrm{GL}(n, \mathbb{C}) \times \mathrm{GL}(n, \mathbb{C})$-action.

To obtain a crystal analog of the see-saw (9.2), we define $\mathrm{GL}(n) \times \mathrm{GL}(n)$ and $\mathrm{GL}(r + s)$ crystal structures on $\mathrm{Mat}_{(r+s)\times n}(\mathbb{N})$. Let $X' \in \mathrm{Mat}_{r \times n}(\mathbb{N})$ and $X'' \in \mathrm{Mat}_{s \times n}(\mathbb{N})$, and let

$$X = \begin{pmatrix} X' \\ X'' \end{pmatrix} \in \mathrm{Mat}_{(r+s) \times n}(\mathbb{N}) \tag{9.3}$$

be the matrix obtained by stacking X' and X''. We have a $\mathrm{GL}(n) \times \mathrm{GL}(r + s)$ crystal structure on $\mathrm{Mat}_{(r+s)\times n}(\mathbb{N})$ as in Section 9.1. On the other hand, we have a $\mathrm{GL}(n) \times \mathrm{GL}(r)$ crystal structure on $\mathrm{Mat}_{r \times n}(\mathbb{N})$ and a $\mathrm{GL}(n) \times \mathrm{GL}(s)$ crystal structure on $\mathrm{Mat}_{s \times n}(\mathbb{N})$, so identifying $X \in \mathrm{Mat}_{(r+s)\times n}(\mathbb{N})$ with

$$X' \boxtimes X'' \in \mathrm{Mat}_{r \times n}(\mathbb{N}) \boxtimes \mathrm{Mat}_{s \times n}(\mathbb{N}),$$

we also obtain a $\mathrm{GL}(n) \times \mathrm{GL}(r) \times \mathrm{GL}(n) \times \mathrm{GL}(s)$ structure. We regard these as commuting $\mathrm{GL}(n) \times \mathrm{GL}(n)$ and $\mathrm{GL}(r) \times \mathrm{GL}(s)$ structures.

We now have $\mathrm{GL}(n) \times \mathrm{GL}(n)$ and $\mathrm{GL}(r+s)$ crystal structures on $\mathrm{Mat}_{(r+s)\times n}(\mathbb{N})$. It must be checked that these two structures are compatible with the tensor product operation on $\mathrm{GL}(n)$ crystals and with Levi branching from $\mathrm{GL}(r + s)$ crystals to $\mathrm{GL}(r) \times \mathrm{GL}(s)$. The compatibility needed is contained in the following result.

Proposition 9.3. *In the above situation, let (P, Q), (P', Q') and (P'', Q'') be the results of applying RSK to X, X' and X'', respectively. Then P is plactically equivalent to $P' \otimes P''$ as $\mathrm{GL}(n)$ matrices, while Q is $\mathrm{GL}(r) \times \mathrm{GL}(s)$ plactically equivalent to $Q' \boxtimes Q''$.*

Proof. As in Theorem 9.1, let R_j denote the row tableau containing $x_{j,i}$ copies of $\boxed{i}$. Then, as in Theorem 9.1, $R_1 \otimes \cdots \otimes R_{r+s}$ is plactically equivalent to P. On the other hand, for the same reason, $R_1 \otimes \cdots \otimes R_r \equiv P'$ and $R_{r+1} \otimes \cdots \otimes R_{r+s} \equiv P''$. Therefore $P \equiv P' \otimes P''$.

Now let $\widetilde{R}_j$ denote the row tableau containing $x_{i,j}$ copies of $\boxed{i}$, for $j = 1, \dots, n$. Remembering that Q bears the same relationship to the transpose ${}^t X$ that P bears to X, Q is plactically equivalent (in terms of the $\mathrm{GL}(r + s)$ crystal structure) to $\widetilde{R}_1 \otimes \cdots \otimes \widetilde{R}_n$.

The row $\widetilde{R}_j$ is obtained by concatenating $\widetilde{R}'_j$ and $\widetilde{R}''_j$, where $\widetilde{R}'_j$ contains $x_{i,j}$ copies of $\boxed{i}$ for $1 \leqslant i \leqslant r$, and $\widetilde{R}''_j$ contains $x_{i,j}$ copies of $\boxed{i}$ for $r + 1 \leqslant i \leqslant$

$r + s$. Now $\widetilde{R}'_i$ only contains entries from $1, \ldots, r$ and $\widetilde{R}''_i$ only contains entries from $r + 1, \ldots, r + s$. This means that on branching to a $\mathrm{GL}(r) \times \mathrm{GL}(s)$ crystal, $\widetilde{R}_i = \widetilde{R}'_i \otimes \widetilde{R}''_i$ is $\mathrm{GL}(r) \times \mathrm{GL}(s)$-plactically equivalent to $\widetilde{R}'_i \boxtimes \widetilde{R}''_i$. (The relation of $\mathrm{GL}(r) \times \mathrm{GL}(s)$-plactic equivalence is coarser than $\mathrm{GL}(r + s)$ plactic equivalence.) Thus as $\mathrm{GL}(r) \times \mathrm{GL}(s)$ crystals, Q is plactically equivalent to

$$(\widetilde{R}'_1 \boxtimes \widetilde{R}''_1) \otimes \cdots \otimes (\widetilde{R}'_n \boxtimes \widetilde{R}''_n).$$

This is $\mathrm{GL}(r) \times \mathrm{GL}(s)$-plactically equivalent to

$$(\widetilde{R}'_1 \otimes \cdots \otimes \widetilde{R}'_n) \boxtimes (\widetilde{R}''_1 \otimes \cdots \otimes \widetilde{R}''_n) \equiv Q' \boxtimes Q''.$$

$\qquad\qquad\qquad\qquad\qquad\qquad\qquad\qquad\qquad\qquad\qquad\qquad\qquad\qquad \square$

Theorem 9.4. *Let λ, μ and ν be partitions. Then the multiplicity of $\mathcal{B}_\lambda^{(n)}$ in $\mathcal{B}_\mu^{(n)} \otimes \mathcal{B}_\nu^{(n)}$ equals the multiplicity of $\mathcal{B}_\mu^{(r)} \boxtimes \mathcal{B}_\nu^{(s)}$ in the $\mathrm{GL}(r) \times \mathrm{GL}(s)$ crystal obtained by branching $\mathcal{B}_\lambda^{(r+s)}$.*

We are assuming that n, r and s are large enough such that the crystals appearing in this statement are all defined. We will use the notation $P(X) = P$, $Q(X) = Q$, etc. for P, Q, P', Q', P'', Q'' as in Proposition 9.3.

Proof. As explained at the beginning of this section, $\mathrm{Mat}_{(r+s) \times n}(\mathbb{N})$ has both a $\mathrm{GL}(n) \times \mathrm{GL}(r+s)$ crystal structure and a $\mathrm{GL}(n) \times \mathrm{GL}(n) \times \mathrm{GL}(r) \times \mathrm{GL}(s)$ structure. These cannot be combined further into a $\mathrm{GL}(n) \times \mathrm{GL}(n) \times \mathrm{GL}(r+s)$ crystal structure because the $\mathrm{GL}(n) \times \mathrm{GL}(n)$ and $\mathrm{GL}(r + s)$ crystal operations do not commute. However, we *can* combine the $\mathrm{GL}(n)$ action from the $\mathrm{GL}(n) \times \mathrm{GL}(r + s)$ structure with the $\mathrm{GL}(r) \times \mathrm{GL}(s)$ structure from the $\mathrm{GL}(n) \times \mathrm{GL}(n) \times \mathrm{GL}(r) \times \mathrm{GL}(s)$ structure and regard $\mathrm{Mat}_{(r+s) \times n}(\mathbb{N})$ as a $\mathrm{GL}(n) \times \mathrm{GL}(r) \times \mathrm{GL}(s)$ crystal.

Let

$$\mathcal{C} = \{X \in \mathrm{Mat}_{(r+s) \times n}(\mathbb{N}) \mid P(X) \in \mathcal{B}_\lambda^{(n)}, \ Q'(X) \in \mathcal{B}_\mu^{(r)}, \ Q''(X) \in \mathcal{B}_\nu^{(s)}\}.$$

This consists of all elements of $\mathrm{Mat}_{(r+s) \times n}(\mathbb{N})$ that are $(\mathrm{GL}(n) \times \mathrm{GL}(r) \times \mathrm{GL}(s))$-plactically equivalent to elements of the $\mathrm{GL}(n) \times \mathrm{GL}(r) \times \mathrm{GL}(s)$ crystal $\mathcal{B}_\lambda^{(n)} \boxtimes \mathcal{B}_\mu^{(r)} \boxtimes \mathcal{B}_\nu^{(s)}$. Therefore $\mathcal{C}$ is a disjoint union of copies of this crystal. We will count these in two different ways.

First we observe that

$$\mathcal{C} \subset \{X \in \mathrm{Mat}_{(r+s) \times n}(\mathbb{N}) \mid P(X) \in \mathcal{B}_\lambda^{(n)}\}.$$

By Corollary 9.2, this is a $\mathrm{GL}(n) \times \mathrm{GL}(r + s)$ crystal isomorphic to $\mathcal{B}_\lambda^{(n)} \boxtimes \mathcal{B}_\lambda^{(r+s)}$. On branching to $\mathrm{GL}(n) \times \mathrm{GL}(r) \times \mathrm{GL}(s)$ the number of subcrystals isomorphic to $\mathcal{B}_\lambda^{(n)} \boxtimes \mathcal{B}_\mu^{(r)} \boxtimes \mathcal{B}_\nu^{(s)}$ is equal to the multiplicity of $\mathcal{B}_\mu^{(r)} \boxtimes \mathcal{B}_\nu^{(s)}$ in the $\mathrm{GL}(r) \times \mathrm{GL}(s)$ crystal obtained by branching $\mathcal{B}_\lambda^{(r+s)}$.

On the other hand,

$$\mathcal{C} \subset \{X \in \mathrm{Mat}_{(r+s) \times n}(\mathbb{N}) \mid Q'(X) \in \mathcal{B}_\mu^{(r)}, Q''(X) \in \mathcal{B}_\nu^{(s)}\}.$$

By Corollary 9.2, this is isomorphic to

$$\mathcal{B}_\mu^{(n)} \boxtimes \mathcal{B}_\nu^{(n)} \boxtimes \mathcal{B}_\mu^{(r)} \boxtimes \mathcal{B}_\nu^{(s)}$$

as a $\mathrm{GL}(n) \times \mathrm{GL}(n) \times \mathrm{GL}(r) \times \mathrm{GL}(s)$ crystal. Remembering that by Proposition 9.3 $P(X) \equiv P'(X) \otimes P''(X)$ as $\mathrm{GL}(n)$ crystal elements, it is clear that the number of subcrystals isomorphic to $\mathcal{B}_\lambda^{(n)} \boxtimes \mathcal{B}_\mu^{(r)} \boxtimes \mathcal{B}_\nu^{(s)}$ equals the multiplicity of $\mathcal{B}_\lambda^{(n)}$ in $\mathcal{B}_\mu^{(n)} \otimes \mathcal{B}_\nu^{(n)}$. Comparing this with our other computation of the multiplicity, we obtain the required equality.   □

Theorem 9.5 (The Littlewood–Richardson Rule). *The multiplicity of $\mathcal{B}_\lambda^{(n)}$ in $\mathcal{B}_\mu^{(n)} \otimes \mathcal{B}_\nu^{(n)}$ equals $c_{\mu\nu}^\lambda$.*

Proof. This follows by combining Theorem 9.4 with Theorem 8.14.   □

Exercise

Exercise 9.1. Compute the Littlewood–Richardson coefficient $c_{(2,1),(2,1)}^{(3,2,1)}$ in three different ways:

(1) Using Definition 8.9.
(2) Using Theorem 8.14.
(3) Using Corollary 9.5.

Chapter 10

Crystals for Stanley Symmetric Functions

As we have seen in (3.3), the Schur polynomial $s_\lambda(x_1, \ldots, x_\ell)$ can be viewed as the character of the $\mathrm{GL}(\ell)$ crystal $\mathcal{B}_\lambda$. Since the Schur polynomials form a basis for the ring of symmetric polynomials, an interesting question is to expand a given symmetric polynomial $f(x)$ in terms of Schur polynomials. Suppose that the Schur expansion is integral and *positive*, meaning that in the expansion

$$f(x) = \sum_\lambda a_\lambda s_\lambda(x)$$

in terms of Schur polynomials, all the coefficients a_λ are nonnegative integers. Then one can try to give a crystal theoretic interpretation of this expansion. More precisely, one can try to find a crystal structure on the combinatorial objects $\mathcal{C}$ defining the symmetric polynomial $f(x)$. Determining the highest weight elements in this crystal $\mathcal{C}$, then yields a combinatorial interpretation for the coefficients a_λ:

$$a_\lambda = |\{c \in \mathcal{C} \mid e_i c = 0 \text{ for all } 1 \leqslant i < \ell, \mathrm{wt}(c) = \lambda\}|.$$

In this chapter, we carry this out for the Stanley symmetric functions F_w following [Morse and Schilling (2016)]. Stanley symmetric functions F_w [Stanley (1984)] are indexed by permutations $w \in S_n$. We begin in Section 10.1 by defining the Stanley symmetric functions and the combinatorial objects (decreasing factorization of w) that govern them. In Section 10.2, we introduce the crystal structure on decreasing factorizations and then use the crystal in Section 10.3 to prove properties of the Stanley symmetric functions. In particular, it turns out that the crystal on decreasing factorizations is related to the tableaux model via the Edelman-Greene insertion [Edelman and Greene (1987)] of Section 7.3.

10.1 Stanley symmetric functions

Stanley symmetric functions are indexed by permutations $w \in S_n$ of the symmetric group. The symmetric group S_n is a Coxeter group (as introduced in Chapter 2) generated by the simple transpositions s_i for $1 \leqslant i < n$, where each s_i interchanges i and $i + 1$. The word $i_1 i_2 \ldots i_m$ of letters $i_j \in \{1, 2, \ldots, n - 1\}$ is called a *reduced word* for w if $w = s_{i_1} s_{i_2} \cdots s_{i_m}$ and there is no shorter word with this property.

(Note that unlike in other chapters, for convenience of notation we do not write reduced words as tuples.) The length $\ell(w)$ of w is equal to m if the word is reduced. We denote by w_0 the longest element in S_n.

An element $v \in S_n$ is called *decreasing* if there is a reduced word $i_1 i_2 \cdots i_m$ for v such that $i_1 > i_2 > \cdots > i_m$. The identity in S_n is considered to be decreasing. The *content* $\mathrm{cont}(v)$ is the set of letters appearing in its reduced word(s). Note that a decreasing element v is completely determined by its content. Given $w \in S_n$, a *decreasing factorization* of w is a factorization $w^k \cdots w^1$ such that $w = w^k \cdots w^1$ with $\ell(w) = \ell(w^1) + \cdots + \ell(w^k)$ and each factor w^i is decreasing. We denote the set of all decreasing factorizations of w by $\mathcal{W}_w$. Then for any $w \in S_n$, the *Stanley symmetric function* $F_w(x)$ is defined as

$$F_w(x) = \sum_{w^k \cdots w^1 \in \mathcal{W}_w} x_1^{\ell(w^1)} \cdots x_k^{\ell(w^k)}. \tag{10.1}$$

One of Stanley's motivations to study these functions was to understand the reduced words for a given w. Let us denote the set of all reduced words for w by $\mathrm{Red}(w)$. For example, since every single letter i by itself is decreasing, the coefficient of the square free term $x_1 x_2 \cdots x_{\ell(w)}$ is precisely the number of reduced words $|\mathrm{Red}(w)|$.

For the next example, it is useful to define the *monomial symmetric functions*

$$m_\lambda(x) = \sum_\alpha x^\alpha$$

where the sum runs over all distinct permutations $\alpha = (\alpha_1, \alpha_2, \ldots)$ of the entries of $\lambda = (\lambda_1, \lambda_2, \ldots)$.

Note that the definition of Stanley symmetric functions, in principle, involves infinitely many variables $x_1, x_2, \ldots$. However, if one restricts to precisely ℓ decreasing factors (some of which might be trivial), then one obtains Stanley symmetric polynomials in ℓ variables $x_1, x_2, \ldots, x_\ell$. Let us denote the set of decreasing factorizations of $w \in S_n$ into ℓ decreasing factors by $\mathcal{W}_w^\ell$.

Example 10.1. *We show how to compute the Stanley symmetric polynomial for the long element* $w_0 = s_1 s_2 s_1 = s_2 s_1 s_2 \in S_3$. *Restricting ourselves to three variables* x_1, x_2, x_3 *or equivalently to* $\mathcal{W}_{w_0}^3$, *we find the following set of decreasing factorizations with three factors for* w_0:

$$(1)(2)(1), \quad ()(1)(21), \quad (1)()(21), \quad (1)(21)(),$$
$$(2)(1)(2), \quad ()(21)(2), \quad (21)()(2), \quad (21)(2)().$$

Therefore,

$$F_{s_1 s_2 s_2}(x_1, x_2, x_3) = 2x_1 x_2 x_3 + x_1^2 x_2 + x_1^2 x_3 + x_2^2 x_3 + x_1 x_2^2 + x_1 x_3^2 + x_2 x_3^2$$
$$= 2m_{(1,1,1)}(x_1, x_2, x_3) + m_{(2,1)}(x_1, x_2, x_3).$$

Note that the square free term is contained in $m_{(1,1,1)}(x)$ *which indeed has coefficient* 2, *the number of reduced words of* $w_0 \in S_3$.

It turns out that the Stanley symmetric functions are indeed symmetric functions.

Theorem 10.2 ([Stanley (1984)]). *The Stanley symmetric functions F_w for $w \in S_n$ satisfy the following properties:*

(1) $F_w(x)$ is a symmetric function in $x = (x_1, x_2, \ldots)$.
(2) Let $a_{w,\lambda} \in \mathbb{Z}$ be the coefficient of the Schur function s_λ in F_w. Then there exist partitions $\lambda(w)$ and $\mu(w)$, so that $a_{w,\lambda(w)} = a_{w,\mu(w)} = 1$ and

$$F_w(x) = \sum_{\lambda(w) \leqslant \lambda \leqslant \mu(w)} a_{w,\lambda} s_\lambda(x).$$

[Edelman and Greene (1987)] and separately [Lascoux and Schützenberger (1985)] showed that the Schur expansion coefficients $a_{w,\lambda}$ are nonnegative.

Theorem 10.3 ([Edelman and Greene (1987); Lascoux and Schützenberger (1985)]). *We have $a_{w,\lambda} \in \mathbb{Z}_{\geqslant 0}$.*

Other interpretations of the coefficients $a_{w,\lambda}$ can be found in [Haiman (1992); Reiner and Shimozono (1998)].

In the next section we provide another proof of the results in Theorems 10.2 and 10.3, based on [Morse and Schilling (2016)]. In particular, we interpret $a_{w,\lambda}$ as the number of highest weight elements of weight λ in a crystal graph.

10.2   Crystal on decreasing factorizations

We are now ready to define a crystal structure $\mathcal{B}(w)$ of type $A_{\ell-1}$ on $\mathcal{W}_w^\ell$ for every $w \in S_n$. As a set $\mathcal{B}(w)$ is the set $\mathcal{W}_w^\ell$, that is, the set of all decreasing factorizations of w into at most ℓ factors. The weight function wt of $w^\ell \cdots w^1 \in \mathcal{B}(w)$ is defined to be $(\ell(w^1), \ell(w^2), \ldots, \ell(w^\ell))$. The Kashiwara raising and lowering operators e_i and f_i only act on the factors $w^{i+1} w^i$. The action is defined by first bracketing certain letters and then moving an unbracketed letter from one factor to the other.

Let us begin by describing the bracketing procedure in analogy to the bracketing procedure in Section 2.4. Start with the largest letter b in $\mathrm{cont}(w^{i+1})$ and pair it with the smallest $a > b$ in $\mathrm{cont}(w^i)$. If there is no such a in $\mathrm{cont}(w^i)$, then b is unpaired. The pairing proceeds in decreasing order on elements of $\mathrm{cont}(w^{i+1})$, and with each iteration, previously paired letters of $\mathrm{cont}(w^i)$ are ignored. Define

$$L_i(w^\ell \cdots w^1) = \{b \in \mathrm{cont}(w^{i+1}) \mid b \text{ is unpaired in the } w^{i+1}w^i\text{-pairing}\}$$

and

$$R_i(w^\ell \cdots w^1) = \{b \in \mathrm{cont}(w^i) \mid b \text{ is unpaired in the } w^{i+1}w^i\text{-pairing}\}.$$

We may now define the crystal operations, beginning with e_i. If $L_i(w^\ell \cdots w^1) = \varnothing$, then we define $e_i(w^\ell \cdots w^1) = 0$. Otherwise, $e_i(w^\ell \cdots w^1)$ is defined by replacing the factors w^{i+1} and w^i by $\widetilde{w}^{i+1}$ and $\widetilde{w}^i$ such that

$$\mathrm{cont}(\widetilde{w}^{i+1}) = \mathrm{cont}(w^{i+1}) \backslash \{b\} \quad \text{and} \quad \mathrm{cont}(\widetilde{w}^i) = \mathrm{cont}(w^i) \cup \{b - t\}$$

for $b = \min\big(L_i(w^\ell \cdots w^1)\big)$ and $t = \min\{j \geqslant 0 \mid b - j - 1 \notin \mathrm{cont}(w^{i+1})\}$. The following Lemma contains the facts needed to see that replacing w^{i+1} and w^i in this way produces another reduced factorization of w.

Lemma 10.4. *The content of w^{i+1} contains $b, b-1, \ldots, b-t$ but not $b-t-1$. The content of w^i contains $b, b-1, \ldots, b-t+1$ but not $b+1$ and not $b-t$. We have*

$$w^{i+1}w^i = \widetilde{w}^{i+1}\widetilde{w}^i. \tag{10.2}$$

Proof. The definition of t implies that $\mathrm{cont}(w^{i+1})$ contains $b, b-1, \ldots, b-t$ but not $b - t - 1$. To see that $\mathrm{cont}(w^i)$ does not contain $b+1$, we note that if it did, then $b + 1$ in $\mathrm{cont}(w^i)$ could not be paired with any element y of $\mathrm{cont}(w^{i+1})$ before b, since y would be strictly greater than b and each element of $\mathrm{cont}(w^i)$ can only be paired with an element of $\mathrm{cont}(w^{i+1})$ that is smaller than it. Thus when we come to pairing $b \in \mathrm{cont}(w^{i+1})$ there will be an unpaired $y > b$ in $\mathrm{cont}(w^i)$ and b will get paired. But this is a contraction since by assumption b is unpaired.

Now by definition, b is the rightmost unpaired element of $\mathrm{cont}(w^{i+1})$, so that $b - 1, \ldots, b - t$ are all paired. Let us ask what elements they may be paired with. The $b-1$ in $\mathrm{cont}(w^{i+1})$ must be paired with an element z of $\mathrm{cont}(w^i)$ that is greater than it. Thus $z > b - 1$. But z cannot be greater than b, because if such a z is unpaired when $b - 1$ is paired, it must be unpaired when b is paired, in which case the b would get paired with z. Therefore $z = b$ and so $\mathrm{cont}(w^i)$ contains b, which is paired with $b - 1$ in $\mathrm{cont}(w^{i+1})$. Next we consider $b - 2$ in $\mathrm{cont}(w^{i+1})$, and similar reasoning shows that it has to be paired with $b - 1$. Continuing in this way, we see that $\mathrm{cont}(w^i)$ must contain $b, \ldots, b - t + 1$ and these are paired with $b - 1, \ldots, b - t$ in $\mathrm{cont}(w^{i+1})$. We will show that $\mathrm{cont}(w^i)$ does not contain $b - t$ at the end of the proof.

Now we may write

$$w^{i+1} = u s_b s_{b-1} \cdots s_{b-t} u' \quad \text{and} \quad w^i = u'' s_b s_{b-1} \cdots s_{b-t+1} u''',$$

where u and u'' are the product of reflections s_k with k with $k > b$, u' is the product of reflections s_k with $k < b - t$, and u''' is the product of reflections s_k with $k < b - t + 1$. We know that u' does not involve s_{b-t-1} and u'' does not involve s_{b+1}, so u' and u'' commute with $s_b, \ldots, s_{b-t}$.

We make note of the identity

$$(s_b s_{b-1} \cdots s_{b-t})(s_b s_{b-1} \cdots s_{b-t+1}) = (s_{b-1} \cdots s_{b-t})(s_b s_{b-1} \cdots s_{b-t})$$

which we leave the reader to check. This, together with the commutations noted above, means that

$$w^{i+1}w^i = (u s_{b-1} \cdots s_{b-t} u)(u'' s_b \cdots s_{b-t} u'''). \tag{10.3}$$

Now we can show that $\mathrm{cont}(w^i)$ does not contain $b - t$. If it did, then s_{b-t} would be the first factor in the decreasing factorization of u'''. So the above factorization would contain two adjacent s_{b-t}, which we could cancel, obtaining $\ell(w^{i+1}w^i) < \ell(w^{i+1}) + \ell(w^i)$, contracting the defintion of a decreasing factorization.

Finally, we note that the two factors in (10.3) are $\widetilde{w}^{i+1}$ and $\widetilde{w}^i$, obtaining (10.2).

$\square$

Note that since $\ell(\widetilde{w}^i) = \ell(w^i) + 1$ and $\ell(\widetilde{w}^{i+1}) = \ell(w^{i+1}) - 1$, we have $\mathrm{wt}(e_i(x)) = \mathrm{wt}(x) + 1$ for $x \in \mathcal{W}_w^\ell$.

Similarly, $f_i(w^\ell \cdots w^1)$ is defined by replacing the factors $w^{i+1} w^i$ by $\widetilde{w}^{i+1} \widetilde{w}^i$ such that

$$\mathrm{cont}(\widetilde{w}^{i+1}) = \mathrm{cont}(w^{i+1}) \cup \{a+s\} \quad \text{and} \quad \mathrm{cont}(\widetilde{w}^i) = \mathrm{cont}(w^i) \backslash \{a\}$$

for $a = \max(R_i(w^\ell \cdots w^1))$ and $s = \min\{j \geqslant 0 \mid a + j + 1 \notin \mathrm{cont}(w^i)\}$. If $R_i(w^\ell \cdots w^1) = \varnothing$, then $f_i(w^\ell \cdots w^1) = 0$. We leave the reader to formulate and prove the analog of Lemma 10.4 for the f_i and to deduce that $\mathcal{B}(w)$ is a crystal; see Exercise 10.1.

Example 10.5. *Let* $(s_3 s_2)(s_3 s_1)(s_2) \in \mathcal{W}_w^3$ *for* $w = s_3 s_2 s_3 s_1 s_2 \in S_4$. *To apply* e_2, *we need to first bracket the letters in* $\mathrm{cont}(w^3) = 32$ *with those in* $\mathrm{cont}(w^2) = 31$. *The letter* 3 *in* $\mathrm{cont}(w^3)$ *is unbracketed since there is no bigger letter in* $\mathrm{cont}(w^2)$, *but the letter* 2 *in* $\mathrm{cont}(w^3)$ *is bracketed with* 3 *in* $\mathrm{cont}(w^2)$. *Hence* $b = \min(L_2(w^3 w^2 w^1)) = 3$ *and* $t = \min\{j \geqslant 0 \mid b - j - 1 \notin \mathrm{cont}(w^3)\} = 1$. *Therefore,* $e_2((s_3 s_2)(s_3 s_1)(s_2)) = (s_2)(s_3 s_2 s_1)(s_2)$. *Similarly,* $f_2((s_3 s_2)(s_3 s_1)(s_2)) = (s_3 s_2 s_1)(s_3)(s_2)$.

Theorem 10.6 ([Morse and Schilling (2016)]). *$\mathcal{B}(w)$ is a Stembridge crystal of type $A_{\ell-1}$.*

A proof of this theorem is given in [Morse and Schilling (2016), Appendix]. See also Exercise 10.2.

Example 10.7. *The crystal $\mathcal{B}(w_0)$ of type A_2 for $w_0 \in S_3$ is provided in Figure 10.1.*

10.3 Applications

As we have seen in Theorem 5.20, normal (and in particular Stembridge) crystals are closed under taking tensor products and connected components in a crystal graph correspond to irreducible components. In addition, the irreducible components are in bijection with the highest weight elements and dominant weights. Therefore, the irreducible components of our crystal $\mathcal{B}(w)$ of type $A_{\ell-1}$ are isomorphic to $\mathcal{B}_\lambda$ for some partition λ. Recall from (3.3) that the Schur function $s_\lambda(x_1, \ldots, x_\ell)$ is the character of the crystal $\mathcal{B}_\lambda$ of type $A_{\ell-1}$

$$s_\lambda(x_1, \ldots, x_\ell) = \sum_{b \in \mathcal{B}_\lambda} x^{\mathrm{wt}(b)}.$$

Denote by $\mathcal{W}_{w,\lambda}^\ell$ all elements in $\mathcal{W}_w^\ell$ of weight λ.

Choosing ℓ sufficiently large, the above arguments immediately yield the following result regarding the Schur expansion of the Stanley symmetric functions.

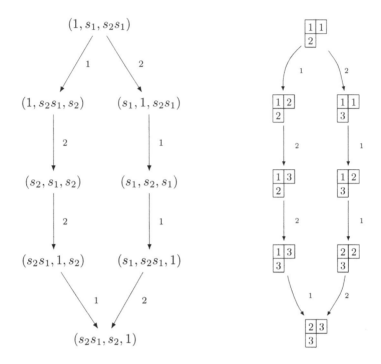

Fig. 10.1   Crystal $\mathcal{B}(w_0)$ of type A_2 for $w_0 = s_1 s_2 s_1 \in S_3$ on the left and the highest weight crystal $\mathcal{B}_{(2,1)}$ of type A_2 in terms of Young tableaux on the right.

Corollary 10.8 ([Morse and Schilling (2016)]). *For any $w \in S_n$, the coefficient $a_{w,\lambda}$ in*

$$F_w(x) = \sum_\lambda a_{w,\lambda}\, s_\lambda(x) \tag{10.4}$$

enumerates the highest weight factorizations in $\mathcal{W}^\ell_{w,\lambda}$, where $\ell > \ell(w)$. That is,

$$a_{w,\lambda} = |\{w^\ell \cdots w^1 \in \mathcal{W}^\ell_{w,\lambda} \mid e_i(w^\ell \cdots w^1) = 0 \text{ for all } 1 \leqslant i < \ell\}|\,.$$

[Edelman and Greene (1987)] (see also [Fomin and Greene (1998), Theorem 1.2]) characterized the coefficients $a_{w,\lambda}$ as the number of semistandard tableaux of shape λ' (the conjugate of λ) whose column-reading word is a reduced word of w.

We can understand this result in terms of a variant of the Edelman–Greene (EG) insertion discussed in Section 7.3. In fact, it can be extended to the full crystal (not just highest weight elements) as follows. Given a decreasing factorization $v^\ell \cdots v^1 \in \mathcal{W}^\ell_w$, consider $\bar{v}^1 \cdots \bar{v}^\ell$ by reversing all factors. In particular, each decreasing factor v^i turns into an increasing factor $\bar{v}^i$. Now successively insert the factors $\bar{v}^i$ for $i = 1, 2, \ldots, \ell$ using the EG insertion of Section 7.3. Recall that, in this insertion, a letter a is inserted into a row by finding the smallest letter $b > a$. If $b = a+1$ and a is also contained in the row, then $a+1$ is inserted into the next row down. Otherwise,

b is replaced by a and then b is inserted into the next row down. In both cases, we consider b to be bumped. For each inserted factor $\bar{v}^i$, the cells in the new shape are recorded by letters i. This yields a correspondence $\varphi_{\mathrm{EG}} \colon v^\ell \cdots v^1 \mapsto (P, Q)$, where P is the EG insertion tableau and Q is the EG recording tableau, where each letter in the factor $\bar{v}^i$ is recorded with the letter i in Q.

Example 10.9. *Take $v^3 v^2 v^1 = (1)(2)(32)$ a factorization of the permutation $s_1 s_2 s_3 s_2 \in S_4$. Then $\bar{v}^1 \bar{v}^2 \bar{v}^3 = (23)(2)(1)$ with insertions for $i = 1, 2, 3$:*

$$\left(\begin{array}{|c|c|} \hline 2 & 3 \\ \hline \end{array}, \begin{array}{|c|c|} \hline 1 & 1 \\ \hline \end{array} \right) \qquad \left(\begin{array}{|c|c|} \hline 2 & 3 \\ \hline 3 \\ \hline \end{array}, \begin{array}{|c|c|} \hline 1 & 1 \\ \hline 2 \\ \hline \end{array} \right) \qquad \left(\begin{array}{|c|c|} \hline 1 & 3 \\ \hline 2 \\ \hline 3 \\ \hline \end{array}, \begin{array}{|c|c|} \hline 1 & 1 \\ \hline 2 \\ \hline 3 \\ \hline \end{array} \right) = (P, Q) .$$

If $v^3 v^2 v^1 = (32)(31)(2)$, then the insertion sequence yields:

$$\left(\begin{array}{|c|} \hline 2 \\ \hline \end{array}, \begin{array}{|c|} \hline 1 \\ \hline \end{array} \right) \qquad \left(\begin{array}{|c|c|} \hline 1 & 3 \\ \hline 2 \\ \hline \end{array}, \begin{array}{|c|c|} \hline 1 & 2 \\ \hline 2 \\ \hline \end{array} \right) \qquad \left(\begin{array}{|c|c|c|} \hline 1 & 2 & 3 \\ \hline 2 & 3 \\ \hline \end{array}, \begin{array}{|c|c|c|} \hline 1 & 2 & 3 \\ \hline 2 & 3 \\ \hline \end{array} \right) = (P, Q) .$$

The map φ_{EG} turns out to be a crystal morphism, when restricted to the recording tableaux as the next theorem shows. We denote the projection of φ_{EG} onto the recording tableau by $\varphi_{\mathrm{EG}}^{\mathrm{rec}}$, that is, $\varphi_{\mathrm{EG}}^{\mathrm{rec}}(v^\ell \cdots v^1) = Q$ if $\varphi_{\mathrm{EG}}(v^\ell \cdots v^1) = (P, Q)$.

Theorem 10.10 ([Morse and Schilling (2016)], Theorem 4.11). *For any permutation $w \in S_n$, the crystal isomorphism*

$$\mathcal{B}(w) \cong \bigoplus_\lambda \mathcal{B}_\lambda^{\oplus a_{w,\lambda}}$$

is explicitly given by $\varphi_{\mathrm{EG}}^{\mathrm{rec}}$. In particular,

$$\varphi_{\mathrm{EG}}^{\mathrm{rec}} \circ e_i = e_i \circ \varphi_{\mathrm{EG}}^{\mathrm{rec}} \qquad \text{and} \qquad \varphi_{\mathrm{EG}}^{\mathrm{rec}} \circ f_i = f_i \circ \varphi_{\mathrm{EG}}^{\mathrm{rec}}.$$

Example 10.11. *Take $v^3 v^2 v^1 = (32)(31)(2)$ from Example 10.9. Then $e_2(v^3 v^2 v^1) = (2)(321)(2)$ and*

$$\varphi_{\mathrm{EG}}^{\mathrm{rec}}((2)(321)(2)) = \begin{array}{|c|c|c|} \hline 1 & 2 & 2 \\ \hline 2 & 3 \\ \hline \end{array},$$

which is the same as $e_2 \circ \varphi_{\mathrm{EG}}^{\mathrm{rec}}((32)(31)(2))$.

Proof of Theorem 10.10. EG-insertion enjoys many of the same properties as RSK-insertion. For example, given that cell c_x is added to a tableau when x is EG-inserted, and cell c_y is added when y is then EG-inserted into the result, c_y lies strictly east of c_x when $y > x$, and c_y lies strictly lower than c_x when $y < x$.

Fix $w \in S_n$. We first note that φ_{EG} is a bijection between $\mathcal{W}_{w,\alpha}^\ell$ and the set of pairs of same-shaped tableaux (P, Q) where the column-reading word of the transpose of P is a reduced expression for w and Q is semistandard of weight α. That is, given $v^\ell \cdots v^1 \in \mathcal{W}_{w,\alpha}^\ell$, let $(P, Q) = \varphi_{\mathrm{EG}}(v^\ell \cdots v^1)$ and recall that $P = P^\ell$ where P^ℓ is defined by inserting the (distinct) letters of $\mathrm{cont}(v^\ell)$ from smallest to largest into $P^{\ell-1}$. By the above remarks about the properties of the EG-insertion, $Q^\ell / Q^{\ell-1}$ is a horizontal $\ell(v^\ell)$-strip and we iteratively find Q to be semistandard of weight α. The column reading word of the transpose of P is a reduced expression for w by Proposition 7.17. It is not difficult to see that the process is invertible by reverse EG-bumping letters from P^i that lie in the positions determined by cells of $\operatorname{shape}(Q^i)/\operatorname{shape}(Q^{i-1})$ taken from right to left.

Denote the letters in $\operatorname{cont}(v^{i+1})$ by $y_{\alpha_{i+1}}\cdots y_1$ and the letters in $\operatorname{cont}(v^i)$ by $x_{\alpha_i}\cdots x_1$ in decreasing order. Let $a = x_j$ be the leftmost unbracketed letter in the pairing in Section 10.2. Inserting the letters $x_1,\ldots,x_{\alpha_i}$ under the EG-insertion yields α_i insertion paths that move strictly to the right in the tableaux P^i by the above remark on the properties of the EG insertion. Since $a = x_j$ is the leftmost unbracketed letter in $\operatorname{cont}(v^i)$, there exists an index $1 \leqslant m \leqslant \alpha_{i+1}$ such that $x_j \leqslant y_m < y_{m+1} < \cdots < y_{\alpha_{i+1}}$ and $y_1 < y_2 < \cdots < y_{m-1} < x_{j-1}$. In addition, all letters $y_1,\ldots,y_{m-1}$ are bracketed under the crystal bracketing which means that the insertion paths of these letters are weakly to the left of the insertion path of x_{j-1} and no letter can bump x_j. Also, the letters $x_{j+1},\ldots,x_{\alpha_i}$ are bracketed under the crystal bracketing so that of the letters i in Q^i corresponding to the insertion paths of $x_j,\ldots,x_{\alpha_i}$ precisely one is not bracketed with an $i+1$ in Q^{i+1}.

Under the application of f_i the letter $a = x_j$ moves from $\operatorname{cont}(v^i)$ to the letter $a + s$ in $\operatorname{cont}(v^{i+1})$. As a result, the insertion paths of $x_{j+1},\ldots,x_{\alpha_i}$ either stay (partially) in their old track or move left (partially) to the insertion of the previously inserted letter. Similarly, the insertion paths of the corresponding y_h move (partially) left. The new letter $a + s$ in v^{i+1} after the application of f_i, then causes the previously unpaired letter i in Q^{i+1} in the insertion to become an $i+1$, possibly by shifting the insertion paths of the subsequent y_h to the right. This proves the claim for f_i.

The proof for e_i is similar. $\qquad\qquad\qquad\qquad\qquad\qquad\qquad\qquad\qquad\square$

Restricting Theorem 10.10 to highest weight elements yields the following corollary.

Corollary 10.12. *For any permutation $w \in S_n$ and partition λ, there is a bijection between the highest weight factorizations,*

$$\{v^\ell \cdots v^1 \in \mathcal{W}^\ell_{w,\lambda} \mid e_i(v^\ell \cdots v^1) = 0 \text{ for all } 1 \leqslant i < \ell\},$$

and the semistandard tableaux of shape λ' whose column-reading word is a reduced word of w. Explicitly, the bijection is given by the conjugate of the insertion tableau $\varphi^{\mathrm{ins}}_{\mathrm{EG}}(v^\ell \cdots v^1) = P$ of the highest weight element $v^\ell \cdots v^1$.

Example 10.13. *Take $v^3 v^2 v^1 = (1)(2)(32)$ of Example 10.9. The element $(1)(2)(32)$ is highest weight of weight $(2,1,1)$ and the column-reading word of the conjugate of P*

$$P' = \begin{array}{|c|c|c|} \hline 1 & 2 & 3 \\ \hline 3 \\ \cline{1-1} \end{array}$$

is 3123, which is indeed a reduced word for $s_1 s_2 s_3 s_2$. This demonstrates the bijective correspondence of Corollary 10.12.

Another immediate outcome of the crystal $\mathcal{B}(w)$ on decreasing factorizations is Stanley's famous result [Stanley (1984)] that the number of reduced expressions for the longest element $w_0 \in S_n$ is equal to the number of standard tableaux of staircase shape $\rho = (n - 1, n - 2, \ldots, 1)$. See Theorem 7.18. Namely, in $\mathcal{B}(w_0)$ there is only one highest weight element given by the factorization $(s_1)(s_2 s_1)(s_3 s_2 s_1) \cdots (s_{n-1} s_{n-2} \cdots s_1)$. Hence $\mathcal{B}(w_0)$ is isomorphic to the highest weight crystal $\mathcal{B}_\rho$. The reduced words of w_0 are precisely given by the factorizations of weight $(1, 1, \ldots, 1)$. In $\mathcal{B}_\rho$ they are the standard tableaux of shape ρ. The bijection between the reduced words of w_0 and standard tableaux of shape ρ induced by the crystal isomorphism is precisely φ_{EG}^Q (which, due to the initial reversal of the factorization, gives the conjugate of the standard tableau from the straight EG insertion). An example of this crystal isomorphism for $\mathcal{B}(s_1 s_2 s_1)$ in S_3 is given in Figure 10.1.

By Theorem 10.10, the crystal $\mathcal{B}(w)$ relates to the crystals on the recording tableaux under the EG correspondence. It was proved in [Edelman and Greene (1987)] that two reduced words EG insert to the same P tableau if and only if they are Coxeter–Knuth equivalent. Two reduced words are Coxeter–Knuth equivalent if one can be obtained from the other by a sequence of Coxeter–Knuth relations on three consecutive letters

$$(a + 1)a(a + 1) \sim a(a + 1)a, \qquad bac \sim bca, \qquad cab \sim acb, \qquad (10.5)$$

where the last two relations only hold when $a < b < c$. The Coxeter–Knuth graph $\mathcal{CK}(w)$ for $w \in S_n$ is a graph on the reduced words for w where two words are connected if they differ by a relation in (10.5).

There is an interesting relation between the crystal $\mathcal{B}(w)$ and its decomposition into irreducible components and the connected components of the Coxeter–Knuth graph.

Proposition 10.14. *Let $w \in S_n$. The connected components of $\mathcal{CK}(w)$ are in one-to-one correspondence with the connected components of $\mathcal{B}(w)$.*

Proof. Every reduced word of w can be viewed as an element of $\mathcal{B}(w)$ by placing each letter in its own factor (assuming that ℓ is bigger than $\ell(w)$). Suppose $\mathbf{i}, \mathbf{j} \in \mathrm{Red}(w)$ differ by a single relation (10.5) with $\mathbf{i}$ having 3 consecutive letters of the left hand side and $\mathbf{j}$ the corresponding letters of the right hand side. Viewing $\mathbf{i}$ and $\mathbf{j}$ as elements of $\mathcal{B}(w)$, it is not hard to check that $f_i f_{i+1} e_i e_{i+1}(\mathbf{i}) = \mathbf{j}$ which proves that two elements in the same component in $\mathcal{CK}(w)$ are also in the same crystal component.

Conversely, suppose $b, b' \in \mathcal{B}(w)$ with $e_i(b) = b'$, so that b and b' lie in the same component in $\mathcal{B}(w)$. We can view b and b' as reduced words of w by disregarding the grouping into factors. By [Morse and Schilling (2016), Lemma 3.8], b' is obtained from b by a sequence of braid and commutation moves. By a close inspection of the proof of [Morse and Schilling (2016), Lemma 3.8] only the Coxeter–Knuth relations (10.5) are used. A similar argument holds for f_i. This implies that if

$b, b' \in \mathcal{B}(w)$ are in the same component, then the corresponding reduced words are in the same component in $\mathcal{CK}(w)$. □

Given that the Edelman–Greene correspondence maps a factorization to a pair of tableaux and Theorem 10.10 relates the crystal on decreasing factorizations to the crystal on the recording tableaux Q, a natural question to ask is whether there is a "dual" crystal on the P-tableaux. By [Hamaker and Young (2014)], two reduced words have the same recording tableau under the EG insertion if and only if they are connected by Little bumps [Little (2005)]. See Exercise 10.3.

In [Morse and Schilling (2016)], the crystal is defined more generally on certain affine permutations into cyclically decreasing factors. Lam [Lam (2006)] defined analogues of the Stanley symmetric functions in terms of cyclically decreasing elements. In [Morse and Schilling (2016)], the crystal on these affine permutations is used to study k-Schur structure coefficients and further applications to flag Gromov–Witten invariances, fusion coefficients, and positroid varieties.

Exercises

Exercise 10.1. Formulate and prove the analog of Lemma 10.4 for the crystal operators f_i. Also prove, for $x, y \in \mathcal{W}_w^\ell$ that $e_i(x) = y$ if and only if $f_i(y) = x$. Deduce that $\mathcal{W}_w^\ell$ is a crystal.

Exercise 10.2. Prove Theorem 10.6 in several steps.

(1) Prove that $\mathcal{B}(w)$ satisfies the Stembridge Axioms S0-S3 of Section 4.2.
(2) Deduce Stembridge Axioms S0'-S3' for $\mathcal{B}(w)$ by appealing to the order reversing map $\star\colon i \mapsto n - i$, which is extended to words by $a_1 \cdots a_h \mapsto a_h^\star \cdots a_1^\star$ and decreasing factorizations $\star\colon w^\ell \cdots w^1 \mapsto w^{1\star} \cdots w^{\ell\star}$, where $w^{i\star}$ is the element in S_n corresponding to $\operatorname{cont}(w_i)^\star$.
    Hint: See [Morse and Schilling (2016), Appendix].

Exercise 10.3. (Open) Describe a "dual" crystal structure on the P-tableaux under the Edelman–Greene correspondence.

Exercise 10.4. (Open) Generalize the results of this chapter to root systems beyond type A.

Chapter 11

Patterns and the Weyl Group Action

A simple and important method of mapping a crystal into the set of lattice points in a polyhedral cone inside $\mathbb{Z}^N$ appears in [Berenstein and Zelevinsky (1996); Kashiwara (1993); Littelmann (1998)]. Here N is the number of positive roots in the root system, or equivalently, the length of the longest element w_0 in the Weyl group. Among other things, this useful parametrization makes it possible to interpret the Littlewood–Richardson coefficients as counting lattice points in polytopes, so that ideas from tropical geometry become relevant. These aspects are explained in [Berenstein and Zelevinsky (2001)], where the relationship with similar but different parametrizations due to [Lusztig (1990b)] is discussed.

We call the patterns from [Berenstein and Zelevinsky (1996)] and [Littelmann (1998)] *string patterns*. They are called *Kashiwara data* in [Kamnitzer (2007)] and *BZL patterns* in [Brubaker, Bump and Friedberg (2011b)]. The idea is quite simple, but depends on the choice of a reduced word $\mathbf{i} = (i_1, \ldots, i_N)$ for w_0, that is, a sequence of minimal length such that $w_0 = s_{i_1} \cdots s_{i_N}$. Let $\mathcal{C}$ be a crystal. To get coordinates for $v \in \mathcal{C}$, we place a "turtle" at the vertex v and let it walk to the highest weight element u_λ as follows. First it goes as far as it can in the α_{i_1} direction. Thus, if a_1 is the largest integer such that $e_{i_1}^{a_1} v \neq 0$, the turtle takes a_1 steps in the α_{i_1} direction, and goes to $e_{i_1}^{a_1} v$. Then it can go no further in that direction, so it goes as far as it can in the α_{i_2} direction, and so forth. At the end, it finds itself at u_λ. (This requires proof, but it is true.) The sequence

$$\text{string}_{\mathbf{i}}(v) = (a_{i_1}, a_{i_2}, \ldots, a_{i_N})$$

is a recipe for retracing its steps, so it determines v.

This works for any reduced word representing w_0. In this way every reduced word $\mathbf{i}$ determines an embedding $\text{string}_{\mathbf{i}}$ of the crystal into a polyhedral cone in $\mathbb{Z}^N$. No matter which $\mathbf{i}$ is used, the set of lattice points in the resulting polyhedral cone can be given the structure of a crystal, and the resulting infinite crystal (which is independent of the choice of $\mathbf{i}$) is one realization of Kashiwara's $\mathcal{B}_\infty$ crystal. The string patterns give a way of embedding each finite normal crystal into $\mathcal{B}_\infty$. We will explain all this in Chapter 12.

[Littelmann (1998)] found that for each of the classical Cartan types A_r, B_r, C_r and D_r, there is one choice of $\mathbf{i}$ that gives a polyhedral cone by particularly simple inequalities. For arbitrary $\mathbf{i}$, it may not be so simple to find the inequalities that describe the cone.

In Section 11.1, we prove the basic facts in type A. In Section 11.2, we introduce Gelfand–Tsetlin patterns and discuss their relation to both tableaux and string patterns, as well as the $\mathrm{GL}(n)$ to $\mathrm{GL}(n-1)$ branching rule. In Section 11.3, we prove that the operators σ_i defined in Definition 2.35 satisfy the braid relations, thereby obtaining an action of the Weyl group on the crystal. The reason for the inclusion of the Weyl group action in this chapter is that a result we need is proved in Section 11.1.

11.1   String patterns

Let W be the Weyl group of the root system Φ with longest element $w_0 \in W$. Let $\mathbf{i} = (i_1, \ldots, i_N)$ be a reduced word for w_0. Let $\mathcal{C}$ be a crystal associated to the root system Φ and $v \in \mathcal{C}$. We define a sequence of integers $a_j = a_j(v, \mathbf{i})$ for $j = 1, \ldots, N$ associated with v as follows. Let $a_1 = a_1(v, \mathbf{i})$ be the maximal integer such that $e_{i_1}^{a_1} v \neq 0$. In other words, $a_1 = \varepsilon_{i_1}(v)$. Then let a_2 be the maximal integer such that $e_{i_2}^{a_2} e_{i_1}^{a_1} v \neq 0$, and so on. We say that $\mathbf{i}$ is a *good word* for the crystal $\mathcal{C}$ if, for every $v \in \mathcal{C}$, the element $e_{i_N}^{a_N} \cdots e_{i_1}^{a_1} v$ is a highest weight element.

Later in Corollary 13.8, we will see that if $\mathcal{C}$ is a normal crystal, then every reduced word for w_0 is a good word for $\mathcal{C}$. In this section, we will prove this fact for type A crystals (see Theorem 11.6).

Assume that $\mathbf{i}$ is a good word for $\mathcal{B}_\lambda$. Let us define a map $\mathrm{string}_{\mathbf{i}}$ from $\mathcal{B}_\lambda$ to the set of N-tuples of nonnegative integers by

$$\mathrm{string}_{\mathbf{i}}(v) = \big(a_1(v, \mathbf{i}), \ldots, a_N(v, \mathbf{i})\big).$$

It is important to note that v is determined by $\mathrm{string}_{\mathbf{i}}(v)$. Indeed, denoting by u_λ the unique highest weight vector in $\mathcal{B}_\lambda$, we have $u_\lambda = e_{i_N}^{a_N} \cdots e_{i_1}^{a_1} v$, and therefore $v = f_{i_1}^{a_1} \cdots f_{i_N}^{a_N} u_\lambda$. Thus we may represent all elements of $\mathcal{B}_\lambda$ by their string patterns.

We may think of the sequence

$$v_0 = v, \quad v_1 = e_{i_1}^{a_1} v, \quad v_2 = e_{i_2}^{a_2} e_{i_1}^{a_1} v, \quad \ldots, \quad v_N = e_{i_N}^{a_N} \cdots e_{i_1}^{a_1} v = u_\lambda \qquad (11.1)$$

as points on a path through the crystal, which we will call the *stations* of v.

Let us illustrate some of these notions with an example.

Example 11.1. *Let $n = 3$ and take $\mathbf{i} = (1, 2, 1)$ for the root system A_2. For this particular word, we will arrange the string pattern in a triangle:* $\begin{pmatrix} a_2 & a_3 \\ & a_1 \end{pmatrix}$. *Choose $\lambda = (4, 2, 0)$. Let us consider the tableau $T = \begin{array}{|c|c|c|c|} \hline 1 & 2 & 2 & 3 \\ \hline 3 & 3 \\ \cline{1-2} \end{array}$. Then*

$$e_1^2 T = \begin{array}{|c|c|c|c|} \hline 1 & 1 & 1 & 3 \\ \hline 3 & 3 \\ \cline{1-2} \end{array}, \qquad (11.2)$$

while $e_1^3 T = 0$. Therefore $a_1 = 2$. We see that the first two stations are T_0 and T_1 where $T_0 = T$ and T_1 is (11.2). Similarly, we find that $a_2 = 3$ and $a_3 = 1$. The next two stations are

$$T_2 = e_2^3 e_1^2 T = \boxed{\begin{array}{cccc}1&1&1&2\\2&2\end{array}} \quad , \quad T_3 = e_1 e_2^3 e_1^2 T = \boxed{\begin{array}{cccc}1&1&1&1\\2&2\end{array}}$$

and so

$$\text{string}(T) = \begin{pmatrix} 3 & 1 \\ & 2 \end{pmatrix}.$$

To summarize what we have seen, going from $T_0 = T$ to T_1 replaces all 2's in the first row by 1's; going from T_1 to T_2 replaces all 3's in the first two rows by 2's; then finally going from T_2 to T_3 replaces all 2's in the first two rows by 1's (again).

Note that the final station T_3 is the highest weight element of $\mathcal{B}_\lambda$. This needs to be true for any tableau T of shape λ for $\mathbf{i} = (1, 2, 1)$ to be a good word. The four stations $T_0 = T, T_1, T_2$ and T_3 in the crystal are shown in Figure 11.1 as points on a path connecting $v = T$ to u_λ.

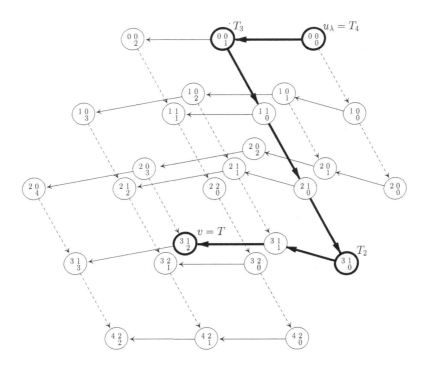

Fig. 11.1 The GL(3) crystal with highest weight $\lambda = (4, 2, 0)$, showing the stations of $v = T$ where T is as in (11.2). Elements of the crystal are labeled by their string patterns. We indicate f_1 by solid arrows and f_2 by dashed ones.

We begin by studying a particular good word for GL(n) crystals. Let

$$\Omega = (1, 2, 1, 3, 2, 1, \ldots, n-1, n-2, \ldots, 3, 2, 1). \qquad (11.3)$$

Since this can be organized as (1), (2, 1), (3, 2, 1), we will follow [Littelmann (1998)] in writing the pattern for this particular word in a triangular array, namely for $v \in \mathcal{B}_\lambda$

$$\text{string}_\Omega(v) = \begin{pmatrix} \ddots & & & \vdots \\ & a_4 & a_5 & a_6 \\ & & a_2 & a_3 \\ & & & a_1 \end{pmatrix}.$$

We also write $a_1 = a_{11}$, $(a_2, a_3) = (a_{21}, a_{22})$, etc. so

$$\text{string}_\Omega(v) = \begin{pmatrix} \ddots & & & \vdots \\ & a_{31} & a_{32} & a_{33} \\ & & a_{21} & a_{22} \\ & & & a_{11} \end{pmatrix}. \tag{11.4}$$

Proposition 11.2. *Let λ be a partition of length $\leqslant n$.*

(i) *Let $T \in \mathcal{B}_\lambda$ and let (11.4) be the string pattern. Regarding T as a tableau, a_{11} equals the number of 2's in the first row, a_{21} equals the number of 3's in the first two rows, a_{22} equals the number of 3's in the top row, and in general a_{ij} equals the number of entries equal to $i + 1$ in the top $i + 1 - j$ rows.*

(ii) *The word Ω is a good word for $\mathcal{B}_\lambda$.*

Proof. Just as we renumbered $(a_1, a_2, a_3, \ldots)$ to $a_{11}, a_{21}, a_{22}, a_{31}, \ldots$, we will renumber the sequence of stations $(T_1, T_2, \ldots)$ (omitting the first station $T_0 = T$) to $T_{11}, T_{21}, T_{22}, T_{31}, \ldots$.

We will check the following description of the T_{ij}. First, T_{11} is obtained from T by replacing all the 2's in the first row by 1's. Then T_{21} is obtained from T_{11} by replacing all the 3's in the first *two* rows of T_{11} by 2's. Then T_{22} is obtained by replacing the 2's in the first row of T_{21} by 1's. In general, T_{ij} is obtained from its predecessor by replacing all the entries equal to $i + 2 - j$ in the first $i + 1 - j$ rows by $i + 1 - j$.

Before we prove that this description is correct, let us show how it implies both (i) and (ii). According to this description, the number of replacements in going to T_{ij} from its predecessor equals the number of entries in that predecessor equal to $i + 2 - j$. However, the locations of these entries were originally (in T) occupied by $i + 1$. Thus this replacement changes a_{ij} entries, where a_{ij} equals the number of entries equal to $i + 1$ in T. This proves (i). It is also clear from the description that in the final tableau $T_{n-1, n-1}$, the entries in the i-th row are all equal to i, so this tableau is the highest weight vector. This shows that Ω is a good word.

It remains to be shown that T_{ij} is obtained from its predecessor by replacing all the entries equal to $i + 2 - j$ in the first $i + 1 - j$ rows by $i + 1 - j$. Let U be the predecessor of T_{ij}. Thus $U = T$ if $i = j = 1$, otherwise $U = T_{i-1, i-1}$ if $i > 1$ and $j = 1$, and otherwise $U = T_{i, j-1}$ if $j > 1$. We will identify $\mathcal{B}_\lambda$ with a subcrystal of $\mathcal{B}_{(\lambda_n)} \otimes \mathcal{B}_{(\lambda_{n-1})} \otimes \cdots \otimes \mathcal{B}_{(\lambda_1)}$, where $\mathcal{B}_{(k)}$ is the crystal of rows of length k. Let $U_1, U_2, \ldots$ be the rows of U, so that $U = U_n \otimes \cdots \otimes U_1$. Now $a_{ij} = \varepsilon_{i+1-j}(U)$ is the number of times e_{i+1-j} can be applied to U, which equals

$$\max\left(\varepsilon_{i+1-j}(U_1), \varepsilon_{i+1-j}(U_1) + \varepsilon_{i+1-j}(U_2) - \varphi_{i+1-j}(U_1), \ldots\right). \tag{11.5}$$

We observe that there are no entries equal to $i+1-j$ in the k-th row of U for any $k \leqslant i-j$, since these were all changed the last time e_{i-j} was applied. Therefore $\varphi_k(U) = 0$ if $k \leqslant i-j$. Thus the first $i+1-j$ entries in (11.5) are $\varepsilon_{i+1-j}(U_1), \varepsilon_{i+1-j}(U_1) + \varepsilon_{i+1-j}(U_2), \ldots$, and the maximum of these is the last term

$$\varepsilon_{i+1-j}(U_1) + \cdots + \varepsilon_{i+1-j}(U_{i+1-j}). \tag{11.6}$$

Subsequent entries in (11.5) are no larger than this. Indeed, the next entry adds $\varepsilon_{i+1-j}(U_{i+2-j})$ and subtracts $\varphi_{i+1-j}(U_{i+1-j})$. We claim that $\varphi_{i+1-j}(U_{i+1-j}) \geqslant \varepsilon_{i+1-j}(U_{i+2-j})$. To see this, note that every entry in U_{i+2-j} equal to $i+2-j$ must be above an entry in the previous row equal to $i+1-j$, because the tableau U has strictly ascending columns. This shows that the next entry is no larger than (11.6). The entries after that are also no larger because $\varepsilon_{i+i-j}(U_k) = 0$ when $k > i+2-j$, also since U has ascending columns.

We have proved that a_{ij} equals (11.6). Applying e_{i+1-j} to U replaces this many $i+2-j$'s by $i+1-j$. But (11.6) is the number of $i+2-j$'s in the first $i+1-j$ rows, so all of these are replaced, as claimed. $\qquad\square$

Before we prove general results for GL(n) we need a couple of specific results for GL(3).

Proposition 11.3. *Let λ be a partition of length $\leqslant 3$ and let $\mathcal{B}_\lambda$ be the GL(3) crystal of tableaux of shape λ. Then every reduced word $\mathbf{i}$ for w_0 is a good word for $\mathcal{B}_\lambda$.*

Proof. There are only two possibilities for $\mathbf{i}$, namely $(1, 2, 1)$ and $(2, 1, 2)$. It follows from Proposition 11.2 that $(1, 2, 1)$ is a good word.

It remains to be proved that the other possibility $\mathbf{i}' = (2, 1, 2)$ is also good. It is not necessary to check this since it may be deduced from the case just proved. The A_2 root system has an "outer automorphism" which interchanges the roots α_1 and α_2. Let $\lambda \mapsto \lambda'$ be the corresponding automorphism of the weight lattice, and let $i \mapsto i'$ be the bijection of the index set $\{1, 2\}$ such that $1' = 2$ and $2' = 1$. If $\mathcal{C}$ is an A_2 crystal we may define another A_2 crystal $\mathcal{C}'$ with the same underlying set and weight function $\mathrm{wt}'(v) = \mathrm{wt}(v)'$, crystal operators $e'_{i'} = e_i$, etc. The fact that $\mathbf{i}$ is a good word for $\mathcal{C}$ implies that $\mathbf{i}'$ is a good word for $\mathcal{C}'$. Now $\mathcal{C}$ clearly satisfies the Stembridge axioms if and only if $\mathcal{C}'$ does, and the conclusion follows. $\qquad\square$

We now recall an important property of the Weyl group W, commonly known as *Matsumoto's Theorem* ([Matsumoto (1964)]). Let s_i be the simple reflections, where $i \in I$, the index set. If i, j are elements of the index set I, let $n(i, j)$ be the order of $s_i s_j$. Thus if the root system is simply-laced, $n(i,j) = 2$ if α_i and α_j are orthogonal and 3 if they are not; for nonsimply-laced types $n(i,j)$ can equal 4 or 6.

The *braid relation* is the relation

$$s_i s_j s_i \cdots = s_j s_i s_j \cdots , \qquad (11.7)$$

where the number of terms on both sides is $n(i,j)$. The intuitive content behind Matsumoto's theorem is that every equality between reduced words can be deduced from the braid relations. To make this rigorous, define a graph structure on the set $\mathrm{Red}(w)$ of reduced words for any element w of the Weyl group, in which we consider $(k_1, \ldots, k_\ell)$ adjacent to $(h_1, \ldots, h_\ell)$ if one is obtained from the other by replacing a subsequence $(i, j, i, \ldots)$ by $(j, i, j, \ldots)$, where the length of the subsequences is $n(i,j)$.

Theorem 11.4 (Matsumoto, Tits). *The graph* $\mathrm{Red}(w)$ *is connected.*

Proof. See [Bump (2013)], Theorem 25.2 or Exercise 13 in Chapter 4, Section 1 of [Bourbaki (2002)]. □

For example, in A_3 two words representing the long element w_0 are $(1,2,1,3,2,1)$ and $(3,2,3,1,2,3)$ and these are connected by edges in $\mathrm{Red}(w)$ as follows:

$$(1,2,1,3,2,1) \text{ —— } (2,1,2,3,2,1) \text{ —— } (2,1,3,2,3,1)$$

$$(2,3,1,2,3,1) \quad \cdot$$

$$(3,2,3,1,2,3) \text{ —— } (2,3,2,1,2,3) \text{ —— } (2,3,1,2,1,3)$$

(We have only drawn a portion of $\mathrm{Red}(w)$; actually, the graph has 14 vertices.)

Proposition 11.5. *Suppose that* $\mathcal{C}$ *is a Stembridge crystal for a simply-laced type. If* $\mathcal{C}$ *has a good word, then every reduced word for the long element* w_0 *is good.*

Proof. By Matsumoto's Theorem 11.4, it is sufficient to show that if two words $\mathbf{i}_1$ and $\mathbf{i}_2$ are adjacent in the graph $\mathrm{Red}(w_0)$, and if $\mathbf{i}_1$ is a good word, then so is $\mathbf{i}_2$. By the definition of the adjacency relation in the graph $\mathrm{Red}(w)$, $\mathbf{i}_2$ is obtained from $\mathbf{i}_1$ by either replacing an occurence of (i,j) with (j,i) (when α_i and α_j are orthogonal roots) or (i,j,i) by (j,i,j) (when they are not).

We will consider the second case first. Thus assume

$$\mathbf{i}_1 = (i_1, \ldots, i_r, i, j, i, i_{r+4}, \ldots, i_N),$$

$$\mathbf{i}_2 = (i_1, \ldots, i_r, j, i, j, i_{r+4}, \ldots, i_N).$$

Let $v \in \mathcal{C}$ and let

$$v_0 = v, \quad v_1 = e_{i_1}^{a_1} v, \quad v_2 = e_{i_2}^{a_2} e_{i_1}^{a_1} v, \quad \ldots, \quad v_N = e_{i_N}^{a_N} \cdots e_{i_1}^{a_1} v$$

be the stations of v with respect to $\mathbf{i}_1$; similarly, let

$$v_0' = v, \quad v_1' = e_{i_1}^{a_1'} v, \quad v_2' = e_{i_2}^{a_2'} e_{i_1}^{a_1'} v, \quad \dots, \quad v_N' = e_{i_N}^{a_N'} \cdots e_{i_1}^{a_1'} v$$

be the stations with respect to $\mathbf{i}_2$. Since $\mathbf{i}_1$ is a good word, $v_N = u_\lambda$, the highest weight vector.

Since the first r entries in $\mathbf{i}_1$ and $\mathbf{i}_2$ agree, we have $v_r' = v_r$. Now we will argue that $v_{r+3}' = v_{r+3}$. Indeed, as explained in Section 2.8, we may branch the crystal to the Levi subsystem A_2 by discarding all edges in the crystal graph except those labeled i or j. Let $\mathcal{D}$ be the connected component of the branched crystal containing v_r. It is a Stembridge crystal. Let u be its highest weight element.

By Proposition 11.3, both (i,j,i) and (j,i,j) are good words for $\mathcal{D}$. It follows that

$$v_{r+3} = e_i^{a_{r+3}} e_j^{a_{r+2}} e_i^{a_{r+1}} v_r, \qquad v_{r+3}' = e_j^{a_{r+3}'} e_i^{a_{r+2}'} e_j^{a_{r+1}'} v_r$$

both equal u. Now the remaining elements of $\mathbf{i}_1$ and $\mathbf{i}_2$ agree, so it follows that the remaining stations agree. Thus $v_N' = v_N$.

The other case is handled similarly except that we branch to an $A_1 \times A_1$ crystal instead of A_2. We see that $\mathbf{i}_2$ is also a good word. $\qquad\square$

Theorem 11.6. *Let $\mathcal{C}$ be a connected Stembridge $GL(n)$ crystal. Then every reduced word for the long Weyl group element is a good word.*

Proof. By Proposition 11.5 it is sufficient to show that one good word exists, and this is accomplished by Proposition 11.2. $\qquad\square$

11.2 Gelfand–Tsetlin patterns

This section can be skipped with no loss of continuity.

Let us consider the branching rule from $GL(n)$ crystals to $GL(n-1)$ crystals. We have a homomorphism from the $GL(n)$ weight lattice $\Lambda_n = \mathbb{Z}^n$ to the $GL(n-1)$ weight lattice Λ_{n-1}, namely

$$(\lambda_1, \dots, \lambda_n) \longmapsto (\lambda_1, \dots, \lambda_{n-1}).$$

Correspondingly, we may branch a $GL(n)$ crystal to $GL(n-1)$. This is almost, but not exactly, the Levi branching of Section 2.8. Indeed, we have a Levi branching to $GL(n-1) \times GL(1)$ crystals, but since the root system of $GL(1)$ is empty, the $GL(n-1) \times GL(1)$ root system is just the $GL(n-1)$ root system. However, the $GL(n-1)$ weight lattice is smaller by one dimension, so the $GL(n-1)$ branched crystal is not exactly the $GL(n-1) \times GL(1)$ branched one. Rather, they are related crystals in the sense of Section 2.7.

Suppose that λ is a partition of length $\leqslant n$ and μ is a partition of length $\leqslant n-1$. We say that λ and μ *interleave* if

$$\lambda_1 \geqslant \mu_1 \geqslant \lambda_2 \geqslant \mu_2 \geqslant \cdots \geqslant \mu_{n-1} \geqslant \lambda_n.$$

Theorem 11.7. *Let λ be a partition of length $\leqslant n$ and let $\mathcal{B}_\lambda$ be the $GL(n)$ crystal of tableaux of highest weight λ. A necessary and sufficient condition for $\mathcal{B}_\lambda$ to have a $GL(n-1)$ subcrystal of weight highest weight μ is that λ and μ interleave. If such a subcrystal exists, it is unique.*

For a representation theoretic analogue of this statement, see Section B.6 in the Appendix.

Proof. An element T of $\mathcal{B}_\lambda$ is a highest weight element of the branched crystal if and only if $e_i T = 0$ for $1 \leqslant i \leqslant n-2$. These may be found exactly as in Theorem 3.2. The condition on T is that the first row can contain only 1's and n's; the second row can contain only 2's and n's, and so forth. In other words, after eliminating all n's from the tableau T, a tableau T' in the alphabet $[n-1]$ remains. If μ is the shape of T', then λ/μ must be a horizonal strip, since T cannot have two n's in the same column. It is easy to see that this is equivalent to assuming that λ and μ interleave.

We have determined the highest weight elements of the branched crystal. Their weights, we see, are exactly the μ such that λ and μ interleave. $\square$

Let λ be a partition of length $\leqslant n$. By a *Gelfand–Tsetlin pattern of size n* we mean a triangular array

$$\Gamma = \left\{ \begin{array}{ccccc} \lambda_{11} & \lambda_{12} & \lambda_{13} & \cdots & \lambda_{1n} \\ & \lambda_{21} & \lambda_{22} & \cdots & \lambda_{2,n-1} \\ & & \lambda_{31} & \cdots & \lambda_{3,n-2} \\ & & & \ddots & \\ & & & \lambda_{n1} & \end{array} \right\}$$

such that each row is a partition, and the rows interleave. We describe a bijection between the Gelfand–Tsetlin patterns Γ whose top row is the partition λ and the crystal $\mathcal{B}_\lambda$ consisting of semistandard Young tableaux T of shape λ in the alphabet $[n]$. Given T, we provide a recipe for Γ is as follows. By removing the boxes labeled $\boxed{n}$ from T gives another tableau T' in the alphabet $[n-1]$. The shape of T' is the second row of Γ. Then removing the boxes labeled $\boxed{n-1}$ gives a tableau T'' whose shape is the next row of Γ, and so forth.

Example 11.8. *There are eight semistandard tableaux of shape* $(2, 1, 0)$ *in the alphabet* [3]. *Here they are with their corresponding Gelfand–Tsetlin patterns:*

$\boxed{\begin{smallmatrix}1&1\\2\end{smallmatrix}}$	$\left\{\begin{smallmatrix}2&&1&&0\\&2&&1\\&&2\end{smallmatrix}\right\}$	$\boxed{\begin{smallmatrix}1&2\\3\end{smallmatrix}}$	$\left\{\begin{smallmatrix}2&&1&&0\\&2&&0\\&&1\end{smallmatrix}\right\}$
$\boxed{\begin{smallmatrix}1&2\\2\end{smallmatrix}}$	$\left\{\begin{smallmatrix}2&&1&&0\\&2&&1\\&&1\end{smallmatrix}\right\}$	$\boxed{\begin{smallmatrix}1&3\\3\end{smallmatrix}}$	$\left\{\begin{smallmatrix}2&&1&&0\\&1&&0\\&&1\end{smallmatrix}\right\}$
$\boxed{\begin{smallmatrix}1&1\\3\end{smallmatrix}}$	$\left\{\begin{smallmatrix}2&&1&&0\\&2&&0\\&&2\end{smallmatrix}\right\}$	$\boxed{\begin{smallmatrix}2&2\\3\end{smallmatrix}}$	$\left\{\begin{smallmatrix}2&&1&&0\\&2&&0\\&&0\end{smallmatrix}\right\}$
$\boxed{\begin{smallmatrix}1&3\\2\end{smallmatrix}}$	$\left\{\begin{smallmatrix}2&&1&&0\\&1&&1\\&&1\end{smallmatrix}\right\}$	$\boxed{\begin{smallmatrix}2&3\\3\end{smallmatrix}}$	$\left\{\begin{smallmatrix}2&&1&&0\\&1&&0\\&&0\end{smallmatrix}\right\}$

In view of Theorem 11.7, we may interpret the rows of the Gelfand–Tsetlin pattern associated with the tableaux T as the highest weights for the sequence of branched crystals under the branchings from the $GL(n)$ crystal $\mathcal{B}_\lambda$ to crystals for $GL(n-1)$, $GL(n-2)$, etc. that contain the given T.

The string pattern can be read off from the Gelfand–Tsetlin pattern. For definiteness, we consider the case $n = 3$, and leave the general case to the reader, see Exercises 11.2.

Proposition 11.9. *Let* $n = 3$. *If the Gelfand–Tsetlin pattern associated to* T *is*

$$\left\{\begin{matrix}\lambda_1 & & \lambda_2 & & \lambda_3\\ & a & & b\\ & & c\end{matrix}\right\},$$

then

$$\mathrm{string}_{(1,2,1)}(T) = \begin{pmatrix}\lambda_1 + \lambda_2 - a - b & \lambda_1 - a\\ & a - c\end{pmatrix}.$$

Proof. This is really a paraphrase of Proposition 11.2 (i). We leave it to the reader to check this. $\qquad\square$

11.3 The Weyl group action

In this section, we describe an action of the Weyl group W on a normal crystal $\mathcal{C}$. The action is to be compatible with the action of W on the weight lattice. Thus, we want

$$\mathrm{wt}(w \cdot v) = w \cdot \mathrm{wt}(v), \qquad w \in W, \, v \in \mathcal{C}.$$

It will *not* be an automorphism of the crystal graph itself, just a permutation of the underlying set. However, the crystal graph underlies the definition of the action.

The action of the symmetric group on tableaux was defined in [Lascoux and Schützenberger (1981, 1990)]. This in effect is the action for type A crystals of tableaux. The useful discussion in [Kirillov and Berenstein (1995)] translates matters into the language of Gelfand–Tsetlin patterns and gives proofs based on piecewise-linear transformations. [Kashiwara (1994)] defined the Weyl group action on arbitrary crystals. An approach based on the Littelmann path model may be found in [Littelmann (1997)]. See also [Kashiwara (1995)].

The group W is generated by simple reflections s_i subject to the braid relations (11.7) together with "quadratic relations" $s_i^2 = 1$. It is a Coxeter group, meaning that these relations give a presentation.

Now we recall the operators σ_i on the crystal defined in Definition 2.35. Recall that $\mathrm{wt}(\sigma_i(v)) = s_i(\mathrm{wt}(v))$, so $\sigma_i(v)$ is the unique element of the root string whose weight is $s_i \mathrm{wt}(v)$. In other words, the effect of σ_i is to reverse each i-root string (see Proposition 2.36).

Since W is a Coxeter group, to obtain an action of W in which the s_i act by σ_i, it is sufficient to check that $\sigma_i^2 = 1$ and that the σ_i satisfy the braid relation. The first point is obvious, but the second requires work. Figure 11.2 illustrates the braid relation by checking it for one particular element of the A_2 crystal $\mathcal{B}_{(5,3,0)}$.

Before setting to work, we outline the method. Since the braid relation only involves two operators at a time, we reduce to rank two crystals, that is, to $A_1 \times A_1$, A_2, C_2 or G_2. Moreover the $A_1 \times A_1$ case is trivial, and by using virtual crystals we will get what we need for C_2 and G_2 from the simply-laced case. Thus we can reduce to checking the braid relation for A_2. Our method (similar to that in [Kirillov and Berenstein (1995)]) is to show that in terms of the string patterns, the effect of the operators σ_i is by means of explicit piecewise-linear transformations.

Proposition 11.10. *Let λ be a partition of length $\leqslant 3$ and let $\mathcal{B}_\lambda$ be the GL(3) crystal of tableaux of shape λ. Let $T \in \mathcal{B}_\lambda$ and let*

$$(a_1, a_2, a_3) = \mathrm{string}_{(1,2,1)}(T), \qquad (b_1, b_2, b_3) = \mathrm{string}_{(2,1,2)}(T).$$

Then

$$(b_1, b_2, b_3) = \big(\max(a_3, a_2 - a_1), a_1 + a_3, \min(a_2 - a_3, a_1)\big).$$

See [Littelmann (1998), Proposition 2.3] for a proof based on the Littelmann path method.

Proof. Let R_1, R_2, R_3 be the rows of T. By Proposition 11.2, we have

$$a_1 = \varphi_2(R_1) = \varepsilon_1(R_1), \qquad a_2 = \varepsilon_2(R_1) + \varepsilon_2(R_2), \qquad a_3 = \varepsilon_2(R_1). \qquad (11.8)$$

Since $v = f_1^{a_1} f_2^{a_2} f_1^{a_3} u_\lambda = f_2^{b_1} f_1^{b_2} f_2^{b_3} u_\lambda$, we have

$$\mathrm{wt}(T) = \lambda - (a_1 + a_3)\alpha_1 - a_2\alpha_2 = \lambda - (b_1 + b_3)\alpha_2 - b_2\alpha_1.$$

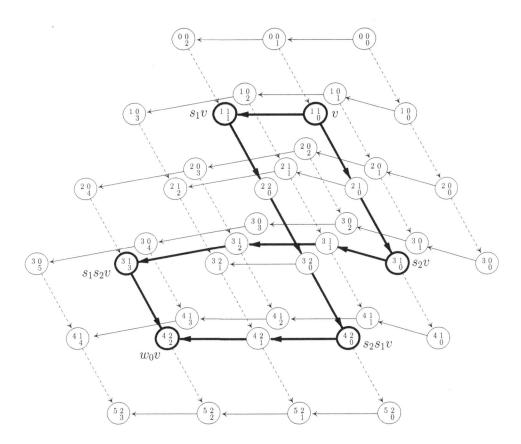

Fig. 11.2 The GL(3) crystal with highest weight $\lambda = (5,3,0)$, illustrating the braid relation for the Weyl group action. If v is the labeled element, this diagram checks that $s_1 s_2 s_1 v = s_2 s_1 s_2 v$.

Thus

$$b_2 = a_1 + a_3, \qquad a_2 = b_1 + b_3.$$

Next,

$$b_1 = \varepsilon_2(T) = \varepsilon_2(R_2 \otimes R_1) = \max(\varepsilon_2(R_1), \varepsilon_2(R_1) + \varepsilon_2(R_2) - \varphi_2(R_1)),$$

so by (11.8) $b_1 = \max(a_3, a_2 - a_1)$, as required. Finally,

$$\max(a_3, a_2 - a_1) + \min(a_2 - a_3, a_1) = a_2$$

and since $b_1 + b_3 = a_2$, we get $b_3 = \min(a_2 - a_3, a_1)$, and we are done. □

Proposition 11.11. *For $N \in \mathbb{Z}$, define maps Σ_N and $\theta: \mathbb{Z}^3 \longrightarrow \mathbb{Z}^3$ by*

$$\Sigma_N(a,b,c) = (N - a + b - 2c, b, c), \qquad \theta(a,b,c) = \big(\max(c, b - a), a + c, \min(b - c, a)\big).$$
(11.9)

Then for any pair N, N' of integers, $\theta \circ \Sigma_{N'} \circ \theta \circ \Sigma_N$ has order 3.

Proof. Let $\Sigma = \Sigma_N$, $\Sigma' = \Sigma_{N'}$. Since $\Sigma^2 = \Sigma'^2 = \theta^2 = 1$, it suffices to show that $\Sigma'\theta\Sigma\theta\Sigma'\theta = \theta\Sigma\theta\Sigma'\theta\Sigma$. We find that

$$(\Sigma\theta)(a,b,c) = \left(N - b + a + c - \min(b-c,a), a+c, \min(b-c,a)\right),$$

$$(\theta\Sigma)(a,b,c) = \left(a + 2c - \min(a+c,N), N + b - a - c, \min(a+c,N) + b - a - 2c\right).$$

Both $\Sigma\theta\Sigma'\theta\Sigma\theta(a,b,c)$ and $\theta\Sigma'\theta\Sigma\theta\Sigma'(a,b,c)$ work out to (X,Y,Z) where

$$X = N' + c + \min(N,b) - \min(a+c,b) - \min(a+c,N'), \qquad Y = N + N' - b,$$

$$Z = N + \min(a+c,b) + \min(a+c,N') - \min(N,b) - a - 2c. \qquad \square$$

Now we may prove the GL(3) braid relation.

Proposition 11.12. *Let λ be a partition of length $\leqslant 3$ and let $\mathcal{B}_\lambda$ be the GL(3) crystal of tableaux of shape λ. Then for $T \in \mathcal{B}_\lambda$*

$$\sigma_1\sigma_2\sigma_1(T) = \sigma_2\sigma_1\sigma_2(T).$$

Proof. Let $\Sigma = \Sigma_N$, $\Sigma' = \Sigma_{N'}$, and θ be the transformations in Proposition 11.11, taking $N = \lambda_1 - \lambda_2$ and $N' = \lambda_2 - \lambda_3$. Let $\text{string}(T)$ denote $\text{string}_{(1,2,1)}(T)$. We will show that

$$\text{string}(\sigma_1(T)) = \Sigma\,\text{string}(T), \qquad \text{string}(\sigma_2(T)) = (\theta\Sigma'\theta)\,\text{string}(T). \qquad (11.10)$$

Note that if this is true, then

$$\text{string}\circ(\sigma_1\sigma_2)^3 = (\theta\Sigma'\theta\Sigma)^3 \circ \text{string} = \text{string},$$

and since the map $\text{string}: \mathcal{B}_\lambda \to \mathbb{Z}^3$ is injective, this implies that $(\sigma_1\sigma_2)^3 = 1$, whence the braid relation.

If $\text{string}(T) = (a,b,c)$, then $e_1^a T$ is the highest weight vector in the 1-root string containing v, regarding the string as an A_1 crystal. Hence b and c will be the same for all elements of the same root string. In particular $\text{string}(\sigma_1 T) = (a',b,c)$ for some a'. We can determine a' by the requirement that $\text{wt}(\sigma_1 T) = s_1(\text{wt}(T)) = \text{wt}(T) - \langle \text{wt}(T), \alpha_1^\vee\rangle\alpha_1$. Now if u_λ is the highest weight element, we have

$$\text{wt}(T) = \text{wt}(f_1^a f_2^b f_1^c u_\lambda) = \lambda - (a+c)\alpha_1 - b\alpha_2,$$

and since $\langle\alpha_1, \alpha_1^\vee\rangle = 2$, $\langle\alpha_2, \alpha_1^\vee\rangle = -1$, this means that

$$\langle\text{wt}(T), \alpha_1^\vee\rangle = \langle\lambda, \alpha_1^\vee\rangle - 2(a+c) + b.$$

We note that $\langle\lambda, \alpha_1^\vee\rangle = \lambda_1 - \lambda_2 = N$, so

$$\text{wt}(\sigma_1 T) = \text{wt}(T) - \langle\text{wt}(T), \alpha_1^\vee\rangle\alpha_1 = \text{wt}(T) - (\langle\lambda, \alpha_1^\vee\rangle - 2(a+c) + b)\alpha_1.$$

This means that $a' = \varepsilon_1(\sigma_1 T) = \varepsilon_1(T) + \langle\lambda, \alpha_1^\vee\rangle - 2(a+c) + b = a + N - 2(a+c) + b$ and

$$(a',b,c) = (N - a + b - 2c, b, c) = \Sigma(a,b,c)$$

which proves the first formula in (11.10).

Similarly string$_{(2,1,2)}(\sigma_2(T)) = \Sigma'$ string$_{(2,1,2)}(T)$. Indeed, this follows from the fact just proved that string$_{(1,2,1)}(\sigma_1(T)) = \Sigma$ string$_{(1,2,1)}(T)$ using the outer automorphism introduced in the proof of Proposition 11.3. Also Proposition 11.10 asserts that

$$\text{string}_{(1,2,1)}(T) = \theta \, \text{string}_{(2,1,2)}(T).$$

Therefore

$$\text{string}_{(1,2,1)}(\sigma_2(T)) = \theta \, \text{string}_{(2,1,2)}(\sigma_2 T) = \theta \Sigma' \, \text{string}_{(2,1,2)}(T) = \theta \Sigma' \theta \, \text{string}_{(1,2,1)}(T),$$

and the second equation in (11.10) is proved. □

Since σ_1 and σ_2 satisfy the braid relation, we may define an action of S_3 on A_2 crystals by $s_i \cdot v = \sigma_i(v)$. This extends to general normal crystals, as we will now show.

Lemma 11.13.

(i) Let $\widehat{s}_i$ with $i \in \{1, 2, 3\}$ be the simple reflections of the Coxeter group of type A_3. Then the elements

$$s_1 = \widehat{s}_1 \widehat{s}_3, \qquad s_2 = \widehat{s}_2$$

satisfy the type C_2 Coxeter relations $s_i^2 = 1$ and $s_1 s_2 s_1 s_2 = s_2 s_1 s_2 s_1$.

(ii) Let $\widehat{s}_i$ with $i \in \{1, 2, 3, 4\}$ be the simple reflections of the Coxeter group of type D_4. Then the elements

$$s_1 = \widehat{s}_1 \widehat{s}_3 \widehat{s}_4, \qquad s_2 = \widehat{s}_2$$

satisfy the type G_2 Coxeter relations $s_i^2 = 1$ and $s_1 s_2 s_1 s_2 s_1 s_2 = s_2 s_1 s_2 s_1 s_2 s_1$.

Although this can easily be checked directly, it is worth noting that the element $s_i s_j$ is a "Coxeter element" in the ambient Weyl group of type A_3 or D_4, and its order h is the Coxeter number. [Coxeter (1934), p. 610] proved that $h = \frac{2N}{r}$ where N is the number of positive roots and r is the rank. Thus $h = 4$ or 6 for A_3 and D_4, respectively, which (combined with $s_i^2 = 1$) implies the required relations.

Proof. The relations can be checked explicitly. See Exercise 11.3. □

Theorem 11.14. *Let C be a normal crystal for a root system with Weyl group W. Then there exists an action of W on C by $s_i \cdot v = \sigma_i(v)$ for $v \in C$, where the operators σ_i are given in Definition 2.35.*

Proof. It is clear that $\sigma_i^2 = 1$. Since W is a Coxeter group, we need only verify that σ_i and σ_j satisfy the braid relation, which means σ_i and σ_j commute if i and j are orthogonal, otherwise $(\sigma_i \sigma_j)^{n(i,j)} = 1$, where $n(i,j)$ is the order of $s_i s_j$ in W.

Since i and j are both contained in a rank two root system, it is sufficient to check the braid relation by branching down to a rank two crystal. See Sections 2.8 and 5.10. In other words, we discard all edges in the crystal graph except those labeled i or j to obtain a normal crystal of type $A_1 \times A_1$, A_2, C_2 or G_2.

Thus we may assume that the root system Φ is one of these types. If the type is $A_1 \times A_1$, it is trivial that σ_i and σ_j commute, and if the type is A_2, the braid relation follows from Proposition 11.12. It remains to handle types C_2 and G_2. In these cases we may assume that $\mathcal{C}$ is a virtual crystal. Now suppose that $\widehat{\mathcal{C}}$ is a crystal of type A_3 or D_4 that is a virtualization of $\mathcal{C}$. Thus we have an embedding of $\mathcal{C}$ into $\widehat{\mathcal{C}}$ as described in Section 5.2. Note that we have already proved the result for simply-laced crystals, so we have a Weyl group action on $\widehat{\mathcal{C}}$. Let $\widehat{\sigma}_i$ be the simple reflections of A_3 or D_4 acting on $\widehat{\mathcal{C}}$.

First suppose that $\mathcal{C}$ is of type C_2 and $\widehat{\mathcal{C}}$ is of type A_3. It follows from (5.2) that $\widehat{\sigma}_1\widehat{\sigma}_3$ induces σ_1 on $\mathcal{C}$, because reversing the root strings labeled 1 and 3 on the virtual crystal will reverse the root strings labeled 1 on $\mathcal{C}$. Similarly $\widehat{\sigma}_2$ induces σ_2. Therefore what we have to check is that $\left((\widehat{\sigma}_1\widehat{\sigma}_3)\widehat{\sigma}_2\right)^4 = 1$, and this follows from Lemma 11.13 Similarly, if $\mathcal{C}$ is of type G_2 and $\widehat{\mathcal{C}}$ is of type D_4, we need to check that in the D_4 Coxeter group we have $c^6 = 1$, where $c = \widehat{s}_1\widehat{s}_3\widehat{s}_4\widehat{s}_2$. This follows from Lemma 11.13 (ii). Thus again, the C_2 or G_2 braid relations follow from what we have already proved in the simply-laced A_3 and D_4 cases.   □

Exercises

Exercise 11.1. Let μ be a partition of length $\leqslant n$ and let λ be a partition of length $\leqslant n - 1$. Show that a necessary and sufficient condition for μ/λ to be a horizontal strip is that μ and λ interleave.

Exercise 11.2. Prove Proposition 11.9, and generalize to arbitrary n.

Exercise 11.3. Prove Lemma 11.13.

Chapter 12

The $\mathcal{B}_\infty$ Crystal

Kashiwara's crystal $\mathcal{B}_\infty$, and the related crystals $\mathcal{T}_\lambda \otimes \mathcal{B}_\infty$, play the role of *Verma modules* in representation theory of Lie groups or Lie algebras. Verma modules are infinite-dimensional "universal highest weight" modules that can have finite-dimensional irreducible modules as quotients. Similarly, $\mathcal{B}_\infty$ is an infinite "universal highest weight crystal" that can be related to normal crystals by morphisms.

To elaborate on this analogy, let G be a complex analytic Lie group with weight lattice Λ, and let $\mathfrak{g}$ be the Lie algebra of G. The *Kostant partition function* P on the weight lattice is defined by letting $P(\mu)$ be the number of ways of writing μ as a sum of positive roots. Thus $P(\mu)$ is the number of representations

$$\mu = \sum_{\alpha \in \Phi^+} k_\alpha \, \alpha,$$

with k_α a nonnegative integer for each positive root α. We call such a decomposition of μ a *root partition*. The generating function is

$$\sum_\mu P(\mu) \, t^{-\mu} = \prod_{\alpha \in \Phi^+} (1 - t^{-\alpha})^{-1}. \tag{12.1}$$

Let $\mathfrak{b} = \mathfrak{t} \oplus \mathfrak{n}_+$ be the Lie algebra of the Borel subgroup B, where $\mathfrak{t}$ and $\mathfrak{n}_+$ are the Lie algebras of the maximal torus $T \subset B$ and the unipotent radical N_+ of B ($\mathfrak{t}$ is also known as the Cartan subalgebra). Similarly, let $\mathfrak{n}_-$ be the "opposite" unipotent subalgebra, generated by the root spaces in $\mathfrak{g}$ in the negative roots. Let $\lambda \in \Lambda$, which we identify with the group of rational characters of T. The differential of λ is a linear functional on $\mathfrak{t}$, which we extend to a functional $\mathfrak{b}$ that is zero on $\mathfrak{n}_+$. We induce λ from $\mathfrak{b}$ to $\mathfrak{g}$ as follows. Let v be a nonzero vector spanning a one-dimensional space $\mathbb{C} \cdot v$ on which $\mathfrak{b}$ acts by λ. Then the *Verma module* $M(\lambda)$ is $U(\mathfrak{g}) \otimes_{U(\mathfrak{b})} (\mathbb{C} \cdot v)$, where $U(\mathfrak{g})$ is the universal enveloping algebra. It has the irreducible representation with highest weight λ as a quotient.

It may be deduced from the Poincaré–Birkhoff–Witt theorem (PBW) that the map $\xi \mapsto \xi \otimes v$ is a vector space isomorphism from $U(\mathfrak{n}_-)$ to $M(\lambda)$. PBW also allows us to compute the character of the T-module $U(\mathfrak{n}_-)$, and we find that (12.1) is the character of $M(0)$. More generally, the character of $M(\lambda)$ is

$$t^\lambda \prod_{\alpha \in \Phi^+} (1 - t^{-\alpha})^{-1} = \sum_\mu P(\mu) \, t^{\lambda - \mu}. \qquad (12.2)$$

The crystal $\mathcal{B}_\infty$ is a crystal basis of the representation $M(0)$. Tensoring it with the crystal $\mathcal{T}_\lambda$ defined in Example 2.28 gives a crystal whose character is (12.2). Among other things, the crystal $\mathcal{B}_\infty$ will give us the tools we need to prove the *refined Demazure character formula* in the next chapter, that will relate the characters of normal crystals to the characters of representations.

In Section 12.1 we begin by introducing the elementary crystals $\mathcal{B}_i$ which serve as building blocks for $\mathcal{B}_\infty$. We proceed in Section 12.2 with the definition and properties of $\mathcal{B}_\infty$ in the simply-laced types. The construction for non-simply-laced types is given in Section 12.3 using the weak virtual crystal construction. We conclude in Section 12.4 with Demazure subcrystals of $\mathcal{B}_\infty$.

12.1   Elementary crystals

We begin our discussion with the *elementary crystals* $\mathcal{B}_i$ that turn out to be important for the construction of $\mathcal{B}_\infty$. Following [Kashiwara (1993)], the crystal $\mathcal{B}_i$ has basis elements $u_i(n)$ for $n \in \mathbb{Z}$, with wt $\big(u_i(n)\big) = n\alpha_i$. Define

$$\varphi_j\big(u_i(n)\big) = \begin{cases} n & \text{if } j = i, \\ -\infty & \text{otherwise,} \end{cases} \qquad \varepsilon_j\big(u_i(n)\big) = \begin{cases} -n & \text{if } j = i, \\ -\infty & \text{otherwise,} \end{cases}$$

and $f_i\big(u_i(n)\big) = u_i(n-1)$, $e_i\big(u_i(n)\big) = u_i(n+1)$, $f_j\big(u_i(n)\big) = e_j\big(u_i(n)\big) = 0$ if $j \neq i$.

Let $\mathcal{T}_\lambda = \{t_\lambda\}$ be the crystal described in Example 2.28. If $\mathcal{C}$ is any crystal, then $x \mapsto t_\lambda \otimes x$ is a bijection $\mathcal{C} \to \mathcal{T}_\lambda \otimes \mathcal{C}$. We have $\text{wt}(t_\lambda \otimes x) = \text{wt}(x) + \lambda$ and $\varepsilon_i(t_\lambda \otimes x) = \varepsilon_i(x)$, though $\varphi_i(t_\lambda \otimes x)$ does not equal $\varphi_i(x)$. It is also true that $x \mapsto x \otimes t_\lambda$ is a bijection $\mathcal{C} \to \mathcal{C} \otimes \mathcal{T}_\lambda$ that shifts the weight by λ. In contrast with $\mathcal{T}_\lambda \otimes \mathcal{C}$, we have $\varphi_i(x \otimes t_\lambda) = \varphi_i(x)$ while $\varepsilon_i(x \otimes t_\lambda)$ and $\varepsilon_i(x)$ differ. Thus $\mathcal{T}_\lambda \otimes \mathcal{C}$ and $\mathcal{C} \otimes \mathcal{T}_\lambda$ are in general not isomorphic.

We will make frequent and implicit use of the isomorphism $\mathcal{T}_\lambda \otimes \mathcal{T}_\mu \cong \mathcal{T}_{\lambda+\mu}$. Also recall the simple reflections s_i acting on the weight lattice by (2.16).

Proposition 12.1. *We have* $\mathcal{B}_i \otimes \mathcal{T}_{s_i\lambda} \cong \mathcal{T}_\lambda \otimes \mathcal{B}_i$.

Proof. It is easy to see that

$$u_i\big(n + \langle \lambda, \alpha_i^\vee \rangle\big) \otimes t_{s_i\lambda} \longleftrightarrow t_\lambda \otimes u_i(n)$$

is an isomorphism. $\qquad\square$

Proposition 12.2. *We have*

$$\mathcal{B}_i \otimes \mathcal{B}_i \cong \bigoplus_{k \in \mathbb{Z}} \mathcal{B}_i \otimes \mathcal{T}_{k\alpha_i} \cong \bigoplus_{k \in \mathbb{Z}} \mathcal{T}_{k\alpha_i} \otimes \mathcal{B}_i.$$

Proof. Since $s_i(\alpha_i) = -\alpha_i$, the two decompositions are equivalent. To prove the second one, the isomorphism takes $u_i(-n) \otimes u_i(-m)$ to $t_{-(n+m-k)\alpha_i} \otimes u_i(-k)$, where $k = \max(n + 2m, m)$. We leave the reader to check that this works. $\qquad\square$

12.2 The crystal $\mathcal{B}_\infty$ for simply-laced types

Our goal in this section is to describe the crystal $\mathcal{B}_\infty$ in the simply-laced case. [Kashiwara (1993)] realized $\mathcal{B}_\infty$ as a subcrystal of the tensor product of simple crystals $\mathcal{B}_i$, and we will proceed similarly. However, our starting point is different from his, since we do not make use of the theory of quantum groups; instead we make use of the weak Stembridge property. In this section we only consider simply-laced Cartan types.

Let $\theta \colon \mathbb{Z}^3 \to \mathbb{Z}^3$ be the map

$$\theta(a,b,c) = \big(\max(c, b-a), a+c, \min(b-c,a)\big). \tag{12.3}$$

It is easy to check that θ, which appeared already in Proposition 11.11, has order 2. Hence it is a bijection.

Proposition 12.3.

(i) If α_i and α_j are orthogonal roots, then

$$\mathcal{B}_i \otimes \mathcal{B}_j \cong \mathcal{B}_j \otimes \mathcal{B}_i$$

under the map $x \otimes y \mapsto y \otimes x$. The maps e_i and f_i (resp. e_j and f_j) do not take the value 0 on $\mathcal{B}_i \otimes \mathcal{B}_j$ and hence are inverse bijections on the crystal.

(ii) If $\langle \alpha_i, \alpha_j^\vee \rangle = -1$, then

$$\mathcal{B}_i \otimes \mathcal{B}_j \otimes \mathcal{B}_i \cong \mathcal{B}_j \otimes \mathcal{B}_i \otimes \mathcal{B}_j$$

under the map Θ defined by

$$\Theta\big(u_i(-a) \otimes u_j(-b) \otimes u_i(-c)\big) = u_j(-a') \otimes u_i(-b') \otimes u_j(-c'),$$

where $(a', b', c') = \theta(a, b, c)$.

Proof. We prove (ii) and leave (i), which is easier, to the reader. Thus assume $\langle \alpha_i, \alpha_j^\vee \rangle = -1$. Since θ is a bijection, so is Θ. We have

$$\mathrm{wt}\big(u_i(-a) \otimes u_j(-b) \otimes u_i(-c)\big) = -(a+c)\alpha_i - b\alpha_j$$
$$= -(a'+c')\alpha_j - b'\alpha_i = \mathrm{wt}\big(u_j(-a') \otimes u_i(-b') \otimes u_j(-c')\big),$$

so Θ is weight-preserving. By (2.13),

$$\varepsilon_i\big(u_i(-a) \otimes u_j(-b) \otimes u_i(-c)\big) = \max\big(c, -\infty, a - \langle -b\alpha_j - c\alpha_i, \alpha_i^\vee \rangle\big).$$

This equals $\max(c, a - b + 2c)$. On the other hand,

$$\varepsilon_i\big(u_j(-a') \otimes u_i(-b') \otimes u_j(-c')\big) = \max\big(-\infty, b' - \langle -c'\alpha_j, \alpha_i^\vee \rangle, -\infty\big),$$

which equals $b' - c'$. Since $\max(c, a - b + 2c) = b' - c'$, we see that Θ preserves ε_i.

We also need to know that Θ commutes with e_i. We have

$$e_i\Theta\big(u_i(-a) \otimes u_j(-b) \otimes u_i(-c)\big) = u_j(-a') \otimes u_i(-(a+c-1)) \otimes u_j(-c'). \tag{12.4}$$

We need to show $\Theta e_i\big(u_i(-a)\otimes u_j(-b)\otimes u_i(-c)\big)$ equals the same thing. If $a+c \geqslant b$, then $(a',b',c') = (c, a+c, b-c)$,

$$e_i\big(u_i(-a)\otimes u_j(-b)\otimes u_i(-c)\big) = e_i\big(u_i(-(a-1))\otimes u_j(-b)\otimes u_i(-c)\big).$$

Applying Θ gives (12.4). The other case $a+c < b$ is similar.

Since $\varphi_i(x) = \langle \mathrm{wt}(x), \alpha_i^\vee\rangle + \varepsilon_i(x)$, the compatibility needed for φ_i follows from those already checked for wt and ε_i. We also need compatibility of Θ with the f_i. To see this, note that e_i and f_i are never zero on $\mathcal{B}_i \otimes \mathcal{B}_j \otimes \mathcal{B}_i$ or $\mathcal{B}_j \otimes \mathcal{B}_i \otimes \mathcal{B}_j$. Thus they are inverse maps, and the f_i compatibility follows from that of the e_i. ☐

We pause to construct $\mathcal{B}_\infty$ for type A_2. We will build it out of the crystal $\mathcal{B}_1 \otimes \mathcal{B}_2 \otimes \mathcal{B}_1$. The latter crystal differs from $\mathcal{B}_\infty$ in that for $v \in \mathcal{B}_1 \otimes \mathcal{B}_2 \otimes \mathcal{B}_1$, $f_i(v)$ and $e_i(v)$ are never zero. This is in contrast with $\mathcal{B}_\infty$, which has a unique highest weight element. We can find a suitable subset of $\mathcal{B}_1 \otimes \mathcal{B}_2 \otimes \mathcal{B}_1$ that models $\mathcal{B}_\infty$.

Let C be the subset of $\mathbb{Z}^3$ consisting of (a,b,c) such that $a \geqslant 0$ and $b \geqslant c \geqslant 0$. It is easy to check that C is stable under the map θ defined in (12.3). If $(a,b,c) \in \mathbb{Z}^3$, let

$$u(a,b,c) = u_1(-a)\otimes u_2(-b)\otimes u_1(-c) \in \mathcal{B}_1 \otimes \mathcal{B}_2 \otimes \mathcal{B}_1. \qquad (12.5)$$

Proposition 12.4. *For Cartan type A_2, let*

$$\mathfrak{C} = \{u(a,b,c) \mid (a,b,c) \in C\}.$$

Then for $v \in \mathfrak{C}$, $f_i(v) \in \mathfrak{C}$ for all i. Moreover $\varepsilon_i(v) \geqslant 0$ for all $i \in I$ and all $v \in \mathfrak{C}$, and $e_i(v) \in \mathfrak{C}$ if and only if $\varepsilon_i(v) > 0$.

Proof. Let $(a',b',c') = \theta(a,b,c)$. We compute

$$\varepsilon_1\big(u(a,b,c)\big) = \max(c, a-b+2c) = b'-c', \qquad \varepsilon_2\big(u(a,b,c)\big) = b-c.$$

From this it is clear that $\varepsilon_1(v), \varepsilon_2(v) \geqslant 0$. Moreover $e_2\big(u(a,b,c)\big) = u(a, b-1, c)$, from which it is clear that if $v \in \mathfrak{C}$ then $e_2(v) \in \mathfrak{C}$ if and only if $\varepsilon_2(v) > 0$. Applying θ and using Proposition 12.3 implies the same statement for e_1. Also

$$f_2\big(u(a,b,c)\big) = u(a, b+1, c),$$

so $f_2(\mathfrak{C}) \subset \mathfrak{C}$, and applying θ and using Proposition 12.3 implies the same statement for f_1. ☐

Now we can describe $\mathcal{B}_\infty$ for type A_2. Indeed take $\mathfrak{C}$ and modify the definition of e_i by redefining $e_i(v) = 0$ if $\varepsilon_i(v) = 0$. It is clear from Proposition 12.4 that this produces a crystal having a unique highest weight vector, and furthermore

$$\varepsilon_i(x) = \max\{k \mid e_i^k(x) \neq 0\}. \qquad (12.6)$$

The resulting crystal is depicted in Figure 12.1. We want to generalize this construction.

We now turn to the construction of $\mathcal{B}_\infty$ for general simply-laced types.

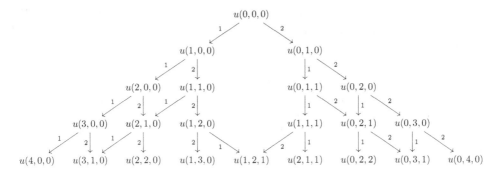

Fig. 12.1 The crystal $\mathcal{B}_\infty$ for A_2. The crystal continues downwards, as $f_i(x)$ is never zero.

Proposition 12.5. *Let w be an element of the Weyl group for a simply-laced Cartan type. Let $w = s_{i_N} \cdots s_{i_2} s_{i_1} = s_{j_N} \cdots s_{j_2} s_{j_1}$ be reduced decompositions. Then there exists an isomorphism*

$$\mathcal{B}_{i_1} \otimes \cdots \otimes \mathcal{B}_{i_N} \to \mathcal{B}_{j_1} \otimes \cdots \otimes \mathcal{B}_{j_N}$$

such that $u_{i_1}(0) \otimes \cdots \otimes u_{i_N}(0) \mapsto u_{j_1}(0) \otimes \cdots \otimes u_{j_N}(0)$.

Proof. By Matsumoto's theorem (Theorem 11.4), we may assume that $(i_1, \ldots, i_N)$ and $(j_1, \ldots, j_N)$ differ by replacing one appearance of (i, j) by (j, i), where i and j are orthogonal roots, or by replacing one appearance of (i, j, i) by (j, i, j), where $\langle \alpha_i, \alpha_j^\vee \rangle = -1$. In these cases, we make use of the isomorphism $\mathcal{B}_i \otimes \mathcal{B}_j \cong \mathcal{B}_j \otimes \mathcal{B}_i$ or $\mathcal{B}_i \otimes \mathcal{B}_j \otimes \mathcal{B}_i \cong \mathcal{B}_j \otimes \mathcal{B}_i \otimes \mathcal{B}_j$ from Proposition 12.3. $\square$

Our immediate concern is with the case when $w = w_0$ is the long element, which we turn to next. If $w_0 = s_{i_N} \cdots s_{i_1}$ is a reduced word, it is easy to see that the crystal $\mathcal{B}_{i_1} \otimes \cdots \otimes \mathcal{B}_{i_N}$ is of finite type. (It is not connected.)

Theorem 12.6. *Let w_0 be the long Weyl group element for a simply-laced Cartan type. Let $w_0 = s_{i_N} \cdots s_{i_1}$ be a reduced decomposition. Then $\mathcal{B}_{i_1} \otimes \cdots \otimes \mathcal{B}_{i_N}$ is a weak Stembridge crystal.*

Proof. First note that e_i never annihilates an element $x \in \mathcal{B}_{i_1} \otimes \cdots \otimes \mathcal{B}_{i_N}$. Hence the condition that $e_i(x) = 0$ implies $\varepsilon_i(x) = 0$ is trivially satisfied. The same is true for f_i.

Fix i and j to verify the Stembridge axioms. We will consider the case where $\langle \alpha_i, \alpha_j^\vee \rangle = -1$, leaving the case where α_i and α_j are orthogonal to the reader.

By Proposition 12.5 we may take any reduced word that we choose. We obtain a reduced word by appending i, j, i to a reduced word for $s_i s_j s_i w_0$, which has length $N - 3$.

We branch to the A_2 root system containing α_i and α_j. We will reduce to the A_2 case by proving that every connected component of $\mathcal{B}_{i_1} \otimes \cdots \otimes \mathcal{B}_{i_N}$ is isomorphic to a connected component of $\mathcal{T}_\lambda \otimes \mathcal{B}_i \otimes \mathcal{B}_j \otimes \mathcal{B}_i$ for some λ.

All of the crystals $\mathcal{B}_{i_k}$ with $i_k \neq i, j$ branch to infinite direct sums of $\mathcal{T}_{n\alpha_k}$. We may move these to the left using Proposition 12.1 and we are left with a sum of crystals of the form

$$\mathcal{T}_\lambda \otimes \mathcal{B}_{i'_1} \otimes \mathcal{B}_{i'_2} \otimes \cdots$$

where $i'_1, i'_2, \ldots$ is the subsequence of the original sequence $i_1, i_2, \ldots$ consisting of elements equal to i or j; here the last three are i, j, i. Now, using Proposition 12.2 we may eliminate adjacent $\mathcal{B}_i \otimes \mathcal{B}_i$ or $\mathcal{B}_j \otimes \mathcal{B}_j$ from this sequence, moving $\mathcal{T}_\mu$'s to the left as they appear. If necessary, we may also replace $\mathcal{B}_j \otimes \mathcal{B}_i \otimes \mathcal{B}_j$ if it occurs by $\mathcal{B}_i \otimes \mathcal{B}_j \otimes \mathcal{B}_i$ and repeat this process, until finally we see that every connected component of $\mathcal{B}_{i_1} \otimes \cdots \otimes \mathcal{B}_{i_N}$ is isomorphic to a connected component of $\mathcal{T}_\lambda \otimes \mathcal{B}_i \otimes \mathcal{B}_j \otimes \mathcal{B}_i$ for some λ.

Since tensoring with $\mathcal{T}_\lambda$ does not affect the Stembridge axioms, we are reduced to the A_2 case. We therefore return to the setting of Proposition 12.3. Consider $x = u(a, b, c)$. We have

$$\varepsilon_i(x) = \max(a - b + 2c, c), \qquad \varepsilon_j(x) = b - c.$$

Furthermore

$$e_i(x) = \begin{cases} u(a - 1, b, c) & \text{if } a + c > b, \\ u(a, b, c - 1) & \text{if } a + c \leqslant b. \end{cases}$$

It follows that if $a + c > b$, then $\varepsilon_j(e_i x) = \varepsilon_j(x)$, $\varepsilon_i(e_j x) = \varepsilon_i(x) + 1$ and $e_i e_j(x) = e_j e_i(x) = u(a - 1, b - 1, c)$. This is consistent with Stembridge Axiom S2. If $a + c < b$ then $\varepsilon_j(e_i x) = \varepsilon_j(x) + 1$, $\varepsilon_i(e_j x) = \varepsilon_i(x)$ and $e_i e_j(x) = e_j e_i(x) = u(a, b - 1, c - 1)$. This is also consistent with Stembridge Axiom S2. Finally if $a + c = b$, then $\varepsilon_j(e_i x) = \varepsilon_j(x) + 1$, $\varepsilon_i(e_j x) = \varepsilon_i(x) + 1$ and $e_i e_j^2 e_i(x) = e_j e_i^2 e_j(x) = u(a - 1, b - 2, c - 1)$. This verifies Axioms S1, S2 and S3. We leave the dual axioms S1', S2' and S3' to the reader. Axioms S0 and S0' are clear since $e_i(x)$ and $f_i(x)$ are never 0 for this crystal. $\qquad\square$

With $w_0 = s_{i_N} \cdots s_{i_1}$ a reduced decomposition as in Theorem 12.6, let $\prec$ be the partial order on $\mathcal{A} = \mathcal{B}_{i_1} \otimes \cdots \otimes \mathcal{B}_{i_N}$ defined in Section 4.4. Let $u_\infty = u_{i_1}(0) \otimes \cdots \otimes u_{i_N}(0)$. Let $\mathfrak{C} = \{x \in \mathcal{A} \mid x \preccurlyeq u_\infty\}$. In other words, $\mathfrak{C}$ consists of all elements that may be obtained from u_∞ by applying the f_i in some order.

The set $\mathfrak{C}$ is independent of the reduced word representing w_0 in the following sense. By Proposition 12.5, if $w_0 = s_{j_N} \cdots s_{j_1}$ is another reduced decomposition, and $\mathfrak{C}'$ is the corresponding subset of $\mathcal{A}' = \mathcal{B}_{j_1} \otimes \cdots \otimes \mathcal{B}_{j_N}$, then the crystal isomorphism $\mathcal{A} \to \mathcal{A}'$ in Proposition 12.5 takes $\mathfrak{C}$ to $\mathfrak{C}'$.

Proposition 12.7. *If $x \in \mathfrak{C}$, then $\varepsilon_i(x) \geqslant 0$ for all $i \in I$. If $\varepsilon_i(x) > 0$, then $e_i(x) \in \mathfrak{C}$.*

Proof. We may choose the reduced word so that $i_N = i$. Now we observe that every element x of $\mathfrak{C}$ is of the form $u_{i_1}(-a_1) \otimes \cdots \otimes u_{i_N}(-a_N)$, where the a_k are nonnegative integers. Indeed, this is clear since applying f_j to such an element increases one of the a_k by one (where $i_k = j$), and every element of $\mathfrak{C}$ is obtained from u_∞ by applying the operators f_j in some sequence. Now remembering that $i_N = i$, it is clear from (2.13) that $\varepsilon_i(x) \geqslant 0$ since the first term in the maximum is $\varepsilon_i(u_{i_N}(-a_N)) = a_N$. This proves the first statement.

For the second statement, assume that $x \in \mathfrak{C}$ such that $\varepsilon_i(x) > 0$. We must show that $e_i(x) \in \mathfrak{C}$. By induction, we may assume that $\varepsilon_i(y) > 0$ implies that $e_i(y) \in \mathfrak{C}$ for all $y \in \mathfrak{C}$ such that $x \prec y$.

Because $x \prec u_\infty$ there exists j such that $e_j(x) \in \mathfrak{C}$. Let $y = e_j(x)$. We have $\varepsilon_i(y) \geqslant \varepsilon_i(x) > 0$. Thus by induction $e_i(y) \in \mathfrak{C}$. By the Stembridge axioms either Axiom S2 or Axiom S3 applies in this situation. That is, if either $\varepsilon_i(y) = \varepsilon_i(x)$ or $\varepsilon_j(e_i(x)) = \varepsilon_j(x)$, then Axiom S2 applies and $e_i(x) = f_j e_i(y)$. Since $e_i(y) \in \mathfrak{C}$, it follows that $x \in \mathfrak{C}$.

The alternative is that $\varepsilon_i(y) = \varepsilon_i(x) + 1$ and $\varepsilon_j(e_i(x)) = \varepsilon_j(x) + 1$, and Axiom S3 applies. In this case, since $\varepsilon_i(x) > 0$, $\varepsilon_i(y) \geqslant 2$, and our induction hypothesis implies that $e_i^2(y) \in \mathfrak{C}$. Moreover $\varepsilon_j(e_i^2(y)) > \varepsilon_j(y) \geqslant 0$, so using the induction hypothesis again $z \in \mathfrak{C}$, where $z = e_j e_i^2(y)$. Now using Axiom S3, we have $e_i(x) = f_i f_j^2(z)$, so $e_i(x) \in \mathfrak{C}$. $\qquad\square$

We may now construct the crystal $\mathcal{B}_\infty$. As a set, $\mathcal{B}_\infty$ consists of the set $\mathfrak{C}$ defined above. We modify the definition of e_i by declaring that $e_i(x) = 0$ if $\varepsilon_i(x) = 0$.

Theorem 12.8. *The crystal $\mathcal{B}_\infty$ is weakly Stembridge and upper seminormal. It has a unique highest weight element u_∞ of weight 0.*

Proof. On defining the crystal this way, the content of Proposition 12.7 is that $\varepsilon_i(x) = \max\{k \mid e_i^k(x) \neq 0\}$. Hence the crystal $\mathcal{B}_\infty$ is upper seminormal.

The ambient crystal $\mathcal{A}$ is weakly Stembridge by Theorem 12.6. It follows from Proposition 12.7 that if $x \in \mathfrak{C}$ is as in Axiom S2 or Axiom S3 then all of the elements whose existence is asserted by those axioms is also in $\mathfrak{C}$. Axiom S0 follows from the fact that $\mathcal{B}_\infty$ is seminormal, and Axiom S0' follows from the fact that $f_i(x)$ is never 0 for this crystal. $\qquad\square$

An important property of $\mathcal{B}_\infty$ is that it is possible to embed other weak Stembridge crystals such as $\mathcal{B}_\lambda$ into it, after twisting by $\mathcal{T}_\lambda$ to adjust the highest weight. That follows from the following more general fact.

Theorem 12.9. *Let C be a connected weak Stembridge crystal with a highest weight element u_λ such that $\mathrm{wt}(u_\lambda) = \lambda$. Then there exists a unique injective crystal morphism $\psi \colon C \longrightarrow \mathcal{T}_\lambda \otimes \mathcal{B}_\infty$ that sends u_∞ to $t_\lambda \otimes u_\infty$.*

The proof is a modification of Theorem 4.13.

Proof. By Theorem 12.8, $\mathcal{B}_\infty$ is a weak Stembridge crystal. Tensoring on the left by the unique element t_λ of $\mathcal{T}_\lambda$ has the effect of changing the weight of each element by

λ, but otherwise leaves e_i, f_i and ε_i unchanged. Thus $\mathcal{T}_\lambda \otimes \mathcal{B}_\infty$ is upper seminormal and so Axiom S0 is satisfied. Such tensoring does change φ_i, but since f_i is never zero on $\mathcal{B}_\infty$ or on $\mathcal{T}_\lambda \otimes \mathcal{B}_\infty$, Axiom S0' is also satisfied and hence $\mathcal{T}_\lambda \otimes \mathcal{B}_\infty$ is a weak Stembridge crystal.

Let Ω be the set of all subsets S of $\mathcal{C}$ with the following properties. First $u_\lambda \in S$. Second, if $x \in S$ and $e_i x \neq 0$, then $e_i x \in S$. Finally, there exists a subset S' of $\mathcal{T}_\lambda \otimes \mathcal{B}_\infty$ and a weight-preserving bijection $x \longmapsto x'$ from $S \longrightarrow S'$ mapping $u_\lambda \longmapsto t_\lambda \otimes u_\infty$ such that if $x \in S$ then $e_i x \neq 0$ if and only if $e_i(x') \neq 0$ and $(e_i x)' = e_i x'$. The set Ω is nonempty since $\{u_\lambda\} \in \Omega$. Let us assume that S is an element of Ω that is maximal with respect to the inclusion of sets. We will prove that $S = \mathcal{C}$. Therefore assume that S is a proper subset. We will obtain a contradiction by extending the map $x \mapsto x'$ to $S \cup \{z\}$, where z is an element of $\mathcal{C} \setminus S$ that is maximal with respect to $\prec$.

The maximality of z implies that if $e_i z \neq 0$, then $e_i z \in S$. We would like to define $z' = f_i(e_i(z)')$, and we have to check that this is independent of i. This is proved exactly as in Theorem 4.13 as a consequence of the Stembridge axioms in both crystals. Depending on whether Axiom S2 or Axiom S3 applies, let $w = e_i e_j(z) = e_j e_i(z)$ or $w = e_j e_i^2 e_j(z) = e_i e_j^2 e_i(z)$, and let w' be the corresponding element of $S' \subset \mathcal{B}_\infty$. Then $z = f_i f_j(w) = f_j f_i(w)$ or $z = f_j f_i^2 f_j(w) = f_i f_j^2 f_i(w)$. For definiteness let us argue the second case. The dual Axiom S3' applies to w and to w', and so $f_j f_i^2 f_j(w') = f_i f_j^2 f_i(w')$. But $f_i^2 f_j(w') = e_j(z)'$ and $f_j^2 f_i(w') = e_i(z)'$, so $f_j(e_j(z)') = f_i(e_i(z)')$, as required.

This shows how to extend $x \mapsto x'$ to a larger subset than S, which is a contradiction unless $S = \mathcal{C}$. It is easy then to see that this map is a morphism of crystals. $\qquad\square$

Corollary 12.10. *Let λ be any dominant weight. Then there exists a unique injective crystal morphism $\psi_\lambda \colon \mathcal{B}_\lambda \longrightarrow \mathcal{T}_\lambda \otimes \mathcal{B}_\infty$ that sends the highest weight element u_λ to $t_\lambda \otimes u_\infty$.*

12.3   The crystal $\mathcal{B}_\infty$ for non-simply-laced types

We now generalize the results of Section 12.2 to non-simply-laced cases by using the construction of weak virtual crystals. Since, as it turns out, we will be able to branch to rank 2 cases, it will suffice to consider types C_2 and G_2. Recall the embeddings of Coxeter groups stated in Lemma 11.13.

We are now going to build the type C_2 crystal $(\mathcal{B}_1 \otimes \mathcal{B}_2)^{\otimes 2}$ for the long word $w_0 = (s_1 s_2)^2$ as a weak virtual crystal inside the ambient crystal $\widehat{\mathcal{V}}^A = (\mathcal{B}_1^A \otimes \mathcal{B}_3^A \otimes \mathcal{B}_2^A)^{\otimes 2}$ of type A_3. Let $\mathcal{V}^C$ be generated by $(u_1(0) \otimes u_3(0) \otimes u_2(0))^{\otimes 2}$ and the virtual crystal operators $f_1 = f_1^A f_3^A$ and $f_2 = f_2^A f_2^A$. Similarly, for the long word $w_0 = (s_2 s_1)^2$ we can realize the type C_2 crystal $(\mathcal{B}_2 \otimes \mathcal{B}_1)^{\otimes 2}$ as a weak virtual crystal inside the ambient crystal $\widehat{\mathcal{V}}'^A = (\mathcal{B}_2^A \otimes \mathcal{B}_1^A \otimes \mathcal{B}_3^A)^{\otimes 2}$ of type A_3. Let $\mathcal{V}'^C$ be generated by $(u_2(0) \otimes u_1(0) \otimes u_3(0))^{\otimes 2}$ and the virtual crystal operators

$f_1 = f_1^A f_3^A$ and $f_2 = f_2^A f_2^A$.

Similarly, we can construct the type G_2 crystal $(\mathcal{B}_1 \otimes \mathcal{B}_2)^{\otimes 3}$ for the long word $w_0 = (s_1 s_2)^3$ as a weak virtual crystal inside the ambient crystal $\widehat{\mathcal{V}}^D = (\mathcal{B}_1^D \otimes \mathcal{B}_3^D \otimes \mathcal{B}_4^D \otimes \mathcal{B}_2^D)^{\otimes 3}$ of type D_4. Let $\mathcal{V}^G$ be generated by $(u_1(0) \otimes u_3(0) \otimes u_4(0) \otimes u_2(0))^{\otimes 3}$ and the virtual crystal operators $f_1 = f_1^D f_3^D f_4^D$ and $f_2 = f_2^D f_2^D f_2^D$. Again we can consider the other reduced word for the long element $w_0 = (s_2 s_1)^3$ and construct $(\mathcal{B}_2 \otimes \mathcal{B}_1)^{\otimes 3}$ as a weak virtual crystal inside the ambient crystal $\widehat{\mathcal{V}}'^D = (\mathcal{B}_2^D \otimes \mathcal{B}_1^D \otimes \mathcal{B}_3^D \otimes \mathcal{B}_4^D)^{\otimes 3}$ of type D_4. Let $\mathcal{V}'^G$ be generated by $(u_2(0) \otimes u_1(0) \otimes u_3(0) \otimes u_4(0))^{\otimes 3}$ and the virtual crystal operators $f_1 = f_1^D f_3^D f_4^D$ and $f_2 = f_2^D f_2^D f_2^D$.

Proposition 12.11.

(i) *Both* $\mathcal{V}^C, \mathcal{V}'^C \subseteq \widehat{\mathcal{V}}^A$ *of type* C_2 *are weak virtual crystals. The elements*

$$u(a_1, \ldots, a_6) := \prod_{i=0}^{1} u_1(-a_{1+3i}) \otimes u_3(-a_{2+3i}) \otimes u_2(-a_{3+3i}) \in \mathcal{V}^C$$

$$u'(a_1, \ldots, a_6) := \prod_{i=0}^{1} u_2(-a_{1+3i}) \otimes u_1(-a_{2+3i}) \otimes u_3(-a_{3+3i}) \in \mathcal{V}'^C$$

are characterized by the conditions

$$\begin{array}{llll} a_1 = a_2, & a_4 = a_5, & a_3, a_6 \in 2\mathbb{Z} & \text{for } \mathcal{V}^C, \\ a_2 = a_3, & a_5 = a_6, & a_1, a_4 \in 2\mathbb{Z} & \text{for } \mathcal{V}'^C. \end{array} \tag{12.7}$$

(ii) *Both* $\mathcal{V}^G, \mathcal{V}'^G \subseteq \widehat{\mathcal{V}}^D$ *of type* G_2 *are weak virtual crystals. The elements*

$$u(a_1, \ldots, a_{12}) := \prod_{i=0}^{2} u_1(-a_{1+4i}) \otimes u_3(-a_{2+4i}) \otimes u_4(-a_{3+4i}) \otimes u_2(-a_{4+4i}) \in \mathcal{V}^G$$

$$u'(a_1, \ldots, a_{12}) := \prod_{i=0}^{2} u_2(-a_{1+4i}) \otimes u_1(-a_{2+4i}) \otimes u_3(-a_{3+4i}) \otimes u_4(-a_{4+4i}) \in \mathcal{V}'^G$$

are characterized by the conditions

$$\begin{array}{llll} a_1 = a_2 = a_3, & a_5 = a_6 = a_7, & a_9 = a_{10} = a_{11}, & a_4, a_8, a_{12} \in 3\mathbb{Z} \quad \text{for } \mathcal{V}, \\ a_2 = a_3 = a_4, & a_6 = a_7 = a_8, & a_{10} = a_{11} = a_{12}, & a_1, a_5, a_9 \in 3\mathbb{Z} \quad \text{for } \mathcal{V}'. \end{array}$$

Proof. We prove the statements for $\mathcal{V}^C$ in part (i) as the other cases are similar. By Theorem 12.6 the ambient crystal $\widehat{\mathcal{V}}^A$ is a weak Stembridge crystal, which confirms Axiom V1 of a weak virtual crystal. We now show that if $u(a_1, \ldots, a_6)$ satisfies the conditions in (12.7), then $f_i u(a_1, \ldots, a_6)$ and $e_i u(a_1, \ldots, a_6)$ also satisfy these conditions. Since the generator $u(0, \ldots, 0)$ satisfies the conditions in (12.7), this shows that $\mathcal{V}^C$ is closed under the virtual crystal operators f_i and e_i, proving Axiom V3 of weak virtual crystals.

By (2.11), it is clear that $\widehat{\varphi}_1(u) = \widehat{\varphi}_3(u)$ and $\widehat{\varphi}_2(u) \in 2\mathbb{Z}$ for $u = u(a_1, \ldots, a_6)$ satisfying (12.7). Also, f_i chooses the rightmost tensor factor where it can act.

Hence $\widehat{f_1}$ and $\widehat{f_3}$ always pick adjacent tensor factors on elements satisfying (12.7). This also shows that both operators in $\widehat{f_2^2}$ pick the same tensor factor if the corresponding parameters are even. The arguments for e_i are similar. This proves both Axioms V2 and V3 of weak virtual crystals. □

We are now in the position to state the analogue of Proposition 12.3.

Proposition 12.12.

(i) *For type C_2,*

$$(\mathcal{B}_1 \otimes \mathcal{B}_2)^{\otimes 2} \cong (\mathcal{B}_2 \otimes \mathcal{B}_1)^{\otimes 2}.$$

(ii) *For type G_2,*

$$(\mathcal{B}_1 \otimes \mathcal{B}_2)^{\otimes 3} \cong (\mathcal{B}_2 \otimes \mathcal{B}_1)^{\otimes 3}.$$

Proof. We prove part (i), as part (ii) is similar. By Proposition 12.11, $(\mathcal{B}_1 \otimes \mathcal{B}_2)^{\otimes 2}$ is isomorphic to the weak virtual crystal inside $(\mathcal{B}_1^A \otimes \mathcal{B}_3^A \otimes \mathcal{B}_2^A)^{\otimes 2}$ and $(\mathcal{B}_2 \otimes \mathcal{B}_1)^{\otimes 2}$ is isomorphic to the weak virtual crystal inside $(\mathcal{B}_2^A \otimes \mathcal{B}_1^A \otimes \mathcal{B}_3^A)^{\otimes 2}$. Note that both $(s_1^A s_3^A s_2^A)^2$ and $(s_2^A s_1^A s_3^A)^2$ are reduced words for w_0^A of type A_3. Hence by Proposition 12.5 we have

$$(\mathcal{B}_1^A \otimes \mathcal{B}_3^A \otimes \mathcal{B}_2^A)^{\otimes 2} \cong (\mathcal{B}_2^A \otimes \mathcal{B}_1^A \otimes \mathcal{B}_3^A)^{\otimes 2}.$$

Since both weak virtual crystals are generated by the same virtual crystal operators, this proves (i). □

We can now generalize Proposition 12.5.

Proposition 12.13. *Let w be an element of the Weyl group for finite Cartan type. Let $w = s_{i_N} \cdots s_{i_2} s_{i_1} = s_{j_N} \cdots s_{j_2} s_{j_1}$ be reduced decompositions. Then there exists an isomorphism*

$$\mathcal{B}_{i_1} \otimes \cdots \otimes \mathcal{B}_{i_N} \longrightarrow \mathcal{B}_{j_1} \otimes \cdots \otimes \mathcal{B}_{j_N}$$

such that $u_{i_1}(0) \otimes \cdots \otimes u_{i_N}(0) \mapsto u_{j_1}(0) \otimes \cdots \otimes u_{j_N}(0)$.

Proof. The proof is the same as for Proposition 12.5 using both Propositions 12.3 and 12.12. □

We may now define $\mathfrak{C}$ as in Section 12.2. Let $w_0 = s_{i_N} \cdots s_{i_1}$ be a reduced decomposition and $\mathcal{A} = \mathcal{B}_{i_1} \otimes \cdots \otimes \mathcal{B}_{i_N}$. With $u_\infty = u_{i_1}(0) \otimes \cdots \otimes u_{i_N}(0)$, let $\mathfrak{C} = \{x \in \mathcal{A} \mid x \preccurlyeq u_\infty\}$ be the set of elements that may be obtained from u_∞ by applying the f_i in some order. As before, the set $\mathfrak{C}$ is independent of the reduced word representing w_0 by Proposition 12.13.

Proposition 12.14. *If $x \in \mathfrak{C}$, then $\varepsilon_i(x) \geqslant 0$ for all $i \in I$. If $\varepsilon_i(x) > 0$, then $e_i(x) \in \mathfrak{C}$.*

Proof. For simply-laced types, this was proven in Proposition 12.7. For non-simply-laced types we can embed the elements of $\mathfrak{C}$ as a virtual crystal into a simply-laced ambient crystal corresponding to the long element by Proposition 12.11. By (5.3a), we have $\varepsilon_i(x) = \frac{1}{\gamma_i}\widehat{\varepsilon}_j(x)$ for all $j \in \sigma(i)$. Since $\widehat{\varepsilon}_j(x) \geqslant 0$ in the simply-laced ambient crystal, this implies $\varepsilon_i(x) \geqslant 0$. If $\varepsilon_i(x) > 0$, this also implies that $\widehat{\varepsilon}_j(x) \geqslant \gamma_i$ for all $j \in \sigma(i)$. By Proposition 12.7, we hence have that $e_i(x) = \prod_{j \in \sigma(i)} \widehat{e}_j^{\gamma_i}(x) \preccurlyeq u_\infty$ proving the statement. $\qquad\square$

As in Section 12.2, we may now define $\mathcal{B}_\infty$ as the set $\mathfrak{C}$. The crystal operators e_i are modified by declaring that $e_i(x) = 0$ if $\varepsilon_i(x) = 0$.

We say that a crystal $\mathcal{C}$ is *weak normal* if it is either a weak Stembridge crystal or a weak virtual crystal.

Theorem 12.15. *The crystal $\mathcal{B}_\infty$ is weak normal and upper seminormal. It has a unique highest weight element u_∞ of weight 0.*

Proof. By Proposition 12.14, the definition of e_i implies that $\varepsilon_i(x) = \max\{k \mid e_i^k(x) \neq 0\}$. Hence the crystal $\mathcal{B}_\infty$ is upper seminormal.

In simply-laced types, $\mathcal{B}_\infty$ is weak Stembridge by Theorem 12.8. In non-simply-laced types, we have shown that it can be constructed as a weak virtual crystal. $\qquad\square$

For types C_2 (resp. G_2), the elements of $\mathcal{B}_\infty$ can be characterized as elements in $(\mathcal{B}_1 \otimes \mathcal{B}_2)^{\otimes 2}$ and $(\mathcal{B}_2 \otimes \mathcal{B}_1)^{\otimes 2}$ (resp. $(\mathcal{B}_1 \otimes \mathcal{B}_2)^{\otimes 3}$ and $(\mathcal{B}_2 \otimes \mathcal{B}_1)^{\otimes 3}$). This is in analogy to the characterization in type A_2 given in Proposition 12.4. See Exercises 12.7 and 12.8.

Corollary 12.10 also generalizes.

Theorem 12.16. *Let λ be any dominant weight. Then there exists a unique injective crystal morphism $\psi_\lambda \colon \mathcal{B}_\lambda \longrightarrow \mathcal{T}_\lambda \otimes \mathcal{B}_\infty$ that sends the highest weight element u_λ to $t_\lambda \otimes u_\infty$.*

Proof. For simply-laced types, the statement was proved in Corollary 12.10. For non-simply-laced types both $\mathcal{B}_\lambda$ and $\mathcal{B}_\infty$ can be realized as virtual crystals. Hence the statement follows from the induced statement for the underlying ambient crystal, which is simply-laced. $\qquad\square$

12.4 Demazure crystals in $\mathcal{B}_\infty$

As we will explain in the next chapter, a *Demazure character* for a Lie group G is not really a character, but part of a character. Similarly, a *Demazure crystal* is not really a crystal as defined in Chapter 2, but part of the crystal. In preparation for the theory in Chapter 13, we will define Demazure crystals in $\mathcal{B}_\infty$ here and prove the properties that we need later.

Let us pick a Weyl group element w. Take a reduced decomposition $w = s_{i_r} \cdots s_{i_1}$ of w and complete it to a reduced decomposition of $w_0 = s_{i_N} \cdots s_{i_1}$.

As in Sections 12.2 and 12.3,

$$\mathcal{C} = \{x \in \mathcal{A} \mid x \preccurlyeq u_\infty\} \quad \text{where} \quad \mathcal{A} = \mathcal{B}_{i_1} \otimes \cdots \otimes \mathcal{B}_{i_N}. \tag{12.8}$$

Let

$$\mathcal{A}(w) = \mathcal{B}_{i_1} \otimes \cdots \otimes \mathcal{B}_{i_r} \otimes u_{i_{r+1}}(0) \otimes \cdots \otimes u_{i_N}(0).$$

Remark 12.17. To justify the notation, we explain why $\mathcal{A}(w)$ is essentially independent of the choice of the reduced word $w = s_{i_r} \cdots s_{i_1}$. If $w = s_{j_r} \cdots s_{j_1}$ is another reduced word, then we may extend the second decomposition to a reduced word for $w_0 = s_{j_N} \cdots s_{j_1}$ by taking $j_k = i_k$ if $k > r$. Then the isomorphisms in Propositions 12.3 and 12.11 take $\mathcal{B}_{i_1} \otimes \cdots \otimes \mathcal{B}_{i_r} \otimes u_{i_{r+1}}(0) \otimes \cdots \otimes u_{i_N}(0)$ to $\mathcal{B}_{j_1} \otimes \cdots \otimes \mathcal{B}_{j_r} \otimes u_{j_{r+1}}(0) \otimes \cdots \otimes u_{j_N}(0)$.

Lemma 12.18. *Let*

$$x = u_{i_1}(-a_1) \otimes \cdots \otimes u_{i_r}(-a_r) \otimes u_{i_{r+1}}(0) \otimes \cdots \otimes u_{i_N}(0) \in \mathcal{A}(w) \cap \mathcal{C}.$$

If $f_i(x) \notin \mathcal{A}(w)$, then $\varepsilon_i(x) = 0$.

Proof. We set $a_k = 0$ if $r + 1 \leqslant k \leqslant N$. Let P be the set of k with $1 \leqslant k \leqslant N$ such that $i_k = i$. By Lemma 2.33, applying f_i to x increments a_{k_0} where $k_0 \in P$ is the first value of k at which

$$-a_k + \left\langle \sum_{j<k} -a_j \alpha_{i_j}, \alpha_i^\vee \right\rangle$$

attains its maximum. Assuming that $f_i(x) \notin \mathcal{A}(w)$ this $k = k_0$ is $> r$. This means that if $k \in P$ and $k \leqslant r$, then

$$-a_k + \left\langle \sum_{j<k} -a_j \alpha_{i_j}, \alpha_i^\vee \right\rangle < -a_{k_0} + \left\langle \sum_{j<k_0} -a_j \alpha_{i_j}, \alpha_i^\vee \right\rangle = \left\langle \sum_{j\leqslant r} -a_j \alpha_{i_j}, \alpha_i^\vee \right\rangle,$$

and so

$$-a_k < \left\langle \sum_{k\leqslant j\leqslant r} -a_j \alpha_{i_j}, \alpha_i^\vee \right\rangle.$$

We separate out the contribution of $j = k$, when $\langle \alpha_{i_j}, \alpha_i^\vee \rangle = \langle \alpha_i, \alpha_i^\vee \rangle = 2$. This proves, for all $k \in P$ such that $k \leqslant r$, the inequality

$$a_k < \left\langle \sum_{k<j\leqslant r} -a_j \alpha_{i_j}, \alpha_i^\vee \right\rangle. \tag{12.9}$$

Now let us consider $\varepsilon_i(x)$. Again using Lemma 2.33,

$$\varepsilon_i(x) = \max_{k\in P} \left(a_k - \sum_{k<j\leqslant r} \langle -a_j \alpha_{i_j}, \alpha_i^\vee \rangle \right).$$

It follows from (12.9) that this is negative for $k \leqslant r$, and if $k > r$, then it is zero. This proves that $\varepsilon_i(x) = 0$. $\qquad\square$

Now let $\mathcal{B}_\infty(w)$ be the subset of $\mathcal{B}_\infty$ defined as follows. If X is a subset of $\mathcal{B}_\infty$, let

$$\mathfrak{D}_i(X) = \{x \in \mathcal{B}_\infty \mid e_i^k(x) \in X \text{ for some } k \geqslant 0\}.$$

With $w = s_{i_r} \cdots s_{i_1}$ a reduced decomposition as before, define

$$\mathcal{B}_\infty(w) = \mathfrak{D}_{i_r} \cdots \mathfrak{D}_{i_1} \{u_\infty\}.$$

We will prove in Theorem 12.19 that this does not depend on the choice of reduced decomposition, justifying the notation.

By an *i-root string* S, we mean the set of all elements of a crystal that may be obtained by applying e_i or f_i from a given element. Every i-root string in $\mathcal{B}_\infty$ has a unique highest weight element, that is, an element u_S such that $e_i(u_S) = 0$. In particular, $S = \mathfrak{D}_i\{u_S\}$.

Theorem 12.19.

(i) *Let $w = s_{i_r} \cdots s_{i_1}$ be a reduced decomposition, completed to a reduced decomposition $w_0 = s_{i_N} \cdots s_{i_1}$ of the long Weyl group element. Identify $\mathcal{B}_\infty$ with the set $\mathfrak{C}$ in $\mathcal{A} = \mathcal{B}_{i_1} \otimes \cdots \otimes \mathcal{B}_{i_N}$. Then $\mathcal{B}_\infty(w) = \mathcal{A}(w) \cap \mathfrak{C}$.*

(ii) *The set $\mathcal{B}_\infty(w)$ is independent of the choice of reduced word $s_{i_r} \cdots s_{i_1}$ representing w.*

(iii) *If S is any root string in $\mathcal{B}_\infty$, then $\mathcal{B}_\infty(w) \cap S$ is one of $\varnothing$, S, or $\{u_S\}$.*

(iv) *When $w = w_0$ is the long Weyl group element, $\mathcal{B}_\infty(w_0) = \mathcal{B}_\infty$.*

We call the set $\mathcal{B}_\infty(w)$ a *Demazure crystal* in $\mathcal{B}_\infty$.

Proof. We will prove (i) by induction on the length of w. If $w = 1$, then there is nothing to prove, so let us write $w = s_{i_r} w'$ where $\ell(w') = r - 1$. By induction,

$$\mathcal{B}_\infty(w') = \left(\mathcal{B}_{i_1} \otimes \cdots \otimes \mathcal{B}_{i_{r-1}} \otimes u_{i_r}(0) \otimes \cdots \otimes u_{i_N}(0)\right) \cap \mathfrak{C}.$$

The set $\mathcal{B}_\infty(w) = \mathfrak{D}_{i_r} \mathcal{B}_\infty(w')$ consists of all elements of $\mathcal{B}_\infty$ that can be obtained by applying f_{i_r} (any number of times) from $\mathcal{B}_\infty(w')$. It is easy to see that applying f_{i_r} to

$$x = u_{i_1}(-a_1) \otimes \cdots \otimes u_{i_r}(-a_r) \otimes u_{i_{r+1}}(0) \otimes \cdots \otimes u_{i_N}(0) \in \mathcal{A}(w) \cap \mathfrak{C} \qquad (12.10)$$

produces an element of the same type, and so $\mathcal{B}_\infty(w) \subset \mathcal{A}(w) \cap \mathfrak{C}$.

To prove the converse inclusion, suppose that x of (12.10) is in $\mathcal{A}(w) \cap \mathfrak{C}$. We will show that $x \in \mathfrak{D}_{i_r} \mathcal{B}_\infty(w')$. If $a_r = 0$ then $x \in \mathcal{B}_\infty(w') \subseteq \mathfrak{D}_{i_r} \mathcal{B}_\infty(w')$. Thus we may assume that $a_r \neq 0$. Using Lemma 2.33, $\varepsilon_{i_r}(x)$ is a maximum of terms, one of which is a_r. Since thus $\varepsilon_{i_r}(x) > 0$, we may consider $e_{i_r}(x)$ which by (12.7) is also in $\mathcal{A}(w) \cap \mathfrak{C}$. By induction on $\sum a_i$, $e_{i_r}(x) \in \mathcal{B}_\infty(w)$ and this implies that $x \in \mathfrak{D}_{i_r} \mathcal{B}_\infty(w') = \mathcal{B}_\infty(w)$. We have proved that $\mathcal{B}_\infty(w) = \mathcal{A}(w) \cap \mathfrak{C}$.

Now Remark 12.17 shows that $\mathcal{B}_\infty(w)$ does not depend on the reduced word representing w. This proves (ii).

Next show that if S is any root string in $\mathcal{B}_\infty$, then $\mathcal{B}_\infty(w) \cap S$ is one of $\varnothing$, S, or $\{u_S\}$. If S is an i-root string, this is equivalent to showing that if $\mathcal{B}_\infty(w) \cap S$

contains an element x such that $\varepsilon_i(x) > 0$, then $e_i(x)$ and $f_i(x)$ are in $\mathcal{B}_\infty(w)$. It is easy to check this for $e_i(x)$, and as for $f_i(x)$, this is Lemma 12.18. This proves (iii). Finally, (iv) is obvious from $\mathcal{B}_\infty(w_0) = \mathcal{A}(w_0) \cap \mathcal{C}$ since $\mathcal{A}(w_0) = \mathcal{A} \supset \mathcal{C}$. □

Exercises

[Cliff (1998); Hong and Lee (2008, 2012)] gave descriptions of the crystal $\mathcal{B}_\infty$ which are useful for their simplicity. The following exercise constructs $\mathcal{B}_\infty$ for type A_n using their method.

Exercise 12.1. Consider $GL(n)$ tableaux of varying shapes λ. Define an equivalence relation on tableaux as follows. Let T and T' be tableaux of shapes λ and λ'. Write $T \equiv T'$ if for each i with $1 \leqslant i \leqslant n$, where R_i and R'_i are the i-th rows, the rows R_i and R'_i only differ by some i's at the beginning of the tableau. For example, the tableaux

$$\begin{array}{|c|c|c|c|}\hline 1 & 2 & 2 & 3 \\\hline 2 & 3 & 4 \\\cline{1-3}\end{array}, \qquad \begin{array}{|c|c|c|c|c|c|c|c|}\hline 1 & 1 & 1 & 1 & 1 & 2 & 2 & 3 \\\hline 2 & 2 & 3 & 4 \\\cline{1-4} 3 \\\cline{1-1}\end{array}$$

are equivalent. We say that a tableau T is *large* if the number of entries in the i-th row that are equal to i is strictly larger than the total number of entries in the $i+1$ row. Thus the second equivalent tableau above is large. If T is a tableau, let $[T]$ denote its equivalence class.

(i) Prove that if T and T' are equivalent large tableaux, then $e_i(T)$ and $e_i(T')$ are equivalent, and that $f_i(T)$ and $f_i(T')$ are equivalent. Therefore give a crystal structure to the set of equivalence classes by defining $e_i([T]) = [e_i(T')]$ where T' is any large tableau equivalent to T. Define $\mathrm{wt}(T)$ in such a way that a Yamanouchi tableau (having only i's in the i-th row) has weight zero. Verify that the set of equivalence classes of $GL(n)$ tableaux is a crystal.

(ii) Prove that this crystal is isomorphic to $\mathcal{B}_\infty$.
   Hint: Use the reduced word $w_0 = (s_{n-1})(s_{n-2}s_{n-1}) \cdots (s_1 \cdots s_{n-1})$. Thus $\mathcal{B}_\infty$ is embedded in $\mathcal{B}(n-1) \otimes \cdots \otimes \mathcal{B}(1)$ where $\mathcal{B}(i) = \mathcal{B}_i \otimes \mathcal{B}_{i+1} \otimes \cdots \otimes \mathcal{B}_{n-1}$. Relate the equivalence classes of i-th rows to elements of the crystal $\mathcal{B}(i)$.

Exercise 12.2. Prove (12.1).

Exercise 12.3. Prove for $GL(n)$ crystals that (12.1) is the character of the crystal $\mathcal{B}_\infty$.

Exercise 12.4. Let λ be a dominant weight. In Corollary 12.10 and Theorem 12.16 we proved that there is an injective crystal morphism $\mathcal{B}_\lambda \longrightarrow \mathcal{T}_\lambda \otimes \mathcal{B}_\infty$ that takes the highest weight element u_λ to $t_\lambda \otimes u_\infty$. Show that there also exists a surjective crystal morphism $\mathcal{T}_\lambda \otimes \mathcal{B}_\infty \longrightarrow \mathcal{B}_\lambda$ that takes $t_\lambda \otimes u_\infty$ to u_λ.

Exercise 12.5. We have seen that for A_2 crystals, a basis of $\mathcal{B}_\infty$ consists of $u(a, b, c)$ defined by (12.5) with $a \geqslant 0$ and $b \geqslant c \geqslant 0$. Compute string$_\mathbf{i}$ $(u(a, b, c))$ with respect to the reduced words $\mathbf{i} = (1, 2, 1)$ and $\mathbf{i} = (2, 1, 2)$ for w_0.

Exercise 12.6. For the A_2 crystal $\mathcal{B}_1 \otimes \mathcal{B}_2 \otimes \mathcal{B}_1$, find all elements $x = u(a, b, c)$ such that $\varepsilon_1(x) = \varepsilon_2(x) = 0$. Prove that $\mathcal{B}_1 \otimes \mathcal{B}_2 \otimes \mathcal{B}_1$ is not connected.

Exercise 12.7. [Kashiwara (1993), Example 2.2.6]

(i) Show that for type C_2 a basis for $\mathcal{B}_\infty$ consists of $u_1(-a) \otimes u_2(-b) \otimes u_1(-c) \otimes u_2(-d) \in \mathcal{B}_1 \otimes \mathcal{B}_2 \otimes \mathcal{B}_1 \otimes \mathcal{B}_2$ with
$$a \geqslant 0 \quad \text{and} \quad d \geqslant c \geqslant b \geqslant 0.$$

(ii) Show that for type C_2 another basis for $\mathcal{B}_\infty$ consists of $u_2(-a) \otimes u_1(-b) \otimes u_2(-c) \otimes u_1(-d) \in \mathcal{B}_2 \otimes \mathcal{B}_1 \otimes \mathcal{B}_2 \otimes \mathcal{B}_1$ with
$$a \geqslant 0, \quad 2b \geqslant c \geqslant 0, \quad \text{and} \quad c \geqslant 2d \geqslant 0.$$

Exercise 12.8. [Kashiwara (1993), Example 2.2.7]

(i) Show that for type G_2 a basis for $\mathcal{B}_\infty$ consists of $u_1(-a) \otimes u_2(-b) \otimes u_1(-c) \otimes u_2(-d) \otimes u_1(-e) \otimes u_2(-f) \in \mathcal{B}_1 \otimes \mathcal{B}_2 \otimes \mathcal{B}_1 \otimes \mathcal{B}_2 \otimes \mathcal{B}_1 \otimes \mathcal{B}_2$ with
$$a \geqslant 0, \quad b \geqslant c \geqslant 0, \quad 2c \geqslant d \geqslant 0, \quad \frac{d}{2} \geqslant e \geqslant 0, \quad e \geqslant f \geqslant 0.$$

(ii) Show that for type G_2 another basis for $\mathcal{B}_\infty$ consists of $u_2(-a) \otimes u_1(-b) \otimes u_2(-c) \otimes u_1(-d) \otimes u_2(-e) \otimes u_1(-f) \in \mathcal{B}_2 \otimes \mathcal{B}_1 \otimes \mathcal{B}_2 \otimes \mathcal{B}_1 \otimes \mathcal{B}_2 \otimes \mathcal{B}_1$ with
$$a \geqslant 0, \quad 3b \geqslant c \geqslant 0, \quad \frac{2c}{3} \geqslant d \geqslant 0, \quad \frac{3d}{2} \geqslant e \geqslant 0, \quad \frac{e}{3} \geqslant f \geqslant 0.$$

Exercise 12.9. In this exercise, we consider $\mathcal{B}_\infty$ for A_2. Explain why the number of elements of weight $-\lambda$ equals the number of ways of writing the weight λ as $(a + c)\alpha_1 + b\alpha_2$ where $a \geqslant 0$ and $b \geqslant c \geqslant 0$. Prove that this equals $P(\lambda)$, where P is the Kostant partition function, by exhibiting a bijection between the set of such a, b, c and the set of root partitions of λ.

Exercise 12.10. Consider the $\mathcal{B}_\infty$ crystal of Cartan type C_2. Use Exercise 12.7 to prove that the number of elements of $\mathcal{B}_\infty$ of weight $-\lambda$ equals the Kostant partition function $P(\lambda)$.

Exercise 12.11. Let λ be a dominant weight for any Cartan type. Using Theorem 12.16, we have an injective morphism $\theta_\lambda : \mathcal{T}_{-\lambda} \otimes \mathcal{B}_\lambda \longrightarrow \mathcal{B}_\infty$ that takes the highest weight $t_{-\lambda} \otimes u_\lambda$ to u_∞. Let $v \in \mathcal{B}_\infty$. Show that there exists a constant N depending on v such that, if $\langle \lambda, \alpha_i^\vee \rangle > N$ for all simple roots α_i, then v is in the image of θ_λ.

Hint: Let $v \in \mathcal{B}_\infty$ be a counterexample. By induction we may assume that v is minimal in the sense that if $e_i(v) \neq 0$, then $e_i(v)$ is in the image of θ_λ whenever $\langle \lambda, \alpha_i^\vee \rangle$ is sufficiently large. Thus $v = f_i(y)$ for some $y \in \mathcal{B}_\infty$ and for every N there exists $x \in \mathcal{B}_\lambda$ with $\langle \lambda, \alpha_i^\vee \rangle > N$ such that $y = \theta_\lambda(t_{-\lambda} \otimes x)$. Argue that $\varphi_i(x) = 0$ and explain how to obtain a contradiction by taking N sufficiently large.

Chapter 13

Demazure Crystals

We have emphasized the analogy between crystals and representations, but now we will prove a firm connection. That is, if λ is a dominant weight, then there is a unique irreducible representation π_λ with highest weight λ, and a unique normal crystal $\mathcal{B}_\lambda$. We will prove that these have the same character.

In the course of proving this, we will encounter *Demazure crystals* which are certain subsets of $\mathcal{B}_\lambda$. They were first conjectured by [Littelmann (1995a)] to generalize Demazure characters; the conjecture was later proven by [Kashiwara (1993)] using $\mathcal{B}_\infty$ and the definition of crystals from $U_q(\mathfrak{g})$-representations. We will make use of $\mathcal{B}_\infty$, but instead of the quantum group $U_q(\mathfrak{g})$, we will give purely combinatorial proofs based on the results in Chapter 12. Thus in place of quantum group methods, we will rely on the properties of Stembridge and virtual crystals.

13.1 Demazure operators and the Demazure character formula

We consider operators on the free abelian group $\mathcal{E}$ spanned by symbols t^μ with μ in the weight lattice Λ. Given a weight $\lambda \in \Lambda$, define the *Demazure operator* on $\mathcal{E}$ by

$$D_i(t^\lambda) := \frac{t^{\lambda+\rho} - t^{s_i(\lambda+\rho)}}{1 - t^{-\alpha_i}} t^{-\rho}$$

$$= \begin{cases} t^\lambda(1 + t^{-\alpha_i} + \cdots + t^{-n\alpha_i}) & \text{if } n = \langle \lambda, \alpha_i^\vee \rangle \geqslant 0, \\ 0 & \text{if } n = \langle \lambda, \alpha_i^\vee \rangle = -1, \\ -t^\lambda(t^{\alpha_i} + \cdots + t^{(-n-1)\alpha_i}) & \text{if } n = \langle \lambda, \alpha_i^\vee \rangle \leqslant -2, \end{cases} \tag{13.1}$$

where ρ is the Weyl vector defined in (2.3) and $s_i(x) = x - \langle x, \alpha_i^\vee \rangle \alpha_i$ is the i-th simple reflection in the Weyl group W. Equation (13.1) follows from $s_i(\lambda + \rho) = \lambda + \rho - (\langle \lambda, \alpha_i^\vee \rangle + 1)\alpha_i$ by summing a finite geometric series.

Lemma 13.1. *The Demazure operators satisfy $D_i^2 = D_i$ and $s_i D_i = D_i$, where s_i is understood as an operator on $\mathcal{E}$ defined by $s_i \cdot t^\mu = t^{s_i \mu}$.*

Proof. Since $s_i(\rho) = \rho - \alpha_i$ due to the fact that $\langle \rho, \alpha_i^\vee \rangle = 1$ for all i by Exercise 2.13, we may also write

$$D_i f(t) = \frac{f(t) - t^{-\alpha_i} f(s_i t)}{1 - t^{-\alpha_i}}. \tag{13.2}$$

Since $s_i \alpha_i = -\alpha_i$, it is easy to check that $D_i f$ is invariant under s_i, that is, $s_i D_i = D_i$. Now if we apply D_i to an s_i-invariant function f of t, it is clear from (13.2) that $D_i f = f$. This means that $D_i^2 = D_i$. □

Proposition 13.2. *The Demazure operators also satisfy the same braid relations as the underlying Weyl group. That is, if $n(i, j)$ is the order of $s_i s_j$, then*

$$D_i D_j D_i \cdots = D_j D_i D_j \cdots \qquad (13.3)$$

where there are $n(i, j)$ factors on each side.

Proof. This is longer to prove than the simple properties in Lemma 13.1, so we will refer to [Bump (2013), Proposition 25.3] for details. Later in this chapter, we will show that these braid relations can be deduced from the results that we will prove for the crystal Demazure operators. □

By Matsumoto's Theorem (Theorem 11.4), the braid relations imply that we may define

$$D_w = D_{i_k} \cdots D_{i_3} D_{i_2} D_{i_1} ,$$

where $w = s_{i_k} \cdots s_{i_3} s_{i_2} s_{i_1}$ is any reduced expression. Let λ be a dominant weight. The *Demazure character* is defined as

$$\mathrm{ch}_{\lambda, w}(t) := D_w(t^\lambda).$$

We will now state the *Demazure character formula*.

Theorem 13.3. *Let w_0 be the long Weyl group element and let λ be a dominant weight. Then $\mathrm{ch}_{\lambda, w_0}(t)$ is the character of the irreducible representation with highest weight λ.*

Proof. See [Bump (2013), Theorem 25.3], where this fact is deduced from the Weyl character formula. □

Remark 13.4. Let G be a complex analytic Lie group with weight lattice Λ, and let $\mathfrak{n}^+$ be the maximal nilpotent Lie subalgebra of its Lie algebra generated by the positive root spaces. Let V_λ be the representation of G having highest weight λ. The Demazure character describes the weight multiplicities in the $U(\mathfrak{n}^+)$-module generated by the extremal weight vector in V_λ with weight $w(\lambda)$. It is also the character of the maximal torus of the Lie group acting on sections of a line bundle over a Schubert or (equivalently) Bott–Samelson variety. See [Demazure (1974); Andersen (1985); Joseph (1985); Kumar (1987); Mathieu (1988)].

13.2   Demazure crystals

Fix a root system Φ. Just as in Section 12.4, we define operators $\mathfrak{D}_i$ on subsets X of a crystal $\mathcal{B}$ defined by

$$\mathfrak{D}_i X = \{x \in \mathcal{B} \mid e_i^k(x) \in X \text{ for some } k \geqslant 0\}. \tag{13.4}$$

Let λ be a dominant weight. Let $w = s_{i_r} \cdots s_{i_1}$ be a reduced word for the Weyl group element w. We define

$$\mathcal{B}_\lambda(w) = \mathfrak{D}_{i_r} \cdots \mathfrak{D}_{i_1} \{u_\lambda\},$$

where u_λ is the highest weight element of the normal crystal $\mathcal{B}_\lambda$ with highest weight λ. We need to check that this is independent of the reduced word representing w, and that is asserted in the following Theorem.

Theorem 13.5.

(i) *The set $\mathcal{B}_\lambda(w)$ is independent of the choice of reduced word $s_{i_r} \cdots s_{i_1}$ representing w.*

(ii) *If S is any root string in $\mathcal{B}_\lambda$, then $\mathcal{B}_\lambda(w) \cap S$ is one of $\varnothing$, S, or $\{u_S\}$.*

(iii) *When $w = w_0$ is the long Weyl group element, $\mathcal{B}_\lambda(w_0) = \mathcal{B}_\lambda$.*

Proof. We recall from Corollary 12.16 that there exists an embedding ψ_λ of $\mathcal{B}_\lambda$ into $\mathcal{T}_\lambda \otimes \mathcal{B}_\infty$ that sends the highest weight element u_λ to $t_\lambda \otimes u_\infty$. It is easy to show by induction on the length of w that $\mathcal{B}_\lambda(w)$ is the inverse image of $\mathcal{B}_\infty(w)$ under the map ψ_λ. So all three parts follow from the corresponding assertions in Theorem 12.19. $\square$

For reference, (ii) of Theorem 13.5 is the following property.

$$X \cap S = \quad \varnothing \quad \text{or} \quad \{u_S\} \quad \text{or} \quad S. \tag{13.5}$$

It is remarkable that there exists a collection of subsets of $\mathcal{B}_\lambda$ that satisfy (ii) and that is closed under the operators $\mathfrak{D}_i$. To see why this requires proof, consider the example in Figure 13.1. Here is a subset X of $\mathcal{B}_\lambda$ that satisfies (ii), but such that $\mathfrak{D}_2 X$ does not. Since Condition (13.5) is not preserved by the $\mathfrak{D}_i$, Theorem 13.5 (ii) seems a deep property of the Demazure crystals that is not easy to prove.

13.3   Crystal Demazure operators

[Littelmann (1995a)] defined crystal analogues of the Demazure operators D_i that were defined in Section 13.1. Let $\mathcal{B}$ be a crystal and $b \in \mathcal{B}$. The crystal Demazure operator acts on the free $\mathbb{Z}$-module $\mathbb{Z}[\mathcal{B}]$ generated by $\mathcal{B}$ by the formula

$$\mathcal{D}_i(b) = \begin{cases} \displaystyle\sum_{0 \leqslant k \leqslant \langle \mathrm{wt}(b), \alpha_i^\vee \rangle} f_i^k b & \text{if } \langle \mathrm{wt}(b), \alpha_i^\vee \rangle \geqslant 0, \\[2ex] -\displaystyle\sum_{1 \leqslant k < -\langle \mathrm{wt}(b), \alpha_i^\vee \rangle} e_i^k b & \text{if } \langle \mathrm{wt}(b), \alpha_i^\vee \rangle < 0. \end{cases} \tag{13.6}$$

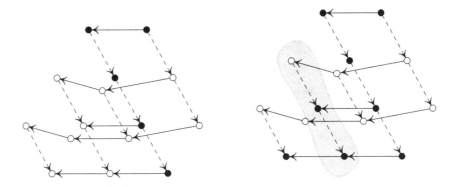

Fig. 13.1 An example showing that Condition (13.5) is not preserved by the $\mathfrak{D}_i$. Left: The subset X of the GL(3) crystal $\mathcal{B}_{(3,1)}$. (The subset is marked in black.) The crystal operators in the solid arrows are f_2 and the operators in the dashed arrows are f_1. Right: The subset $\mathfrak{D}_2 X$. Observe that X satisfies Condition (13.5) while $\mathfrak{D}_2 X$ does not, since the circled root string does not satisfy (13.5).

This lifts the formula (13.1) in the following sense. Extend the map $b \mapsto t^{\mathrm{wt}(b)}$ on $\mathcal{B}$ by linearity to a map $\mathrm{wt}_t \colon \mathbb{Z}[\mathcal{B}] \longrightarrow \mathcal{E}$. Comparing (13.6) with (13.1), the diagram

$$
\begin{array}{ccc}
\mathbb{Z}[\mathcal{B}] & \xrightarrow{\;\mathcal{D}_i\;} & \mathbb{Z}[\mathcal{B}] \\
{\scriptstyle \mathrm{wt}_t}\downarrow & & \downarrow{\scriptstyle \mathrm{wt}_t} \\
\mathcal{E} & \xrightarrow{\;\mathcal{D}_i\;} & \mathcal{E}
\end{array}
\tag{13.7}
$$

commutes. Parallel to Lemma 13.1, the reader will easily check that

$$
\mathcal{D}_i^2 = \mathcal{D}_i. \tag{13.8}
$$

Unlike the D_i, the $\mathcal{D}_i$ do not satisfy the braid relations.

The importance of Condition (13.5) may be seen from the following result. If X is a subset of $\mathcal{B}$, let $\Sigma(X) \in \mathbb{Z}[\mathcal{B}]$ be the sum of the elements of X. Recall that $\mathfrak{D}_i$ acts on sets via (13.4).

Lemma 13.6. *Let X be a subset of $\mathcal{B}$, and suppose that for every i-string S of $\mathcal{B}$, (13.5) is satisfied. Then*

$$
\mathcal{D}_i \Sigma(X) = \Sigma(\mathfrak{D}_i X).
$$

Proof. We may clearly assume that X is nonempty and contained in single i-string S. So if $X = \{u_S\}$ or $X = S$, then we need to show that $\mathcal{D}_i \Sigma(X) = \Sigma(S)$. If $X = \{u_S\}$, then the statement means

$$
\mathcal{D}_i u_S = \Sigma(S) \tag{13.9}
$$

which follows from the first case of (13.6) since $\langle \mathrm{wt}(u_S), \alpha_i^\vee \rangle \geqslant 0$ and $S = \{u_S, f_i(u_S), \ldots, f_i^{\langle \mathrm{wt}(u_S), \alpha_i^\vee \rangle}(u_S)\}$. To check the case where $X = S$, apply $\mathcal{D}_i$ to (13.9). Using (13.8), this gives us $\Sigma(S) = \mathcal{D}_i u_S = \mathcal{D}_i^2 u_S = \mathcal{D}_i \Sigma(S)$. $\qquad\square$

The following result is the *refined Demazure character formula* that was proved by [Littelmann (1995a); Kashiwara (1993)].

Theorem 13.7. *Let λ be a dominant weight. Then*

$$\sum_{v \in \mathcal{B}_\lambda(w)} t^{\mathrm{wt}(v)} = D_{i_r} \cdots D_{i_1} t^\lambda. \tag{13.10}$$

Proof. Since by Theorem 13.5 the $\mathcal{B}_\lambda(w)$ satisfy Condition (13.5), we may use Lemma 13.6 repeatedly to see that

$$\Sigma(\mathcal{B}_\lambda(w)) = \mathcal{D}_{i_r} \cdots \mathcal{D}_{i_1} u_\lambda.$$

Applying wt_t and using the commutative diagram (13.7) gives (13.10).   □

Corollary 13.8. *Let $\mathcal{C} = \mathcal{B}_\infty$ or the normal crystal $\mathcal{B}_\lambda$, where λ is a dominant weight. Then every reduced word for w_0 is a good word for $\mathcal{C}$.*

Proof. Let $w_0 = s_{i_1} \cdots s_{i_N}$ and let $x \in \mathcal{C}$ and let $v \in \mathcal{C}$, and define the stations (11.1). Since $v = v_0 \in \mathcal{C}(w_0) = \mathcal{D}_{i_1} \cdots \mathcal{D}_{i_N}\{u_\infty\}$, we have $v_1 = e_{i_1}^{a_1} v_0 \in \mathcal{D}_{i_2} \cdots \mathcal{D}_{i_N}\{u_\infty\}$. Here $a_1 = \varepsilon_{i_1}(v_0)$. Repeating this process, we learn that $v_k \in \mathcal{D}_{i_{k+1}} \cdots \mathcal{D}_{i_N}\{u_0\}$. Now taking $k = N$, $v_N = u_\infty$. This means that $(i_1, \ldots, i_N)$ is a good word.   □

Corollary 13.9. *The character of the irreducible representation with highest weight λ equals the character of the connected normal crystal $\mathcal{B}_\lambda$ with highest weight λ.*

Proof. We take $w = w_0$. The statement follows from Theorem 13.3 and Theorem 13.5 (iii).   □

Exercises

Exercise 13.1. Consider the A_2 crystal with highest weight $\lambda = (3,1,0)$. This crystal has 15 elements. Its crystal graph appears in Figure 13.1. For all 6 elements of the A_2 Weyl group, compute the Demazure crystal $\mathcal{B}_\lambda(w)$, and thus confirm Theorem 13.5 for this crystal.

Exercise 13.2. Consider the C_2 crystal with highest weight $\lambda = (2,1)$. This crystal has 16 elements. For all 8 elements of the C_2 Weyl group, compute the Demazure crystal $\mathcal{B}_\lambda(w)$, and thus confirm Theorem 13.5 for this crystal.

Exercise 13.3. Consider the G_2 crystal with highest weight $\lambda = \varpi_2$. One realization of this crystal is as the adjoint crystal in Figure 5.7. The crystal has 14 elements. For all 12 elements of the G_2 Weyl group, compute the Demazure crystal $\mathcal{B}_\lambda(w)$, and thus confirm Theorem 13.5 for this crystal.

Exercise 13.4.

(i) Let P be the Kostant partition function defined in Chapter 12, and let λ be a dominant weight. Let μ be another weight, and let $m(\mu)$ be the multiplicity of μ in the character of $\mathcal{B}_\lambda$, that is, the number of $v \in \mathcal{B}_\lambda$ such that $\mathrm{wt}(v) = \mu$. Prove the *Kostant multiplicity formula*

$$m(\mu) = \sum_{w \in W} (-1)^{\ell(w)} P\big(w(\lambda + \rho) - \rho - \mu\big),$$

where ρ is the Weyl vector, half the sum of the positive roots.

Hint: By Corollary 13.9, the character of $\mathcal{B}_\lambda$ is given by the Weyl character formula, which we may write in the form

$$\chi_\lambda(t) = \prod_{\alpha \in \Phi^+} (1 - t^{-\alpha})^{-1} \sum_{w \in W} (-1)^{\ell(w)} t^{w(\alpha + \rho) - \rho}.$$

Now make use of (12.1).

(ii) Make use of Exercise 12.11 to show that the number of $v \in \mathcal{B}_\infty$ of weight μ equals $P(\mu)$.

(iii) Prove that the character of $\mathcal{B}_\infty$ is $\prod_{\alpha \in \Phi^+} (1 - t^{-\alpha})^{-1}$.

Chapter 14

The $\star$-Involution of $\mathcal{B}_\infty$

The "$\star$-involution" of $\mathcal{B}_\infty$ appeared in [Lusztig (1990a,b)]. It played an important role in the proof of [Kashiwara (1993)] of the Demazure character formula. See also [Kashiwara (1994)]. For both Lusztig and Kashiwara, the involution derived from an anti-automorphism of the quantum group that induces the permutation $\alpha_i \mapsto -w_0\alpha_i$ of the simple roots. We will return to this point in Chapter 15.

In this chapter, we will give combinatorial proofs of the basic facts about this involution. [Joseph (2012)] was useful in preparing this chapter.

Let us fix a reduced word $\mathbf{i} = (i_1, \ldots, i_N)$ for the long Weyl group element w_0. Then there exist two different embeddings of $\mathcal{B}_\infty$ into $\mathbb{Z}^N$. First, we have seen that $\mathcal{B}_\infty$ is isomorphic to a subcrystal of $\mathcal{B}_{i_1} \otimes \cdots \otimes \mathcal{B}_{i_N}$; for $x = u_{i_1}(-a_1) \otimes \cdots \otimes u_{i_N}(-a_N) \in \mathcal{B}_\infty$, we define $\iota_1(x) = (a_1, \ldots, a_N)$. Second, let $\iota_2(x) = \text{string}_{\mathbf{i}}(x)$, which was defined in Chapter 11 for $\mathcal{B}_\lambda$, but which may be defined the same way for $\mathcal{B}_\infty$. Thus $\iota_2(x) = (b_1, \ldots, b_N)$ such that $b_1 = \varepsilon_{i_1}(x)$, $b_2 = \varepsilon_{i_2}(e_{i_1}^{b_1}x)$, $b_3(x) = \varepsilon_{i_3}(e_{i_2}^{b_2}e_{i_1}^{b_1}x)$, etc.

[Kashiwara (1993)] proved that the two embeddings ι_1 and ι_2 differ by a weight preserving bijection $\star\colon \mathcal{B}_\infty \longrightarrow \mathcal{B}_\infty$ of order two with remarkable properties. Although the embeddings ι_1 and ι_2 depend on the choice of reduced word $\mathbf{i}$, the bijection $\star$ does not. Using the bijection $\star$, we may define a new crystal structure on $\mathcal{B}_\infty$ with crystal operators $e_i^\star$ and $f_i^\star$ by transportation of structure, that is,

$$e_i^\star(x) = e_i(x^\star)^\star, \qquad f_i^\star(x) = f_i(x^\star)^\star. \tag{14.1}$$

Also

$$\varepsilon_i^\star(x) = \varepsilon_i(x^\star), \qquad \varphi_i^\star(x) = \varphi_i(x^\star). \tag{14.2}$$

In this chapter we will prove:

Theorem 14.1. *There exists a weight preserving map $\mathcal{B}_\infty \longrightarrow \mathcal{B}_\infty$ of order two to be denoted $x \mapsto x^\star$ such that $\iota_1(x^\star) = \iota_2(x)$. Although the maps ι_1 and ι_2 depend on the choice of reduced word $\mathbf{i}$, the map $\star$ does not. Defining a new crystal structure by (14.1) and (14.2), the operators e_i and f_i commute with $e_j^\star$ and $f_j^\star$ if $i \neq j$.*

If $i = j$, the operators e_i and f_i do *not* in general commute with $e_i^\star$ and $f_i^\star$.

14.1   The A_2 case

We will give a quick proof for type A_2 crystals. This section will not be required for the proofs in the general case.

Define maps $\tau, \theta \colon \mathbb{R}^3 \longrightarrow \mathbb{R}^3$ by

$$\tau(a, b, c) = \big(\max(a - b + 2c, c), b, \min(a, b - c)\big),$$

$$\theta(a, b, c) = \big(\max(c, b - a), a + c, \min(a, b - c)\big).$$

We have already encountered the map θ twice, in seemingly different ways: in (11.9) and the proof of Proposition 11.12, and in Proposition 12.3.

Lemma 14.2. *The maps τ and θ both have order two and they commute. They both preserve the cone $C = \{(a, b, c) \mid a \geqslant 0, b \geqslant c \geqslant 0\}$.*

Proof. We have already seen in Section 12.2 that θ has order 2 and preserves the cone C. It is not hard to check that τ also has order two and that

$$\theta\tau(a, b, c) = \tau\theta(a, b, c) = (b - c, a + c, c).$$

Let us divide the cone C into two parts, C_1 being the part where $a + c \geqslant b$ and C_2 being the part where $a + c < b$. The two parts C_1 and C_2 are preserved separately by τ, and interchanged by θ. The statement follows. $\square$

As was explained after Proposition 12.4, we constructed $\mathcal{B}_\infty$ in the A_2 case as the subcrystal of

$$u(a, b, c) = u_1(-a) \otimes u_2(-b) \otimes u_1(-c) \in \mathcal{B}_1 \otimes \mathcal{B}_2 \otimes \mathcal{B}_1$$

such that $(a, b, c) \in C$. This requires redefining $e_i(x) = 0$ if $\varepsilon_i(x) = 0$.

Proposition 14.3. *With this identification of $\mathcal{B}_\infty$ as a subcrystal of $\mathcal{B}_1 \otimes \mathcal{B}_2 \otimes \mathcal{B}_1$, we have*

$$\mathrm{string}_{(1,2,1)}\big(u(a, b, c)\big) = \tau(a, b, c). \tag{14.3}$$

Theorem 14.1 is satisfied in this case.

Proof. We will show that for $(a, b, c) \in C$, $\mathrm{string}_{(1,2,1)}(x) = \tau(a, b, c)$. Indeed, clearly $\mathrm{wt}(x) = -(a+c)\alpha_1 - b\alpha_2$. On the other hand, let $\mathrm{string}_{(1,2,1)}\big(u(a, b, c)\big) = (a', b', c')$. Then $\mathrm{wt}(x)$ also equals $-(a'+c')\alpha_1 - b'\alpha_2$ since by Proposition 11.2 or Corollary 13.8 $u_\infty = e_1^{c'} e_2^{b'} e_1^{a'} u(a, b, c)$. Thus $a' + c' = a + c$ and $b' = b$. Since by definition $a' = \varepsilon_1(x)$, it follows from (2.13) that $a' = \max(a - b + 2c, c)$. This proves (14.3).

Now define $x \mapsto x^\star$, where if $x\grave{} = u(a, b, c)$ then $x^\star = u(a', b', c')$, with $(a', b', c') = \tau(a, b, c)$. By Lemma 14.2, this is a bijective map of order 2 such that in the notation of Theorem 14.1, we have $\iota_1(x^\star) = \iota_2(x)$.

We must check that the map $\star$ has the same property with respect to the other reduced word $(2, 1, 2)$ for w_0. By Proposition 12.3, x is mapped to $u\big(\theta(a, b, c)\big)$ under the isomorphism $\mathcal{B}_1 \otimes \mathcal{B}_2 \otimes \mathcal{B}_1 \cong \mathcal{B}_2 \otimes \mathcal{B}_1 \otimes \mathcal{B}_2$. Under the same map $\star$, this is mapped to $u\big(\tau\theta(a, b, c)\big)$, so that $\mathrm{string}_{(2,1,2)}(x) = \tau\theta(a, b, c)$. By Lemma 14.2, τ and θ commute, so that $\mathrm{string}_{(2,1,2)}(x) = \theta\tau(a, b, c) = \theta\,\mathrm{string}_{(1,2,1)}(x)$. This is indeed true by Proposition 11.10.

We leave it to the reader to check that e_i and f_i commute with $e_j^\star$ and $f_j^\star$ when $i = 1$ and $j = 2$ or vice versa. $\qquad\square$

14.2 The general case

Assume for a moment that Theorem 14.1 is known. Conjugating e_i and f_i by $\star$ gives new crystal operators, which we denote $e_i^\star$ and $f_i^\star$. We also obtain new $\varepsilon_i^\star$ and $\varphi_i^\star$. (There is no new weight function since the involution is weight preserving.) Thus $\mathcal{B}_\infty$ has a second, complementary crystal structure. One approach to constructing the involution is to first construct $e_i^\star$ and $f_i^\star$, and to deduce the existence of the involution.

Proposition 14.4. *There exists a unique morphism of crystals $\psi_i \colon \mathcal{B}_\infty \longrightarrow \mathcal{B}_i \otimes \mathcal{B}_\infty$ such that $\psi_i(u_\infty) = u_i(0) \otimes u_\infty$. Choose a reduced word $(i_1, \ldots, i_N)$ for w_0 such that $i_1 = i$ and identify $\mathcal{B}_\infty$ with a subcrystal of $\mathcal{B}_{i_1} \otimes \cdots \otimes \mathcal{B}_{i_N}$ as in Section 12.2. Then if $x = u_{i_1}(-a_1) \otimes \cdots \otimes u_{i_N}(-a_N) \in \mathcal{B}_\infty$*

$$\psi_i(x) = u_i(-a_1) \otimes u_{i_1}(0) \otimes u_{i_2}(-a_2) \otimes \cdots \otimes u_{i_N}(-a_N). \tag{14.4}$$

Proof. Since every element of $\mathcal{B}_\infty$ is obtained by applying the f_i to u_∞ in some order, the embedding ψ_i, if it exists, is unique. We note that $\mathcal{B}_i$ is embedded in $\mathcal{B}_i \otimes \mathcal{B}_i$ by

$$u_i(-a) \mapsto \begin{cases} u_i(-a) \otimes u_i(0) & \text{if } a \geqslant 0, \\ u_i(0) \otimes u_i(-a) & \text{otherwise.} \end{cases}$$

On the other hand, $\mathcal{B}_\infty$ is embedded as the set of elements

$$x = u_{i_1}(-a_1) \otimes \cdots \otimes u_{i_N}(-a_N) \in \mathcal{B}_{i_1} \otimes \cdots \otimes \mathcal{B}_{i_N}$$

such that $x \preccurlyeq u_\infty = u_{i_1}(0) \otimes \cdots \otimes u_{i_N}(0)$. Thus the embedding $\mathcal{B}_i \hookrightarrow \mathcal{B}_i \otimes \mathcal{B}_i$ gives rise to the embedding $\psi_i \colon \mathcal{B}_\infty \hookrightarrow \mathcal{B}_i \otimes \mathcal{B}_\infty$ in which (14.4) is satisfied. We used the fact that our chosen reduced word for w_0 has $i_1 = i$. But apart from that condition, the embedding does not depend on the choice of reduced word since as we have already noted, there can be at most one such embedding. $\qquad\square$

Lemma 14.5. *Suppose that $i_1 = i$ and that*

$$x = u_{i_1}(-a_1) \otimes u_{i_2}(-a_2) \otimes \cdots \otimes u_{i_N}(-a_N) \in \mathcal{B}_\infty.$$

Then

$$y = u_{i_1}(0) \otimes u_{i_2}(-a_2) \otimes \cdots \otimes u_{i_N}(-a_N) \in \mathcal{B}_\infty.$$

Moreover $\psi_i(x) = u_i(-a_1) \otimes y$.

Proof. By (14.4) we have

$$u_{i_1}(-a_1) \otimes u_{i_1}(0) \otimes u_{i_2}(-a_2) \otimes \cdots \otimes u_{i_N}(-a_N) = \psi_i(x) \in \mathcal{B}_i \otimes \mathcal{B}_\infty.$$

So discarding the first entry must give an element of $\mathcal{B}_\infty$.   □

Let us define the subset $\mathcal{B}^i$ of $\mathcal{B}_\infty$ to be the set of $x \in \mathcal{B}_\infty$ such that $\psi_i(x) = u_i(0) \otimes x$.

Proposition 14.6.

(i) *If $x \in \mathcal{B}^i$ and $e_j(x) \neq 0$ for any j, then $e_j(x) \in \mathcal{B}^i$.*
(ii) *If $x \in \mathcal{B}^i$, then $\varphi_i(x) \geq 0$. Moreover, $\varphi_i(x) > 0$ if and only if $f_i(x) \in \mathcal{B}^i$.*

Proof. Let us prove (i). Since $x \in \mathcal{B}^i$, we have $\psi_i(x) = u_i(0) \otimes x$. Then $\psi_i(e_j(x)) = e_j(u_i(0) \otimes x)$ since by Proposition 14.4 the map ψ_i is a crystal morphism. This cannot equal $e_j(u_i(0)) \otimes x$. Indeed, the latter is $u_i(1) \otimes x$ if $j = i$ and 0 otherwise; both are impossible since $\psi_i(e_j(x))$ must be $\preccurlyeq \psi_i(u_\infty) = u_i(0) \otimes u_\infty$. Thus $\psi_i(e_j(x)) = u_i(0) \otimes e_j(x)$. This means that $e_j(x) \in \mathcal{B}^i$.

We prove (ii). By Lemma 14.5, we have

$$x = u_{i_1}(0) \otimes z, \qquad z = u_{i_2}(-a_2) \otimes \cdots \otimes u_{i_N}(-a_N).$$

Hence by (2.14), $\varphi_i(x) = \max(0, \varphi_i(z))$. It is now clear that $\varphi_i(x) \geq 0$. Suppose that $\varphi_i(x) > 0$. Then $\varphi_i(z) > 0$ and by Proposition 2.33, $f_i(x) = u_i(0) \otimes f_i(z)$. Since $f_i(x) \preccurlyeq x \preccurlyeq u_\infty$, $f_i(x) \in \mathcal{B}_\infty$ and clearly $f_i(x) \in \mathcal{B}^i$. On the other hand, if $\varphi_i(x) = 0$, then $f_i(x) = u_i(-1) \otimes z$ is not in $\mathcal{B}^i$.   □

In view of Proposition 14.6, we make $\mathcal{B}^i$ into a crystal by redefining $f_i(x) = 0$ if $\varphi_i(x) = 0$ for $x \in \mathcal{B}^i$. Also, let $\mathcal{B}_i^+ = \{u_i(-a) \mid a \geq 0\}$. Here we redefine $e_i(u_i(0)) = 0$ to make $\mathcal{B}_i^+$ into a crystal.

We say that a crystal if i-*seminormal* if (2.6) is satisfied.

Proposition 14.7.

(i) *The crystal $\mathcal{B}^i$ is upper seminormal and i-seminormal.*
(ii) *The inclusion map $\mathcal{B}_i^+ \otimes \mathcal{B}^i \hookrightarrow \mathcal{B}_i \otimes \mathcal{B}_\infty$ is a strict morphism of crystals.*
(iii) *The map ψ_i induces an isomorphism between $\mathcal{B}_\infty$ and $\mathcal{B}_i^+ \otimes \mathcal{B}^i$.*

Proof. For (i), the fact that $\mathcal{B}^i$ is upper seminormal follows from the fact that $\mathcal{B}_\infty$ is upper seminormal, and the fact that $\mathcal{B}^i \sqcup \{0\}$ is stable under the e_i. Now with $f_i(x)$ redefined, it is i-seminormal by Proposition 14.6 (ii).

As for (ii), this verification is needed because we have two different definitions of $f_i(x)$ for $x \in \mathcal{B}^i$: If $\varphi_i(x) = 0$, then $f_i(x) = 0$ in $\mathcal{B}^i$ but not in $\mathcal{B}_\infty$. So we must check that this does not affect the computation of $f_i(u_i(-a) \otimes x)$. Indeed, $f_i(u_i(-a) \otimes x) = u_i(-(a+1)) \otimes x$ unless $\varphi_i(x) > \varepsilon_i(u_i(-a)) = a$. So either f_i applies to the first component or $\varphi_i(x) > 0$ so that $f_i(x)$ has the same value whether x is regarded as an element of $\mathcal{B}_\infty$ or $\mathcal{B}^i$. In either case, $f_i(u_i(-a) \otimes x)$ has the same value whether computed in $\mathcal{B}_i^+ \otimes \mathcal{B}^i$ or in $\mathcal{B}_i \otimes \mathcal{B}_\infty$. We see that the inclusion $\mathcal{B}_i^+ \otimes \mathcal{B}^i \longrightarrow \mathcal{B}_i \otimes \mathcal{B}_\infty$ is a strict morphism.

For (iii), it is clear that the image ψ_i is contained in $\mathcal{B}_i^+ \otimes \mathcal{B}^i$. To show that this map is surjective, we will show that $\mathcal{B}_i^+ \otimes \mathcal{B}^i$ has a unique highest weight element $u_i(0) \otimes u_\infty$. This is sufficient since if we know this, then any element of $\mathcal{B}_i^+ \otimes \mathcal{B}^i$ can be obtained by applying $f_{j_1} \cdots f_{j_r}$ to $u_i(0) \otimes u_\infty$ for some sequence $j_1, \ldots, j_r$, and then that element is equal to $\psi_i(f_{j_1} \cdots f_{j_r} u_\infty)$.

Thus let $u_i(-a) \otimes x$ be a highest weight element of $\mathcal{B}_i^+ \otimes \mathcal{B}^i$. This means that $e_j(u_i(-a) \otimes x) = 0$ for all j. We have $e_j(u_i(-a) \otimes x) = u_i(-a) \otimes e_j(x)$ for all j except that if $j = i$, in which case

$$
e_i(u_i(-a) \otimes x) = \begin{cases} u_i(-a) \otimes e_i(x) & \text{if } a \leqslant \varphi_i(x), \\ u_i(-(a-1)) \otimes x & \text{if } a > \varphi_i(x). \end{cases}
$$

Now $a > \varphi_i(x)$ is impossible since $\varphi_i(x) \geqslant 0$, so $e_i(u_i(-a) \otimes x) = 0$ would fail. It follows that $e_j(x) = 0$ for all j, and so $x = u_\infty$. Now $e_i(u_i(-a) \otimes x) = 0$ implies $a = 0$, and so we have proved that $u_i(0) \otimes u_\infty$ is the unique highest weight element of $\mathcal{B}_i^+ \otimes \mathcal{B}^i$. $\qquad \square$

We emphasize that although we used a particular realization of $\mathcal{B}_\infty$ as a subcrystal of $\mathcal{B}_{i_1} \otimes \cdots \otimes \mathcal{B}_{i_N}$ for a fixed reduced word $(i_1, \ldots, i_N)$ of $w_0 = s_{i_1} \cdots s_{i_N}$ in the proof of Lemma 14.5 (subject to the constraint $i_1 = i$), the crystals $\mathcal{B}^i$, $\mathcal{B}_i^+$ and the isomorphism ψ_i are all independent of this choice.

We may now describe the modified crystal operators.

Proposition 14.8. *There exists an alternative crystal structure on $\mathcal{B}_\infty$ with crystal operators $e_i^\star$ and $f_i^\star$ and functions $\varepsilon_i^\star$ and $\varphi_i^\star$, defined as follows. Let $x \in \mathcal{B}_\infty$ and let $\psi_i(x) = u_i(-a) \otimes y$, where $y \in \mathcal{B}^i$. We define $\varepsilon_i^\star(x) = a$ and $\varphi_i^\star(x) = a + \langle \mathrm{wt}(x), \alpha_i^\vee \rangle$. We define $e_i^\star(x)$ and $f_i^\star(x)$ by requiring that*

$$
\psi_i(e_i^\star(x)) = \begin{cases} u_i(-(a-1)) \otimes y & \text{if } a > 0, \\ 0 & \text{if } a = 0, \end{cases} \tag{14.5}
$$

$$
\psi_i(f_i^\star(x)) = u_i(-(a+1)) \otimes y.
$$

Proof. It follows from Proposition 14.7 that (14.5) defines maps $\mathcal{B}_\infty \longrightarrow \mathcal{B}_\infty \sqcup \{0\}$, and the crystal definition is easily checked. $\qquad \square$

When we want to emphasize that we consider $\mathcal{B}_\infty$ but with the crystal operators $e_i^\star$ and $f_i^\star$ instead of e_i and f_i, we write $\mathcal{B}_\infty^\star$.

Proposition 14.9. *If $i \neq j$, then the operators $e_i^\star$ and $f_i^\star$ commute with e_j and f_j. For $x \in \mathcal{B}_\infty$, we have $\varepsilon_j\big(f_i^\star(x)\big) = \varepsilon_j(x)$.*

Proof. This follows from the fact that ψ_i induces an isomorphism $\mathcal{B}_\infty \longrightarrow \mathcal{B}_i^+ \otimes \mathcal{B}^i$. Since $\varepsilon_j\big(u_i(-a)\big) = -\infty$, we have

$$e_j\big(u_i(-a) \otimes x\big) = u_i(-a) \otimes e_j(x) \qquad \text{for} \quad u_i(-a) \otimes x \in \mathcal{B}_i^+ \otimes \mathcal{B}^i,$$

and similarly for f_j. On the other hand, $e_i^\star$ and $f_i^\star$ simply decrement or increment a without affecting x. It is thus clear that they commute with e_j and f_j. The fact that $\varepsilon_j\big(f_i^\star(x)\big) = \varepsilon_j(x)$ then follows from the fact that $\mathcal{B}_\infty$ is upper seminormal. $\qquad\square$

Proposition 14.10. *If $x \in \mathcal{B}_\infty^\star$ is a highest weight element, then $x = u_\infty$.*

Proof. Let Y be the set of highest weight elements of $\mathcal{B}_\infty^\star$. Thus

$$Y = \{x \in \mathcal{B}_\infty \mid e_i^\star(x) = 0 \text{ for all } i\}.$$

Note that $e_i^\star(x) = 0$ means that $x \in \mathcal{B}^i$, so $Y = \bigcap_i \mathcal{B}^i$. By Proposition 14.6 (i), if $x \in \mathcal{B}^i$ and $x \preccurlyeq y$ for $y \in \mathcal{B}_\infty$, then $y \in \mathcal{B}^i$. Therefore if $x \in Y$ and $x \preccurlyeq y$, then $y \in Y$.

Suppose that $x \in Y$ and $x \neq u_\infty$. It follows from the last observation that, if x is maximal, then $x = f_i u_\infty$ for some i. Therefore $\psi_i(x) = f_i\big(u_i(0) \otimes u_\infty\big) = u_i(-1) \otimes u_\infty$. Thus $\varepsilon_i^\star(x) = 1$, which is a contradiction. $\qquad\square$

Let $\mathcal{B}^{\star i} = \{x \in \mathcal{B}_\infty \mid e_i(x) = 0\}$.

Proposition 14.11.

(i) *If $x \in \mathcal{B}^{\star i}$ and $e_j^\star(x) \neq 0$ for any j, then $e_j^\star(x) \in \mathcal{B}^{\star i}$.*
(ii) *If $x \in \mathcal{B}^{\star i}$, then $\varphi_i^\star(x) \geq 0$. Moreover $\varphi_i^\star(x) > 0$ if and only if $f_i^\star(x) \in \mathcal{B}^{\star i}$.*

Proof. We prove (i). If $j \neq i$, then this is clear since e_i and $e_j^\star$ commute, so we may assume that $i = j$. Write $\psi_i(x) = u_i(-a) \otimes y$ where $y \in \mathcal{B}^i$. Then $a = \varepsilon_i^\star(x) > 0$ by assumption, and $\psi_i\big(e_i^\star(x)\big) = u_i\big(-(a - 1)\big) \otimes y$. We have

$$0 = \varepsilon_i(x) = \varepsilon_i(u_i(-a) \otimes y) = \max\big\{\varepsilon_i(y), a - \langle \mathrm{wt}(y), \alpha_i^\vee \rangle\big\}.$$

From this we learn that $\varepsilon_i(y) = 0$ and $\langle \mathrm{wt}(y), \alpha_i^\vee \rangle \geq a$. Consequently,

$$0 = \max\big\{\varepsilon_i(y), (a - 1) - \langle \mathrm{wt}(y), \alpha_i^\vee \rangle\big\} = \varepsilon_i\big(\psi_i(e_i^\star(x))\big) = \varepsilon_i\big(e_i^\star(x)\big).$$

Thus $e_i(e_i^\star(x)) = 0$, as required.

Next we prove (ii). Since $x \in \mathcal{B}^{\star i}$, we have $e_i(x) = 0$ and we will prove that $\varphi_i^\star(x) \geq 0$. Let $\psi_i(x) = u_i(-a) \otimes y \in \mathcal{B}_i \otimes \mathcal{B}_\infty$. Then

$$\varphi_i(x) = \langle \mathrm{wt}(x), \alpha_i^\vee \rangle + \varepsilon_i(x) = \langle \mathrm{wt}(x), \alpha_i^\vee \rangle, \qquad \varphi_i^\star(x) = \langle \mathrm{wt}(x), \alpha_i^\vee \rangle + \varepsilon_i^\star(x)$$

and since $\varepsilon_i^\star(x) = a$ we obtain

$$\varphi_i^\star(x) = \varphi_i(x) + a. \qquad (14.6)$$

On the other hand,

$$0 = \varepsilon_i(x) = \varepsilon_i(\psi_i(x)) = \max\{\varepsilon_i(y), a - \langle \mathrm{wt}(y), \alpha_i^\vee \rangle\}$$

from which we learn that $\varepsilon_i(y) = 0$ and

$$a \leqslant \langle \mathrm{wt}(y), \alpha_i^\vee \rangle = \langle \mathrm{wt}(x) + a\alpha_i, \alpha_i^\vee \rangle = \langle \mathrm{wt}(x), \alpha_i^\vee \rangle + 2a = \varphi_i(x) + 2a \ .$$

Hence using (14.6) we obtain $\varphi_i^\star(x) \geqslant 0$. Furthermore if $\varphi_i^\star(x) > 0$ then the preceding calculations show that $\langle \mathrm{wt}(y), \alpha_i^\vee \rangle > a$. Therefore

$$\varepsilon_i\left(f_i^\star(x)\right) = \varepsilon_i\left(\psi_i(f_i^\star x)\right) = \varepsilon_i\left(u_i(-(a+1)) \otimes y\right)$$
$$= \max\{\varepsilon_i(y), a + 1 - \langle \mathrm{wt}(y), \alpha_i^\vee \rangle\} = 0.$$

This shows that if $\varphi_i^\star(x) > 0$, then $f_i^\star(x) \in \mathcal{B}^{\star i}$. The converse is also true, since if $\varphi_i^\star(x) = 0$ then $\varphi_i^\star(f_i^\star(x)) = -1$ and by what we have already shown, $f_i^\star(x)$ cannot be in $\mathcal{B}^{\star i}$. $\qquad \square$

Now we may make $\mathcal{B}^{\star i}$ into a crystal with crystal operators $e_i^\star$ and $f_i^\star$, analogous to $\mathcal{B}^i$. (Note that Proposition 14.11 is analogous to Proposition 14.6.) As with $\mathcal{B}^i$ we redefine $f_i^\star(x) = 0$ if $\varphi_i^\star(x) = 0$, and $\mathcal{B}^{\star i}$ becomes a subcrystal of $\mathcal{B}_\infty^\star$. By Proposition 14.11, $\mathcal{B}_\infty^\star$ is upper seminormal and i-seminormal.

Let $\mathcal{B}_i^\star$ be the same as the crystal $\mathcal{B}_i$ with a notational difference: we will denote the crystal operations as $\varepsilon_i^\star$, $\varphi_i^\star$, $e_i^\star$ and $f_i^\star$, and its elements $u_i^\star(-a)$ $(a \in \mathbb{Z})$. The subcrystal $\mathcal{B}_i^{\star+}$ consists of the elements $u_i^\star(-a)$ with $a \geqslant 0$.

We define a map $\psi_i^\star : \mathcal{B}_\infty^\star \to \mathcal{B}_i^{\star+} \otimes \mathcal{B}^{\star i}$ as follows. Let $x \in \mathcal{B}_\infty^\star$. Let $a = \varepsilon_i(x)$ and $y = e_i^a(x)$. Since the crystal $\mathcal{B}_\infty$ is upper seminormal, $e_i(y) = 0$, that is, $y \in \mathcal{B}^{\star i}$. We define

$$\psi_i^\star(x) = u_i^\star(-a) \otimes y.$$

Proposition 14.12. *Let $x \in \mathcal{B}_\infty$ be such that $\psi_i(x) = u_i(-t) \otimes y$, $\psi_i^\star(x) = u_i^\star(-v) \otimes z$. Then*

$$\varepsilon_i^\star(x) - \varphi_i(y) = \varepsilon_i(x) - \varphi_i^\star(z). \qquad (14.7)$$

To prove this we may use that ψ_i is a morphism, but not $\psi_i^\star$, since we have not established that yet.

Proof. First assume that the root system is simply-laced. Observe that $t = \varepsilon_i^\star(x)$ and $y = e_i^{\star t}(x)$ while $v = \varepsilon_i(x)$ and $z = e_i^v(x)$.

We assume by induction that x is a counterexample that is maximal with respect to $\preccurlyeq$. Since both sides of (14.7) would vanish if $x = u_\infty$, we may write $x' = \varepsilon_j(x)$ for some j and so by induction

$$\varepsilon_i^\star(x') - \varphi_i(y') = \varepsilon_i(x') - \varphi_i^\star(z') \ , \qquad (14.8)$$

where $\psi_i(x') = u_i(-t') \otimes y'$ and $\psi_i^\star(x') = u_i(-v')^\star \otimes z'$.

Assume first that $j = i$. Then we may argue as follows. First note that $\varepsilon_i(x') = \varepsilon_i(x) - 1 = v - 1$ and $z' = \varepsilon_i^{v-1}(x') = z$. Thus

$$\varepsilon_i(x) - \varphi_i^\star(z) = \varepsilon_i(x') - \varphi_i^\star(z') + 1. \tag{14.9}$$

We will prove that

$$\varepsilon_i^\star(x) - \varphi_i(y) = \varepsilon_i^\star(x') - \varphi_i(y') + 1. \tag{14.10}$$

This is sufficient since then combining (14.8), (14.9) and (14.10) we obtain (14.7).

Since ψ_i is a homomorphism we have, remembering that $\varepsilon_i(u_i(-t)) = t = \varepsilon_i^\star(x)$

$$\psi_i(x') = e_i\psi_i(x) = \begin{cases} u_i(-t) \otimes e_i(y) & \text{if } \varphi_i(y) \geq \varepsilon_i^\star(x), \\ u_i(-(t-1)) \otimes y & \text{otherwise.} \end{cases} \tag{14.11}$$

First let us consider the case where $\varphi_i(y) \geq \varepsilon_i^\star(x)$. Then $y' = e_i(y)$ and $t' = t$, that is, $\varepsilon_i^\star(x') = \varepsilon_i^\star(x)$. And since $y' = e_i(y)$ we have $\varphi_i(y) = \varphi_i(y') - 1$. Thus we obtain (14.10) in this case. On the other hand, if $\varphi_i(y) < \varepsilon_i^\star(x)$, then (14.11) implies that $t' = t - 1$ and $y' = y$. Thus $\varepsilon_i^\star(x) = t = t' + 1 = \varepsilon_i^\star(x') + 1$, and again we have (14.10).

This settles the case where $j = i$. If $j \neq i$, we may as well assume that $e_i(x) = 0$, because if $e_i(x) \neq 0$, then we could take $j = i$ and we have already settled that case. Also, because $\mathcal{B}_\infty$ is a weak Stembridge crystal, we have either $\varepsilon_i(x') = \varepsilon_i(x)$ or $\varepsilon_i(x') = \varepsilon_i(x) + 1$. In either case, because $\varepsilon_j(u_i(-t)) = -\infty$, we have

$$\psi_i(x') = e_j\psi_i(x) = u_i(-t) \otimes e_j y,$$

and therefore $y' = e_j y$ and $t' = t$, that is, $\varepsilon_i^\star(x) = \varepsilon_i^\star(x')$.

We consider the first case $\varepsilon_i(x') = \varepsilon_i(x) = 0$ now. We have

$$0 = \varepsilon_i(x) = \varepsilon_i\psi_i(x) = \max\{\varepsilon_i(y), t + \varepsilon_i(y) - \varphi_i(y)\},$$

so $\varepsilon_i(y) = 0$ and $\varepsilon_i^\star(x) + \varepsilon_i(y) - \varphi_i(y) = 0$, that is, $\varphi_i(y) = \varepsilon_i^\star(x)$. Thus $\varepsilon_i^\star(x) - \varphi_i(y) = 0$ and similarly $\varepsilon_i^\star(x') - \varphi_i(y') = 0$. And since $\varepsilon_i(x') = \varepsilon_i(x) = 0$ and $z' = z$, we also have

$$\varepsilon_i(x') - \varphi_i^\star(z') = \varepsilon_i(x) - \varphi_i^\star(z).$$

Therefore $\varepsilon_i^\star(x') - \varphi_i(y') = \varepsilon_i(x') - \varphi_i(z')$ implies $\varepsilon_i^\star(x) - \varphi_i(y) = \varepsilon_i(x) - \varphi_i(z)$ and we are done in this case.

We are left with the case where $\varepsilon_i(x) = 0$ and $\varepsilon_i(x') = 1$. Because $\mathcal{B}_\infty$ is a Stembridge crystal, this can only happen if $\langle \alpha_j, \alpha_i^\vee \rangle = -1$. We have $z = x$ while $z' = e_i(x')$. Since $\varepsilon_i\psi_i(x) = \varepsilon_i(x) = 0$ we have $\max\{\varepsilon_i(y), t + \varepsilon_i(y) - \varphi_i(y)\} = 0$. Thus

$$\varepsilon_i(y) = 0, \qquad \varepsilon_i^\star(x) \leq \varphi_i(y).$$

On the other hand since $\varepsilon_i\psi_i(x') = \varepsilon_i(x') = 1$ we have $\max\{\varepsilon_i(y'), \varepsilon_i^\star(x') + \varepsilon_i(y') - \varphi_i(y')\} = 1$, where we remind the reader that $\varepsilon_i^\star(x') = t = \varepsilon_i^\star(x)$. We must now consider two cases. Either $\varepsilon_i(y') = 1$ and $\varepsilon_i^\star(x) \leqslant \varphi_i(y')$, or $\varepsilon_i(y') = 0$ and $\varepsilon_i^\star(x) - \varphi_i(y') = 1$.

Assume that $\varepsilon_i(y') = 0$ and $\varepsilon_i^\star(x) - \varphi_i(y') = 1$. Since $\mathcal{B}_\infty$ is a Stembridge crystal we have either $\varphi_i(y') = \varphi_i(y)$ or $\varphi_i(y') = \varphi_i(y) - 1$ but because $\varepsilon_i^\star(x) \leqslant \varphi_i(y)$ and $\varepsilon_i^\star(x) - \varphi_i(y') = 1$, we must have $\varphi_i(y) = \varepsilon_i^\star(x)$. In particular, we have $\varphi_i(y') = \varepsilon_i^\star(x) - 1 = \varphi_i(y) - 1$. Next, $t - \varphi_i(y') = 1$ implies that

$$\psi_i(z') = e_i\psi_i(x') = e_i(u_i(-t) \otimes y') = u_i(-(t-1)) \otimes y'$$

and therefore $\varepsilon_i^\star(z') = \varepsilon_i^\star(x) - 1 = \varepsilon_i^\star(z) - 1$. Since $\mathrm{wt}(z') = \mathrm{wt}(z) + \alpha_i + \alpha_j$, this implies that $\varphi_i^\star(z') = \varphi_i^\star(z) - 1 + \langle \alpha_i + \alpha_j, \alpha_i^\vee \rangle = \varphi_i^\star(z)$. We have seen that

$$\varepsilon_i(x') = \varepsilon_i(x) + 1, \qquad \varphi_i(y') = \varphi_i(y) - 1, \qquad \varepsilon_i^\star(x') = \varepsilon_i^\star(x), \qquad \varphi_i^\star(z') = \varphi_i^\star(z).$$

Therefore (14.8) implies (14.7) and we are done in this case.

Finally we must consider the case when $\varepsilon_i(y') = 1$. Of course, we are still assuming that $\varepsilon_i(x) = 0$ and $\varepsilon_i(x') = 1$ with $x' = e_j(x)$ and $j \neq i$. As before, $z = x$, $z' = e_i(x')$, and applying e_j to $\psi_i(x) = u_i(-t) \otimes y$ gives $\psi_i(x') = u_i(-t) \otimes y'$ with $y' = e_j(x)$. So $\varepsilon_i^\star(x') = t = \varepsilon_i^\star(x)$ while $\varphi_i(y') = \varphi_i(y)$ by Proposition 4.5, which implies $\varepsilon_i^\star(x) \leqslant \varphi_i(y')$. Therefore when we apply e_i to $\psi_i(x') = u_i(-t) \otimes y'$ we obtain $\psi_i(z') = u_i(-t) \otimes e_i(y')$. This means that $\varepsilon_i^\star(z') = t = \varepsilon_i^\star(x) = \varepsilon_i^\star(z)$. Now

$$\varphi_i^\star(z') - \varepsilon_i^\star(z') = \varphi_i^\star(z) - \varepsilon_i^\star(z) + \langle \mathrm{wt}(z') - \mathrm{wt}(z), \alpha_i^\vee \rangle,$$

where $\langle \mathrm{wt}(z') - \mathrm{wt}(z), \alpha_i^\vee \rangle = \langle \alpha_i + \alpha_j, \alpha_i^\vee \rangle = 1$. It follows that $\varphi_i^\star(z') = \varphi_i^\star(z) + 1$. Hence

$$\varepsilon_i(x') = \varepsilon_i(x) + 1, \qquad \varphi_i(y') = \varphi_i(y), \qquad \varepsilon_i^\star(x') = \varepsilon_i^\star(x), \qquad \varphi_i^\star(z') = \varphi_i^\star(z) + 1.$$

Therefore (14.8) implies (14.7) and we are done in the simply-laced case.

Next we consider the non-simply-laced case. As shown in Section 12.3, $\mathcal{B}_\infty$ for non-simply-laced type X can be realized as a virtual crystal $\mathcal{V}_\infty$ inside a simply-laced crystal $\widehat{\mathcal{V}}_\infty$ of type Y. In particular, by (5.3a) and (5.3b) this implies that $\varepsilon_i(x) = \frac{1}{\gamma_i}\widehat{\varepsilon}_j(x)$ and $\varphi_i(x) = \frac{1}{\gamma_i}\widehat{\varphi}_j(x)$ for all $j \in \sigma(i)$. We are going to show that $\mathcal{B}_\infty^\star$ can also be realized as a virtual crystal. This would imply that $\varepsilon_i^\star(x) = \frac{1}{\gamma_i}\widehat{\varepsilon}_j^\star(x)$ and $\varphi_i^\star(x) = \frac{1}{\gamma_i}\widehat{\varphi}_j^\star(x)$ for all $j \in \sigma(i)$. Making use of the simply-laced case which is already established,

$$\widehat{\varepsilon}_j^\star(x) - \widehat{\varphi}_j(y) = \widehat{\varepsilon}_j(x) - \widehat{\varphi}_j^\star(z),$$

which implies

$$\varepsilon_i^\star(x) - \varphi_i(y) = \varepsilon_i(x) - \varphi_i^\star(z)$$

as desired.

It remains to show that $\mathcal{B}_\infty^\star$ can be realized as a virtual crystal. Recall the definition of the starred crystal operators (14.5). By (14.4) and the virtual crystal embedding of Section 12.3, in $\widehat{\mathcal{V}}_\infty$ we have

$$\widehat{\psi}_i \colon \widehat{\mathcal{V}}_\infty \longrightarrow \bigotimes_{j \in \sigma(i)} \mathcal{B}_j^Y \otimes \widehat{\mathcal{V}}_\infty$$

$$x \longmapsto \bigotimes_{j \in \sigma(i)} u_j(-a) \otimes y.$$

We define the starred virtual crystal operators as

$$\widehat{\psi}_i\big(e_i^\star(x)\big) = \begin{cases} \displaystyle\bigotimes_{j\in\sigma(i)} u_j\big(-(a-\gamma_i)\big)\otimes y & \text{if } a > 0, \\[2mm] 0 & \text{if } a = 0, \end{cases} \tag{14.12}$$

$$\widehat{\psi}_i\big(f_i^\star(x)\big) = \bigotimes_{j\in\sigma(i)} u_j\big(-(a+\gamma_i)\big)\otimes y.$$

The ambient crystal of type Y is a weak Stembridge crystal (confirming Axiom V1). Furthermore, $\widehat{\varepsilon}_j^\star(x) = \widehat{\varepsilon}_i^\star(x) = a \in \gamma_i\mathbb{Z}$ and $\widehat{\varphi}_j^\star(x) = \widehat{\varphi}_i^\star(x) = a + \langle\widehat{\mathrm{wt}}(x),\widehat{\alpha}_i^\vee\rangle \in \gamma_i\mathbb{Z}$ for all $j \in \sigma(i)$ (confirming Axiom V2). Comparing with the characterization of $\mathcal{V}_\infty$ provided by Proposition 12.11, it is also clear from (14.12) that $\mathcal{V}_\infty = \mathcal{V}_\infty^\star$ is closed under $e_i^\star$ and $f_i^\star$ (confirming Axiom V3). This proves that $\mathcal{B}_\infty^\star$ can be realized as a virtual crystal. $\qquad\square$

Proposition 14.13. *The map* $\psi_i^\star: \mathcal{B}_\infty^\star \longrightarrow \mathcal{B}_i^{\star+} \otimes \mathcal{B}^{\star i}$ *is a morphism for the* $\star$ *crystal structure.*

Proof. It is sufficient to show that

$$\varepsilon_i^\star(x) = \varepsilon_i^\star\psi_i^\star(x) \qquad\text{and}\qquad e_i^\star\psi_i^\star(x) = \psi_i^\star e_i^\star(x), \tag{14.13}$$

since the corresponding statements for the $\varphi_i^\star$ and $f_i^\star$ follow by the definition of a crystal. The fact that $\varepsilon_j^\star(x) = \varepsilon_j^\star\psi_i^\star(x)$ and $e_j^\star\psi_i^\star(x) = \psi_i^\star e_j^\star(x)$ when $j \neq i$ is an easy consequence of (14.9).

We use the notation of Proposition 14.12 and denote $\psi_i(x) = u_i(-t)\otimes y$, $\psi_i^\star(x) = u_i(-v)\otimes z$, where $t = \varepsilon_i^\star(x)$ and $v = \varepsilon_i(x)$, $y = e_i^{\star\,t}(x)$ and $z = e_i^v(x)$.

There are two cases. First assume that $t = \varepsilon_i^\star(x) \leqslant \varphi_i(y)$. By Proposition 14.12, we have $v = \varepsilon_i(x) \leqslant \varphi_i^\star(z)$. We have $v = \varepsilon_i\psi_i(x) = \max\{\varepsilon_i(y), t+\varepsilon_i(y)-\varphi_i(y)\} = \varepsilon_i(y)$. Because $t \leqslant \varphi_i(y)$, each time we apply e_i to $\psi_i(x) = u_i(-t)\otimes y$ it applies to the second component, and hence

$$\psi_i(z) = \psi_i(e_i^v x) = e_i^v\psi_i(x) = u_i(-t)\otimes e_i^v(y). \tag{14.14}$$

This means that $\varepsilon_i^\star(z) = t = \varepsilon_i^\star(x)$. On the other hand, since $\varepsilon_i(x) \leqslant \varphi_i^\star(z)$

$$\varepsilon_i^\star\psi_i^\star(x) = \max\{\varepsilon_i^\star(z),\varepsilon_i(x)+\varepsilon_i^\star(z)-\varphi_i^\star(z)\} = \varepsilon_i^\star(z). \tag{14.15}$$

Combining these facts, $\varepsilon_i^\star\big(\psi_i^\star(x)\big) = \varepsilon_i^\star(x)$. This proves the first identity in (14.13) in this case. To prove the second identity, we will assume that $t > 0$; if $t = 0$ we leave it to the reader to check that both sides of the second identity in (14.13) are 0. We note that

$$\psi_i\big(e_i^\star(x)\big) = u_i\big(-(t-1)\big)\otimes y$$

so

$$\varepsilon_i\big(e_i^\star(x)\big) = \varepsilon_i\psi_i\big(e_i^\star(x)\big) = \max\{\varepsilon_i(y), t-1+\varepsilon_i(y)-\varphi_i(y)\} = \varepsilon_i(y) = v.$$

Thus $\psi_i^\star(e_i^\star(x)) = u_i^\star(-v) \otimes e_i^v(e_i^\star(x))$. We will prove that $e_i^v(e_i^\star(x)) = e_i^\star(z)$. Indeed, note that

$$\psi_i(e_i^v(e_i^\star(x))) = e_i^v(\psi_i(e_i^\star(x))) = e_i^v(u_i(-(t-1)) \otimes y).$$

Since $t - 1 = \varepsilon_i^\star(x) - 1 < \varphi_i(y)$, each time we apply e_i^v it goes onto the y and so

$$\psi_i(e_i^v(e_i^\star(x))) = u_i(-(t-1)) \otimes e_i^v y = \psi_i(e_i^\star(z)),$$

where we have used (14.14) and the definition (14.5) of $e_i^\star$. This proves $e_i^v(e_i^\star(x)) = e_i^\star(z)$. Now by the definition of $\psi_i^\star$ and the fact that $\varepsilon_i(e_i^\star(x)) = v$ we have

$$\psi_i^\star(e_i^\star(x)) = u_i^\star(-v) \otimes e_i^v(e_i^\star(x)) = u_i^\star(-v) \otimes e_i^\star(z).$$

On the other hand

$$e_i^\star(\psi_i^\star(x)) = e_i^\star(u_i^\star(-v) \otimes z) = u_i^\star(-v) \otimes e_i^\star(z)$$

also which proves the second identity in (14.13) in this case.

In the other case, $t = \varepsilon_i^\star(x) > \varphi_i(y)$ and $v = \varepsilon_i(x) > \varphi_i^\star(z)$. We note that

$$v = \varepsilon_i(x) = \varepsilon_i \psi_i(x) = \max\{\varepsilon_i(y), t + \varepsilon_i(y) - \varphi_i(y)\} = t + \varepsilon_i(y) - \varphi_i(y).$$

Now applying e_i repeatedly to $\psi_i(x) = u_i(-t) \otimes y$, the first $t - \varphi_i(y)$ applications go on the $u_i(-t)$. Thus

$$e_i^{t-\varphi_i(y)} \psi_i(x) = u_i(-\varphi_i(y)) \otimes y.$$

After this, each application of e_i goes to the second component and we have

$$\psi_i(z) = e_i^v \psi_i(x) = u_i(-\varphi_i(y)) \otimes e_i^{\varepsilon_i(y)} y. \tag{14.16}$$

Therefore $\varepsilon_i^\star(z) = \varphi_i(y)$. By the first equality in (14.15) and Proposition 14.12 we have

$$\varepsilon_i^\star \psi_i^\star(x) = \varepsilon_i(x) + \varepsilon_i^\star(z) - \varphi_i^\star(z) = \varepsilon_i^\star(x) + \varepsilon_i^\star(z) - \varphi_i(y) = \varepsilon_i^\star(x)$$

which proves first identity in (14.13) in this case. We note that since $y \in \mathcal{B}^i$ we have $\varphi_i(y) \geq 0$ and hence $t = \varepsilon_i^\star(x) > \varphi_i(y) \geq 0$. We have

$$\varepsilon_i(e_i^\star(x)) = \varepsilon_i \psi_i(e_i^\star(x)) = \varepsilon_i(u_i(-(t-1)) \otimes y) = t - 1 + \varepsilon_i(y) - \varphi_i(y),$$

that is, $\varepsilon_i(e_i^\star(x)) = v - 1$. Also

$$\psi_i(e_i^{v-1}(e_i^\star(x))) = e_i^{v-1}(u_i(-(t-1)) \otimes y)$$

which we may compute the same way we computed $\psi_i(z) = \psi_i(e_i^v(x))$ above. We obtain

$$\psi_i(e_i^{v-1}(e_i^\star(x))) = u_i(-\varphi_i(y)) \otimes e_i^{\varepsilon_i(y)}(y) = \psi_i(z)$$

and therefore $e_i^{v-1}(e_i^\star(x)) = z$.

Now by the definition of $\psi_i^\star$ and (14.16) we have

$$\psi_i^\star(e_i^\star(x)) = u_i^\star(-(v-1)) \otimes z = e_i^\star(\psi_i^\star(x))$$

which proves the second identity in (14.13) in this case. $\square$

Theorem 14.14. *The crystal* $\mathcal{B}_\infty^\star$ *is isomorphic to* $\mathcal{B}_\infty$. *That is, there exists a weight-preserving bijection* $\vartheta\colon \mathcal{B}_\infty \longrightarrow \mathcal{B}_\infty$ *such that for every* i *we have*

$$\varepsilon_i^\star \circ \vartheta = \varepsilon_i, \qquad \varphi_i^\star \circ \vartheta = \varphi_i, \qquad e_i^\star \circ \vartheta = \vartheta \circ e_i, \qquad f_i^\star \circ \vartheta = \vartheta \circ f_i. \qquad (14.17)$$

Proof. Let $w_0 = s_{i_1} \cdots s_{i_N}$ be a reduced decomposition of the long Weyl group element w_0. Since we have embeddings $\mathcal{B}_\infty \hookrightarrow \mathcal{B}_i \otimes \mathcal{B}_\infty$ we have an embedding

$$\mathcal{B}_\infty \hookrightarrow \mathcal{B}_{i_1} \otimes \cdots \otimes \mathcal{B}_{i_N} \otimes \mathcal{B}_\infty.$$

In Chapter 12 we constructed $\mathcal{B}_\infty$ as the subcrystal of $\mathcal{B}_{i_1} \otimes \cdots \otimes \mathcal{B}_{i_N}$ generated by the highest weight element $u_{i_1}(0) \otimes \cdots \otimes u_{i_N}(0)$. Thus this embedding may be understood as an embedding $\mathcal{B}_\infty \hookrightarrow \mathcal{B}_\infty \otimes \mathcal{B}_\infty$. The latter map is just $x \mapsto x \otimes u_\infty$. Indeed, it is sufficient to check that when we apply f_i to $x \otimes u_\infty$ the f_i goes to the x, in other words, that $\varphi_i(u_\infty) \leqslant \varepsilon_i(x)$. This follows from the upper seminormality of $\mathcal{B}_\infty$. Thus we have identified $\mathcal{B}_\infty$ as the subcrystal of $\mathcal{B}_{i_1} \otimes \cdots \otimes \mathcal{B}_{i_N} \otimes \mathcal{B}_\infty$ generated by $u_{i_1}(0) \otimes \cdots \otimes u_{i_N}(0) \otimes u_\infty$ and we see that it is contained in $\mathcal{B}_{i_1} \otimes \cdots \otimes \mathcal{B}_{i_N} \otimes u_\infty$.

On the other hand, we also have embeddings $\mathcal{B}_\infty^\star \hookrightarrow \mathcal{B}_i^\star \otimes \mathcal{B}_\infty^\star$ and since $\mathcal{B}_i^\star$ is upper seminormal we similarly obtain the embedding of $\mathcal{B}_\infty^\star$ into $\mathcal{B}_{i_1}^\star \otimes \cdots \otimes \mathcal{B}_{i_N}^\star \otimes \mathcal{B}_\infty^\star$ as the subcrystal generated by the highest weight element $u_{i_1}^\star(0) \otimes \cdots \otimes u_{i_N}^\star(0) \otimes u_\infty^\star$. As in the previous case, $\mathcal{B}_\infty^\star$ is contained in $\mathcal{B}_{i_1}^\star \otimes \cdots \otimes \mathcal{B}_{i_N}^\star \otimes u_\infty^\star$. This crystal is now clearly isomorphic to the corresponding subcrystal in $\mathcal{B}_{i_1} \otimes \cdots \otimes \mathcal{B}_{i_N} \otimes u_\infty$, since we can map

$$\vartheta\big(u_{i_1}(-a_1) \otimes \cdots \otimes u_{i_N}(-a_N) \otimes u_\infty\big) = u_{i_1}^\star(-a_1) \otimes \cdots \otimes u_{i_N}^\star(-a_N) \otimes u_\infty^\star.$$

$\qquad\qquad\qquad\qquad\qquad\qquad\qquad\qquad\qquad\qquad\qquad\qquad\qquad\qquad\qquad\qquad\qquad\square$

Proposition 14.15. *The map* ϑ *has order two.*

Proof. In this proof we will not distinguish between $\mathcal{B}_i$ and $\mathcal{B}_i^\star$. Thus we will identify $u_i^\star(-t)$ with $u_i(-t)$. The identity (14.5) may be written

$$\psi_i e_i^\star = (e_i \otimes 1)\psi_i, \qquad \psi_i f_i^\star = (f_i \otimes 1)\psi_i. \qquad (14.18)$$

We also have

$$\psi_i^\star e_i = (e_i \otimes 1)\psi_i^\star, \qquad \psi_i^\star f_i = (f_i \otimes 1)\psi_i^\star. \qquad (14.19)$$

Indeed, this follows easily from the definition of $\psi_i^\star$ in which we defined $\psi_i^\star(x) = u_i(-a) \otimes y$ where $a = \varepsilon_i(x)$ and $y = e_i^a x$, so that

$$\psi_i^\star(f_i x) = u_i\big(-(a+1)\big) \otimes e_i^{a+1} f_i x = u_i\big(-(a+1)\big) \otimes y = (f_i \otimes 1)\psi_i^\star(x).$$

Next we argue that

$$(1 \otimes \vartheta)\psi_i = \psi_i^\star \vartheta. \qquad (14.20)$$

Indeed, the map ϑ is a morphism from $\mathcal{B}_\infty$ (with the e_i and f_i crystal operations) to $\mathcal{B}_\infty^\star$ (with the starred operations) so $(1 \otimes \vartheta)\psi_i \vartheta^{-1}$ is a morphism from $\mathcal{B}_\infty^\star$ into $\mathcal{B}_i \otimes \mathcal{B}_\infty^\star$. Since $\mathcal{B}_\infty^\star$ is connected with highest weight element u_∞, there can only be one such morphism so $(1 \otimes \vartheta)\psi_i \vartheta^{-1} = \psi_i^\star$ by Proposition 14.13.

Now using (14.17), (14.18), (14.19) and (14.20) we have

$$\psi_i^\star \vartheta^2 f_i = \psi_i^\star \vartheta f_i^\star \vartheta = (1 \otimes \vartheta)\psi_i f_i^\star \vartheta = (f_i \otimes \vartheta)\psi_i \vartheta = (f_i \otimes 1)\psi_i^\star \vartheta^2 = \psi_i^\star f_i \vartheta^2.$$

Since $\psi_i^\star$ is injective, this proves that $\vartheta^2 f_i = f_i \vartheta^2$. Thus $\vartheta^2 \colon \mathcal{B}_\infty \longrightarrow \mathcal{B}_\infty$ is a bijective crystal morphism. This implies that $\vartheta^2 = 1$. □

Let us use the notation $\star$ instead of ϑ for the map. Theorem 14.1 is now proved in the simply-laced case.

14.3 Properties of the involution

14.3.1 *Relation to Demazure crystals*

The first result of this section justifies the statement in the introduction to this Chapter, that the two embeddings ι_1 and ι_2 of $\mathcal{B}_\infty$ into $\mathbb{Z}^N$ differ by the involution. Let $\mathbf{i} = (i_1, \ldots, i_N)$ be a reduced word for the long Weyl group element $w_0 = s_{i_1} \cdots s_{i_N}$. Let us identify $\mathcal{B}_\infty$ with a subcrystal of $\mathcal{B}_{i_1} \otimes \cdots \otimes \mathcal{B}_{i_N}$ as explained in Section 12.4.

Theorem 14.16. *Suppose that $x \in \mathcal{B}_\infty$ and that*

$$x^\star = u_{i_1}(-a_1) \otimes \cdots \otimes u_{i_N}(-a_N).$$

Then

$$\mathrm{string}_{\mathbf{i}}(x) = (a_1, \ldots, a_N). \tag{14.21}$$

Proof. Instead of computing the image of $x^\star$ in $\mathcal{B}_{i_1} \otimes \cdots \otimes \mathcal{B}_{i_N}$ we may compute the a_i by computing the image of x, regarded as an element of the crystal $\mathcal{B}_\infty^\star$ (whose underlying set is $\mathcal{B}_\infty$ but with the $e_i^\star$ and $f_i^\star$ crystal operations) under the embedding of $\mathcal{B}_\infty^\star$ in $\mathcal{B}_{i_1}^\star \otimes \cdots \otimes \mathcal{B}_{i_N}^\star$. This embedding is obtained by the composition

$$\mathcal{B}_\infty^\star \xrightarrow{\ \psi_{i_1}^\star\ } \mathcal{B}_{i_1}^\star \otimes \mathcal{B}_\infty^\star \xrightarrow{\ 1 \otimes \psi_{i_2}^\star\ } \mathcal{B}_{i_1}^\star \otimes \mathcal{B}_{i_2}^\star \otimes \mathcal{B}_\infty^\star \xrightarrow{\ 1 \otimes 1 \otimes \psi_{i_3}^\star\ } \cdots$$

ending in $\mathcal{B}_{i_1}^\star \otimes \cdots \otimes \mathcal{B}_{i_N}^\star \otimes \{u_\infty^\star\} \subset \mathcal{B}_{i_1}^\star \otimes \cdots \otimes \mathcal{B}_{i_N}^\star \otimes \mathcal{B}_\infty^\star$. We have $\psi_{i_1}^\star(x) = u_{i_1}^\star(-a_1) \otimes y$, where $y = e_{i_1}^{a_1} x$. Thus $a_1 = \varepsilon_{i_1}(x)$. Then applying $1 \otimes \psi_{i_2}^\star$ gives $u_{i_1}^\star(-a_1) \otimes u_{i_2}^\star(-a_2) \otimes z$, where $u_{i_2}^\star(-a_2) \otimes z = \psi_{i_2}^\star(y)$ and so we learn that $a_2 = \varepsilon_{i_2}(e_{i_1}^{a_1} x)$ and $z = e_{i_2}^{a_2} e_{i_1}^{a_1} x$. Continuing in this fashion, we obtain (14.21). □

Let $w \in W$, and let $\mathcal{B}_\infty(w)$ be the Demazure crystal defined in Section 12.4. Kashiwara showed that the involution interchanges the Demazure crystals $\mathcal{B}_\infty(w)$ and $\mathcal{B}_\infty(w^{-1})$. We prove this next.

Theorem 14.17 ([Kashiwara (1993)]). *We have $\mathcal{B}_\infty(w^{-1})^\star = \mathcal{B}_\infty(w)$.*

Proof. We start with a reduced word $(i_1, \ldots, i_r)$ for $w = s_{i_1} \cdots s_{i_r}$ and complete it to a reduced word for the long element $w_0 = s_{i_1} \cdots s_{i_N}$. We want to apply Theorem 12.19 (i). Note that, in the notation of Section 12.4, the reduced word for w is read backwards. To compensate for this difference in notation, we consider $\mathcal{B}_\infty(w^{-1})$ and we see that this Demazure crystal is identified with $\mathfrak{C} \cap \mathcal{A}(w^{-1})$, where $\mathfrak{C}$ is defined in (12.8) and

$$\mathcal{A}(w^{-1}) = \mathcal{B}_{i_1} \otimes \cdots \otimes \mathcal{B}_{i_r} \otimes u_{i_{r+1}}(0) \otimes \cdots \otimes u_{i_N}(0).$$

Applying $\star$ and using Theorem 14.16, we see that $\mathcal{B}_\infty(w^{-1})^\star$ is contained in the set of crystal elements whose string patterns terminate after r steps. In the notation of Section 12.4, it is obvious that such an element is contained in $\mathcal{B}_\infty(w) = \mathfrak{D}_{i_1} \cdots \mathfrak{D}_{i_r} u_\infty$.

We have proved that $\mathcal{B}_\infty(w^{-1})^\star \subseteq \mathcal{B}_\infty(w)$. Similarly, $\mathcal{B}_\infty(w)^\star \subseteq \mathcal{B}_\infty(w^{-1})$ and applying $\star$ we obtain $\mathcal{B}_\infty(w) \subseteq \mathcal{B}_\infty(w^{-1})^\star$. Therefore $\mathcal{B}_\infty(w^{-1})^\star = \mathcal{B}_\infty(w)$. $\square$

14.3.2 *Characterization of highest weight crystals*

As we have seen in Section 14.2, the image of $\psi_i \colon \mathcal{B}_\infty \hookrightarrow \mathcal{B}_i \otimes \mathcal{B}_\infty$ is characterized by $\{u_i(-a) \otimes y \in \mathcal{B}_i \otimes \mathcal{B}_\infty \mid \varepsilon_i^\star(y) = 0, a \geq 0\}$. Next we are going to characterize the image of the embedding $\mathcal{B}_\lambda \hookrightarrow \mathcal{T}_\lambda \otimes \mathcal{B}_\infty$.

Lemma 14.18. *Let*

$$\Sigma = \{t_\lambda \otimes x \mid \varepsilon_i^\star(x) \leqslant \langle \lambda, \alpha_i^\vee \rangle \text{ for all } i\} \subseteq \mathcal{T}_\lambda \otimes \mathcal{B}_\infty.$$

If $t_\lambda \otimes x \in \Sigma$, then $\varphi_i(t_\lambda \otimes x) \geqslant 0$. Furthermore, $\varphi_i(t_\lambda \otimes x) > 0$ if and only if $f_i(t_\lambda \otimes x) \in \Sigma$.

Proof. Choose a reduced word $(i_1, i_2, \ldots, i_N)$ for w_0 with $i_1 = i$. We recall that when we identify $\mathcal{B}_\infty$ with a subcrystal of $\mathcal{B}_{i_1} \otimes \cdots \otimes \mathcal{B}_{i_N}$ we may write

$$t_\lambda \otimes x = t_\lambda \otimes u_i(-a) \otimes u_{i_2}(-a_2) \otimes \cdots \otimes u_{i_N}(-a_N) \tag{14.22}$$

with $a = \varepsilon_i^\star(x)$. We apply Lemma 2.33 to this and obtain $\varphi_i(t_\lambda \otimes x)$ as a maximum of terms as in (2.11). The first, corresponding to t_λ, is $-\infty$, and the second, corresponding to $u_i(-a)$, is $\langle \lambda, \alpha_i^\vee \rangle - \varepsilon_i^\star(x)$. Therefore, Lemma 2.33 tells us that

$$\varphi_i(t_\lambda \otimes x) \geqslant \langle \lambda, \alpha_i^\vee \rangle - \varepsilon_i^\star(x). \tag{14.23}$$

Moreover by Lemma 2.33, we have equality in (14.23) if and only if when we apply f_i to (14.22), the operator f_i is applied to the factor $u_i(-a)$. In other words, we have equality in (14.23) if and only if $\varepsilon_i^\star(f_i(x)) = \varepsilon_i^\star(x) + 1$; otherwise we have $\varepsilon_i^\star(f_i(x)) = \varepsilon_i^\star(x)$. Also, since $\varepsilon_j^\star$ and f_i commute if $i \neq j$ and since $\mathcal{B}_\infty^\star$ is upper seminormal, we always have $\varepsilon_j^\star(f_i(x)) = \varepsilon_j^\star(x)$ if $j \neq i$ and $f_i(x) \neq 0$. It is now clear that if $t_\lambda \otimes x \in \Sigma$, then $\varphi_i(t_\lambda \otimes x) \geqslant 0$. Furthermore, if $\varphi_i(t_\lambda \otimes x) > 0$, then $f_i(t_\lambda \otimes x) = t_\lambda \otimes f_i(x) \in \Sigma$. Moreover, if $\varphi_i(t_\lambda \otimes x) = 0$ for $t_\lambda \otimes x \in \Sigma$, we must have $\langle \lambda, \alpha_i^\vee \rangle - \varepsilon_i^\star(x) = 0$ and in this case what we have shown implies that applying f_i increases $\varepsilon_i^\star(x)$, so $f_i(t_\lambda \otimes x) \notin \Sigma$. $\square$

In Theorems 12.10 and 12.16 we defined a crystal embedding $\psi_\lambda \colon \mathcal{B}_\lambda \hookrightarrow \mathcal{T}_\lambda \otimes \mathcal{B}_\infty$ that maps the highest weight element u_λ to $t_\lambda \otimes u_\infty$.

Theorem 14.19. *The image of ψ_λ is*

$$\{t_\lambda \otimes x \in \mathcal{T}_\lambda \otimes \mathcal{B}_\infty \mid \varepsilon_i^\star(x) \leqslant \langle \lambda, \alpha_i^\vee \rangle \text{ for all } i\}.$$

Note that this theorem provides an answer to the following question: What is the smallest dominant weight λ such that $t_\lambda \otimes x \in \mathcal{B}_\infty$ is in $\mathcal{B}_\lambda$. The answer is

$$\lambda = \sum_{i \in I} \varepsilon_i^\star(x) \varpi_i.$$

Proof of Theorem 14.19. It follows from Lemma 14.18 that if we redefine $f_i(t_\lambda \otimes x) = 0$ when $\varphi_i(t_\lambda \otimes x) = 0$, the set Σ becomes a crystal that is both upper seminormal (since $\mathcal{B}_\infty$ is) and lower seminormal (by Lemma 14.18).

We claim that the image of ψ_λ is contained in Σ. Indeed, the highest weight element u_λ is taken to $t_\lambda \otimes u_\infty$. If $v \in \mathcal{B}_\lambda$ is not a highest weight element, write $v = f_i(y)$ for some $y \in \mathcal{B}_\lambda$. By induction $\psi_\lambda(y) \in \Sigma$, and $\varphi_i(\psi_\lambda(y)) = \varphi_i(y) > 0$. Since Σ is seminormal, it follows that $v = f_i(\psi_\lambda(y)) \in \Sigma$.

Now the crystal Σ has a unique highest weight element $t_\lambda \otimes u_\infty$, so it is a connected crystal. Therefore the image of ψ_λ is precisely Σ. $\qquad\square$

Another curious fact that follows from Theorem 14.19 is a reformulation of the highest weight condition for a tensor product of two highest weight crystals $\mathcal{B}_\mu \otimes \mathcal{B}_\lambda$. Recall from the tensor product rule, that $b \otimes c \in \mathcal{B}_\mu \otimes \mathcal{B}_\lambda$ is highest weight if $c = u_\lambda$ and $\varepsilon_i(b) \leqslant \varphi_i(u_\lambda)$ for all $i \in I$. Since for the highest weight element u_λ we have $\varphi_i(u_\lambda) = \langle \lambda, \alpha_i^\vee \rangle$, this means that $b^\star \in \mathcal{B}_\lambda$, where $b^\star = \psi_\lambda^{-1}(\psi_\mu(b)^\star)$.

14.3.3 *Commutor*

The $\star$-involution is also related to the commutor introduced by [Henriques and Kamnitzer (2006)] following (they say) a suggestion of Berenstein. Recall from Exercise 5.1 the Lusztig (or Schützenberger in type A) involution $S \colon \mathcal{B}_\lambda \longrightarrow \mathcal{B}_\lambda$ which maps the highest weight vector to the lowest weight vector and interchanges f_i with $e_{i'}$, where $w_0(\alpha_i) = -\alpha_{i'}$. This can be extended to reducible crystals by acting as above on each highest weight component. Then the *commutor* is the map

$$C \colon \mathcal{B}_\mu \otimes \mathcal{B}_\lambda \longrightarrow \mathcal{B}_\lambda \otimes \mathcal{B}_\mu$$
$$b \otimes c \longmapsto S\big(S(c) \otimes S(b)\big).$$

We use the notation $C_{\mathcal{A},\mathcal{B}} \colon \mathcal{A} \otimes \mathcal{B} \longrightarrow \mathcal{B} \otimes \mathcal{A}$ if we want to emphasize the crystals involved.

Proposition 14.20. *The commutor is an involution and satisfies*

$$C_{\mathcal{B} \otimes \mathcal{A}, \mathcal{C}} \circ (C_{\mathcal{A}, \mathcal{B}} \otimes 1) = C_{\mathcal{A}, \mathcal{C} \otimes \mathcal{B}} \circ (1 \otimes C_{\mathcal{B}, \mathcal{C}}) \tag{14.24}$$

on $\mathcal{A} \otimes \mathcal{B} \otimes \mathcal{C}$.

Proof. Both assertions are easy. To check (14.24), applying either composition to $a \otimes b \otimes c$ gives $S\big(S(c) \otimes S(b) \otimes S(a)\big)$.   □

Equation (14.24) is called the *cactus relation*. It implies that the category of crystals is a *coboundary category* in the sense of [Drinfeld (1989)]. See [Henriques and Kamnitzer (2006); Kamnitzer and Tingley (2009b,a); Savage (2009)]. As explained in these references, the cactus relation imposes the action of the *cactus group* on the tensor product of finite-dimensional highest weight crystals, which can also be identified as the fundamental group of the moduli space of marked, real, genus-zero stable curves.

Note that the commutor is not a braiding, which would mean that

$$(1_\mathcal{B} \otimes C_{\mathcal{A},\mathcal{C}}) \circ (C_{\mathcal{A},\mathcal{B}} \otimes 1_\mathcal{C}) = C_{\mathcal{A},\mathcal{B}\otimes\mathcal{C}},$$
$$(C_{\mathcal{A},\mathcal{C}} \otimes 1_\mathcal{B}) \circ (1_\mathcal{C} \otimes C_{\mathcal{B},\mathcal{C}}) = C_{\mathcal{A}\otimes\mathcal{B},\mathcal{C}}.$$

A counterexample would be to take $\mathcal{A} = \mathcal{B} = \mathcal{C} = \mathcal{B}_{(1)}$ of type A_2 and $a \otimes b \otimes c =$ $\boxed{2} \otimes \boxed{3} \otimes \boxed{1}$. Then

$$(1_\mathcal{B} \otimes C_{\mathcal{B}\otimes\mathcal{B}}) \circ (C_{\mathcal{B}\otimes\mathcal{B}} \otimes 1_\mathcal{B}) \left(\boxed{2} \otimes \boxed{3} \otimes \boxed{1}\right) = \boxed{2} \otimes \boxed{3} \otimes \boxed{1},$$
$$\text{whereas} \quad C_{\mathcal{B},\mathcal{B}\otimes\mathcal{B}} \left(\boxed{2} \otimes \boxed{3} \otimes \boxed{1}\right) = \boxed{2} \otimes \boxed{1} \otimes \boxed{3}.$$

For highest weight vectors, the commutor can be expressed in terms of $\star$.

Theorem 14.21 ([Kamnitzer and Tingley (2009b)]). *For a highest weight element* $b \otimes u_\lambda \in \mathcal{B}_\mu \otimes \mathcal{B}_\lambda$, *we have*

$$C(b \otimes u_\lambda) = b^\star \otimes u_\mu,$$

where $b^\star = \psi_\lambda^{-1}\left(\psi_\mu(b)^\star\right)$.

We leave the proof as an exercise to the reader in Exercise 15.11.

Exercises

Exercise 14.1. Let s_1 and s_2 be the simple reflections for the A_2 Weyl group. Compute $\mathcal{B}_\infty(s_1 s_2)$ and $\mathcal{B}_\infty(s_2 s_1)$ explicitly for type A_2 and verify that they are interchanged by $\star$ as in Theorem 14.17.

Exercise 14.2.

(a) [Kashiwara (2002), Lemma 7.1.2] Show that for dominant weights λ and μ, the map $\mathcal{T}_\mu \otimes \mathcal{B}_\lambda \longrightarrow \mathcal{B}_{\lambda+\mu}$ defined by $t_\mu \otimes b \mapsto u_\mu \otimes b$ is a crystal morphism that commutes with the operators e_i.

(b) Deduce from part (a) that there exists a crystal morphism

$$I_{\lambda,\lambda+\mu} \colon \mathcal{T}_{-\lambda} \otimes \mathcal{B}_\lambda \longrightarrow \mathcal{T}_{-\lambda-\mu} \otimes \mathcal{B}_{\lambda+\mu}$$

which commutes with the operators e_i and maps $t_{-\lambda} \otimes u_\lambda$ to $t_{-\lambda-\mu} \otimes u_{\lambda+\mu}$.

(c) [Kashiwara (2002)] For dominant weights λ, μ, ξ, show that the following diagram commutes

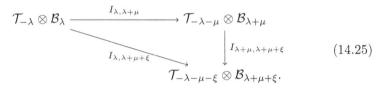

$$(14.25)$$

(d) Using the order that $\lambda \geqslant \mu$ if $\lambda - \mu$ is dominant, show that

$$\mathcal{B}_\infty = \varinjlim_{\lambda \in P^+} \mathcal{T}_{-\lambda} \otimes \mathcal{B}_\lambda.$$

Chapter 15

Crystals and Tropical Geometry

In this chapter, we will discuss the combinatorial theory of the Lusztig parametrization of $\mathcal{B}_\infty$, and the closely related combinatorial theories of MV polytopes and geometric crystals. The topics in this chapter have, besides their connections with the Lusztig canonical basis, another common theme, which is *tropical geometry*. This term refers to a profound parallel between piecewise linear combinatorics and algebraic geometry. See [Maclagan and Sturmfels (2015)] for a current introduction to this field. Unlike previous chapters, in this one we will content ourselves to only consider the simply-laced case.

The *Lusztig parametrization* of $\mathcal{B}_\infty$ was introduced in [Lusztig (1990a)] as a *canonical basis* of the quantized enveloping algebra $U_q(\mathfrak{n})$ of the maximal nilpotent Lie algebra $\mathfrak{n}$ of a semisimple Lie algebra $\mathfrak{g}$ of simply-laced type. Later, Lusztig extended the construction to the non-simply-laced cases in [Lusztig (1992, 2011)]. See [Tingley (2016)] for an efficient treatment of the canonical basis in the simply-laced case. As Lusztig noted, this topic has deep connections with the subject of *total positivity*. See [Lusztig (1994, 1997); Berenstein, Fomin and Zelevinsky (1996); Berenstein and Zelevinsky (2001)].

One ingredient of Lusztig's description of the canonical basis is purely combinatorial, and we can explain it without reference to quantum groups. It works as follows. For every reduced word $\mathbf{i} = (i_1, \ldots, i_N)$ for the long Weyl group element w_0, we have a map $v \mapsto v_{\mathbf{i}}$ of $\mathcal{B}_\infty$ into $\mathbb{N}^N$, where $\mathbb{N}$ is the set $\{0, 1, 2, \ldots\}$ of non-negative integers, and N is the length of w_0. The different maps $v \mapsto v_{\mathbf{i}}$ may be thought of as different "views" of the crystal. They are related by piecewise-linear transformations of $\mathbb{N}^N$, which are "tropicalizations" of certain algebraic maps of the maximal unipotent subgroup N^+ of the Lie group that preserve the set of totally positive matrices $N^+(\mathbb{R})$. The crystal operations may be described quite easily: to calculate $e_i(v)$ or $f_i(v)$ we simply choose a reduced word $\mathbf{i}$ such that $i_1 = i$, and then we decrement or increment the first component of $v_{\mathbf{i}}$.

After we discuss the Lusztig parametrization of $\mathcal{B}_\infty$, we will turn our attention to combinatorics of *MV polytopes*, which are closely connected with the Lusztig parametrization. Although we will restrict ourselves to the combinatorial theory of these polytopes, let us briefly explain how they arose in the geometric Langlands theory.

Let G be a split reductive group over $\mathbb{C}$, and let $\widehat{G}$ be its Langland dual group. Let $\mathcal{O} = \mathbb{C}[[\tau]]$ be the ring of formal power series and $\mathcal{K} = \mathbb{C}((\tau))$ its field of fractions, which is the field of formal Laurent series. The quotient $\mathcal{G}r = G((\mathcal{K}))/G((\mathcal{O}))$ is called the *affine* or *loop Grassmannian* $\mathcal{G}r$. It is called a Grassmannian because if $G = \mathrm{GL}(r)$, it may be regarded as a parameter space for the set of lattices in $\mathcal{K}^n$, that is, the finitely generated $\mathcal{O}$-submodules that span this vector space. The affine Grassmannian is a central object in the geometric Langlands theory, because [Ginzburg (1995)] and [Mirković and Vilonen (2000, 2007)], building on [Lusztig (1983)] and ideas of Drinfeld, proved the *geometric Satake isomorphism*, which is an equivalence of categories between the finite-dimensional representations of $\widehat{G}$ and the perverse sheaves on $\mathcal{G}r$. In [Mirković and Vilonen (2000, 2007)] certain cycles, now called *MV-cycles*, are important, since their images in the intersection homology of a perverse sheaf form a basis of the corresponding representation of $\widehat{G}$ under the geometric Satake isomorphism.

It was shown in [Anderson (2000, 2003); Anderson and Kogan (2004); Kamnitzer (2010, 2007)] that the MV-cycles could be fruitfully studied by their *moment maps* (see [Atiyah and Pressley (1983)]) which represent each cycle by a convex polytope in the weight lattice of $\widehat{G}$. Moreover, these polytopes are organized into a crystal base for the representation. Alternatively, these *MV-polytopes* give a model of $\mathcal{B}_\infty$ in which each element is represented by a polytope. Some of the key ideas, particularly the *tropical Plücker relations*, are due to [Berenstein and Zelevinsky (2001)].

MV polytopes and their crystal structure have a significance beyond their origins in the geometric Langlands theory. In [McNamara (2011)], MV cycles are used to parametrize contributions to certain important p-adic integrals, explaining the appearance of crystals in p-adic Whittaker functions. In [Tingley and Webster (2012)], the crystal structure on MV polytopes is connected to Khovanov-Lauda-Rouqier (KLR) algebras which appear in the categorification of quantum groups. As we will see, the shape of the MV polytope corresponding to $v \in \mathcal{B}_\infty$ is closely connected with the Lusztig parametrization of $\mathcal{B}_\infty$. Indeed, the components of the view v_i are simply the lengths of the edges of the polytope as we follow a path from highest to lowest weight, depending on the choice of reduced word $\mathbf{i}$. So there are many reasons to study MV polytopes.

An interesting aspect of these theories is that the piecewise linear maps that appear throughout are *tropicalizations* of certain corresponding algebraic maps. We define the *tropical semi-ring* $\mathbb{T}$ as the set $\mathbb{R} \cup \{\infty\}$ with operations of addition, multiplication and division (but no subtraction). These operations differ from the usual $+$, $\times$ and $\div$, so we will denote them by $\oplus$, $\otimes$ and $\oslash$. These tropical operations may be expressed in terms of the usual ones by

$$x \oplus y = \min(x, y), \qquad x \otimes y = x + y, \qquad x \oslash y = x - y.$$

The usual 0 element in $\mathbb{R}$ serves as a multiplicative identity in $\mathbb{T}$ so we denote $\mathbb{1} = 0$, while ∞ serves as an additive identity, so we denote $\mathbb{0} = \infty$. Now a polynomial map can be interpreted using the tropical relations, as long as it can be expressed by a formula that does not involve subtraction. For example, the tropicalization of the map $f(x, y, z) = (x + y)/z$ is the piecewise linear map $f(x, y, z) = \min(x, y) - z$.

Conversely, given a piecewise linear map, we can try to interpret it as the tropicalization of an algebraic map, called a *geometric* (or *algebraic*) *lifting*. The geometric lifting is not uniquely determined, but if the right lifting can be found, some structure present in the set of piecewise linear maps may be reflected in the lifted maps. [Kirillov (2001); Morier-Genoud (2008)] consider the geometric lifting of the Schützenberger–Lusztig involution and Weyl group action while [Kirillov (2001); Noumi and Yamada (2004); Danilov and Koshevoy (2005/07)] consider the geometricalization of the RSK correspondence. The most important development in this direction is the theory of *geometric crystals*. In this theory, [Berenstein and Kazhdan (2000, 2007)] showed that the crystal base itself has a geometric lifting. That is, they exhibit algebraic varieties called *geometric crystals* with algebraic maps whose tropicalizations are the weight map, crystal operators, and Weyl group action.

Remark 15.1. In this chapter, we continue to denote the roots in the weight lattice Λ as α_i and the coroots as $\alpha_i^\vee$. This means that our notation is the opposite of that in [Kamnitzer (2007, 2010)], where the MV polytopes are polytopes in the coweight lattice. That is, our α_i are Kamnitzer's $\alpha_i^\vee$. The group G, when it appears, will be the same as Kamnitzer's G, and the weight lattice Λ will be the weight lattice of the Langlands dual group $\widehat{G}$. This is a compromise between the conflicting goals of being consistent with the notation in the papers of [Anderson (2003)] and [Kamnitzer (2007, 2010)], and with the notations in the rest of the book.

15.1 Lusztig parametrization: The A_2 case

Let R be a commutative ring. If $a \in R$, let $x_1(a)$ and $x_2(a)$ be 3×3 matrices over R defined by

$$x_1(a) = \begin{pmatrix} 1 & a & \\ & 1 & \\ & & 1 \end{pmatrix}, \qquad x_2(a) = \begin{pmatrix} 1 & & \\ & 1 & a \\ & & 1 \end{pmatrix},$$

where unspecified entries are zero. We compute

$$x_1(a)x_2(b)x_1(c) = \begin{pmatrix} 1 & a+c & ab \\ & 1 & b \\ & & 1 \end{pmatrix}, \qquad x_2(a)x_1(b)x_2(c) = \begin{pmatrix} 1 & b & bc \\ & 1 & a+c \\ & & 1 \end{pmatrix}.$$

Therefore $x_1(a)x_2(b)x_1(c) = x_2(a')x_1(b')x_2(c')$, where $(a', b', c') = \vartheta_{\mathrm{alg}}(a, b, c)$ with

$$\vartheta_{\mathrm{alg}}(a, b, c) = \left(\frac{bc}{a+c}, a+c, \frac{ab}{a+c} \right). \tag{15.1}$$

The algebraic map ϑ_{alg} is defined on

$$\{(a, b, c) \mid a + c \neq 0, \ b \neq 0\}.$$

It maps this set to itself since if $(a', b', c') = \vartheta(a, b, c)$, then $b' = a+c$ and $a'+c' = b$. The map ϑ_{alg} has order 2.

We now consider the *tropicalization* of the algebraic map ϑ_{alg}. Since the definition of ϑ_{alg} involves only addition, multiplication and division, but no subtraction, we may reinterpret the operations in (15.1) as the corresponding tropical operations. Thus the tropicalization of the algebraic map ϑ_{alg} is the piecewise linear map ϑ on $\mathbb{R}^3$:

$$\vartheta(a, b, c) = \big(b + c - \min(a, c), \min(a, c), a + b - \min(a, c)\big). \qquad (15.2)$$

We may now construct a crystal isomorphic to $\mathcal{B}_\infty$ in the A_2 case by a different method than in Chapter 12. Let $\mathcal{L}$ be the set $\mathbb{N}^3$, where as usual $\mathbb{N}$ is the set of nonnegative integers. We define $\mathrm{wt} \colon \mathbb{N}^3 \to \Lambda$, where Λ is the weight lattice, by

$$\mathrm{wt}(x) = -(a + b)\alpha_1 - (b + c)\alpha_2 \qquad \text{for} \quad x = (a, b, c) . \qquad (15.3)$$

We also define

$$\varepsilon_1(x) = a, \qquad e_1(x) = \begin{cases} (a - 1, b, c) & \text{if } a > 0, \\ 0 & \text{if } a = 0. \end{cases}$$

The other crystal operator e_2 is obtained by conjugating e_1 by ϑ, which is a symmetry that interchanges the two simple roots of A_2. Thus $e_2 = \vartheta \circ e_1 \circ \vartheta$ and $\varepsilon_2 = \varepsilon_1 \circ \vartheta$. In addition, $f_1(x)$ is obtained by adding 1 to the first component of x, and $f_2 = \vartheta \circ f_1 \circ \vartheta$. We define φ_i by requiring that $\varphi_i(x) - \varepsilon_i(x) = \langle \mathrm{wt}(x), \alpha_i^\vee \rangle$.

Proposition 15.2. *The crystal $\mathcal{L}$ is a weak Stembridge crystal with a unique highest weight element $(0, 0, 0)$. It is isomorphic to $\mathcal{B}_\infty$.*

Proof. It is easy to check that $\mathcal{L}$ is a crystal which is upper seminormal. Let us consider the Stembridge axioms. Let $x = (a, b, c)$. Assume that $\varepsilon_1(x) = a > 0$. Then it is easy to check that

$$\varepsilon_2\big(e_1(x)\big) = \begin{cases} \varepsilon_2(x) & \text{if } a > c, \\ \varepsilon_2(x) + 1 & \text{if } a \leqslant c. \end{cases}$$

Replacing x by $\vartheta(x)$, this implies that

$$\varepsilon_1\big(e_2(x)\big) = \begin{cases} \varepsilon_1(x) & \text{if } a < c, \\ \varepsilon_1(x) + 1 & \text{if } a \geqslant c. \end{cases}$$

Axioms S0 and S1 are now clear. We check Axiom S3 leaving Axiom S2 to the reader. From the preceding, we see that the situation in Axiom S3 can only arise if

$a = c$ with $a, b > 0$. Thus let $x = (a, b, a)$. Then we compute the following:

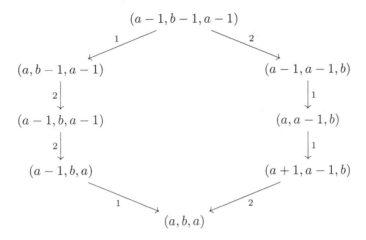

This proves Axiom S3. We leave the dual axioms to the reader.

Now since $\mathcal{L}$ is thus an upper seminormal weak Stembridge crystal with a unique highest weight vector $(0, 0, 0)$, it is isomorphic to $\mathcal{B}_\infty$, by an argument similar to the type used in Corollary 12.10.

$\square$

15.2 Geometric preparations

In this section and the next we will generalize the results of the last section to the simply-laced case. The required piecewise linear maps are tropicalizations of geometric ones which we will also consider. See [Lusztig (1993), Section 14.2] for further discussion of how the piecewise linear transition maps are related to the geometry of N^+.

Let us reconsider the decomposition

$$x_1(a)x_2(b)x_1(c) = \begin{pmatrix} 1 & a+c & ab \\ & 1 & b \\ & & 1 \end{pmatrix}$$

for GL(3) that we made use of in the previous section. Most upper triangular unipotent matrices can be written in this form. However we note that the matrix $\begin{pmatrix} 1 & x & z \\ & 1 & y \\ & & 1 \end{pmatrix}$ cannot be written as $x_1(a)\,x_2(b)\,x_1(c)$ if $y = 0$ and $z \neq 0$. Nevertheless, the matrices of the form $x_1(a)\,x_2(b)\,x_1(c)$ are Zariski dense in the group N^+ of upper triangular unipotent matrices.

We generalize this fact about GL(3) to an arbitrary reductive group G. We denote by $\mathbb{G}_a$ the "additive group," which is the affine algebraic group such that $\mathbb{G}_a(R)$ is the additive group of R for any commutative ring R. If $\alpha \in \Phi$ is a root, let

$x_\alpha \colon \mathbb{G}_a \longrightarrow G$ be the one-parameter subgroup tangent to α, and also let $x_i = x_{\alpha_i}$. This is consistent with the notation we used for GL(3) in Section 15.1.

Let $\mathbf{i} = (i_1, \ldots, i_N)$ be a reduced decomposition of $w_0 = s_{i_1} \cdots s_{i_N}$. Consider the following sequence of roots

$$\gamma(\mathbf{i}) = (\gamma_1(\mathbf{i}), \ldots, \gamma_N(\mathbf{i})) = (\alpha_{i_1}, s_{i_1}(\alpha_{i_2}), s_{i_1} s_{i_2}(\alpha_{i_3}), \ldots). \tag{15.4}$$

For example, in the A_2 case, if $\mathbf{i} = (1, 2, 1)$, then $\gamma(\mathbf{i}) = (\alpha_1, \alpha_1 + \alpha_2, \alpha_2)$.

Proposition 15.3. *Each positive root appears exactly once in the sequence $\gamma(\mathbf{i})$.*

Proof. This follows from [Bump (2013), Proposition 20.10] or [Bourbaki (2002), Corollary 2, Section IV.1.6]. □

Let V and W be irreducible algebraic varieties. A *rational map* $f \colon V \longrightarrow W$ is a morphism defined on a dense Zariski-open subset U of V. If $f' \colon V \longrightarrow W$ is another rational map defined on U', we consider f and f' to be the same if $f = f'$ on $U \cap U'$, which is still dense in the Zariski topology. A *birational map* is then one that has an isomorphism in the category of irreducible varieties with rational maps.

For example, let N^+ (resp. N^-) be the maximal unipotent subgroup generated by the x_α with $\alpha \in \Phi^+$ (resp. Φ^-).[1] Let B (resp. B^-) be the normalizer of N^+ (resp. N^-). Thus B and B^- are opposite Borel subgroups of G. The *flag variety* $X = G/B^-$ is a complete projective variety. It follows from the Bruhat decomposition [Borel (1991), Corollary 14.14] that

$$G = \bigcup_{w \in W} B w B^- \qquad \text{(disjoint).}$$

This induces a decomposition of X into relatively open subsets $B w B^-/B^-$. There is a unique open cell $B B^-/B^-$ corresponding to $w = 1$, and all the other cells have strictly smaller dimension. Using the Bruhat decomposition, the morphism $j \colon N^+ \longrightarrow X$ defined by $j(n) = n B^-$ is injective, and its image is a dense open set. Hence the morphism j is a birational equivalence.

For i in the index set, the subgroup of G generated by x_{α_i} and $x_{-\alpha_i}$ is isomorphic to SL(2). We denote this subgroup as SL(2)$_i$. We also denote by B_i^- the intersection of B^- with SL(2)$_i$. Let P_i^- be the minimal parabolic subgroup generated by SL(2)$_i$ and B^-.

The group $(B^-)^N$ acts on the product $P_{i_1}^- \times \cdots \times P_{i_N}^-$ on the right, where $(b_1, \ldots, b_N)$ sends

$$(p_1, \ldots, p_N) \longrightarrow (p_1 b_1, b_1^{-1} p_2 b_2, \ldots, b_{N-1}^{-1} p_N b_N).$$

The *Bott–Samelson variety* $X_{\mathbf{i}}$ is the quotient of $P_{i_1}^- \times \cdots \times P_{i_N}^-$ by this action. It is a projective variety of dimension N. Since P_i^-/B^- is a projective line, the variety $X_{\mathbf{i}}$ is built up from scratch by successive fiberings by $\mathbb{P}^1$. It comes with a morphism $\pi_{\mathbf{i}} \colon X_{\mathbf{i}} \longrightarrow X$ in which the orbit of $(p_1, \ldots, p_N)$ is sent to $p_1 \cdots p_N B^-$.

---

[1] We use the notation N to denote the number of positive roots. We will therefore avoid using N to denote the group N^+.

Proposition 15.4. *The map* $\pi_{\mathbf{i}}$ *is a birational equivalence.*

Proof. Consider the Bruhat decomposition $G = \bigcup_{w \in W} BwB^-$. The image BB^-/B^- of the big Bruhat cell BB^- in X is an affine space that may be identified with N^+ since the map $n \mapsto nB^-$ from N^+ to BB^-/B^- is an isomorphism.

Let $\mathbb{A}^N$ be affine N-space. We may map $\mathbb{A}^N$ into $X_{\mathbf{i}}$ by sending $a = (a_1, \ldots, a_N) \in \mathbb{A}^N$ to the orbit of

$$(x_{i_1}(a_1)s_{i_1}, \ldots, x_{i_N}(a_N)s_{i_N}) \in \mathrm{SL}(2)_{i_1} \times \cdots \times \mathrm{SL}(2)_{i_N}.$$

The elements of this form comprise an open set U of $X_{\mathbf{i}}$. Under $\pi_{\mathbf{i}}$ the above element maps to $\xi_1 \cdots \xi_N$ where

$$\xi_k = (s_{i_1} \cdots s_{i_{k-1}}) x_{i_k}(a_i)(s_{i_1} \cdots s_{i_{k-1}})^{-1} \in x_{\gamma_k}(\mathbb{G}_a),$$

with γ_k as in (15.4). Since the γ_k are precisely the positive roots of G, every element of N^+ may be written uniquely in this form. Hence $\pi_{\mathbf{i}}$ maps the open set U in $X_{\mathbf{i}}$ bijectively onto the big cell of X, proving that $\pi_{\mathbf{i}}$ is a birational equivalence. □

Let $\mathbb{A}^N$ be N-dimensional affine space. Let us now define a morphism $\phi_{\mathbf{i}} \colon \mathbb{A}^N \to X$ by

$$\phi_{\mathbf{i}}(a_1, \ldots, a_N) = x_{i_1}(a_1) \cdots x_{i_N}(a_N).$$

Theorem 15.5. *The morphism* $\phi_{\mathbf{i}} \colon \mathbb{A}^N \to X$ *is a birational equivalence.*

Proof. Let $\widetilde{\phi}_{\mathbf{i}} \colon \mathbb{A}^N \to X_{\mathbf{i}}$ be the map that sends $(a_1, \ldots, a_N)$ to the orbit of $(x_{i_1}(a_1), \ldots, x_{i_N}(a_N))$. This map lifts $\phi_{\mathbf{i}}$ to the Bott–Samelson variety since $\pi_{\mathbf{i}} \circ \widetilde{\phi}_{\mathbf{i}} = \phi_{\mathbf{i}}$. Now $\widetilde{\phi}_{\mathbf{i}}$ is a birational morphism. Indeed, remembering that $X_{\mathbf{i}}$ is built up from scratch by successive fiberings of $\mathbb{P}^1$, this fact boils down to the fact that the inclusion of $\mathbb{A}^1$ into $\mathbb{P}^1$ is birational. Since $\pi_{\mathbf{i}}$ is a birational equivalence, the statement follows. □

15.3 The Lusztig parametrization in the simply-laced case

The combinatorial part of Lusztig's description of the canonical basis depends on Matsumoto's Theorem 11.4, which we now review. As in Chapter 11, let $\mathrm{Red}(w)$ be the set of reduced words $\mathbf{i} = (i_1, \ldots, i_N)$ for a Weyl group element $w = s_{i_1} \cdots s_{i_N}$. We may define a graph structure on $\mathrm{Red}(w)$ in which $\mathbf{i}$ and $\mathbf{j}$ are adjacent in the following two cases. (Since we are in the simply-laced case, these are the only possible ones.)

(M1) For some k, the word $\mathbf{j}$ is obtained from $\mathbf{i}$ by replacing $(i_k, i_{k+1}) = (i, j)$ by (j, i), where s_i and s_j commute; or

(M2) For some k, the word $\mathbf{j}$ is obtained from $\mathbf{i}$ by replacing $(i_k, i_{k+1}, i_{k+2}) = (i, j, i)$ by (j, i, j), where $s_i s_j$ has order 3.

The content of Matsumoto's Theorem 11.4 is that $\mathrm{Red}(w)$ is a connected graph.

Lusztig's parametrization depends on a family of piecewise linear maps indexed by pairs of elements of $\mathrm{Red}(w)$. Kashiwara's description of $\mathcal{B}_\infty$, which we took up in Chapter 12 may be understood the same way. This similarity is worth pointing out, so we digress to reconsider Kashiwara's coordinatization of $\mathcal{B}_\infty$ before turning to the similar Lusztig parametrization.

Given $v \in \mathcal{B}_\infty$ and $\mathbf{i} \in \mathrm{Red}(w_0)$, we associate an element of $\mathbb{Z}^N$ by a procedure that we will now explain. To avoid confusion with the Lusztig parametrization, we denote this element of $\mathbb{Z}^N$ by $\widehat{v}_{\mathbf{i}}$. When we get to the Lusztig parametrization, there will be another family of elements of $\mathbb{Z}^N$ indexed by $\mathbf{i} \in \mathrm{Red}(w_0)$, that we simply write as $v_{\mathbf{i}}$.

To define $\widehat{v}_{\mathbf{i}}$, embed $\mathcal{B}_\infty$ into $\mathcal{B}_{i_1} \otimes \cdots \otimes \mathcal{B}_{i_N}$ as was done in Chapter 12. Let $u_{i_1}(-a_1) \otimes \cdots \otimes u_{i_N}(-a_N)$ be the image of v under this embedding, and define $\widehat{v}_{\mathbf{i}} = (a_1, \ldots, a_N)$.

We may think of the map $v \mapsto \widehat{v}_{\mathbf{i}} \in \mathbb{Z}^N$ as a "view" of the crystal. Let us ask how the different views vary. We make use of two maps $\theta_2 \colon \mathbb{Z}^2 \longrightarrow \mathbb{Z}^2$ and $\theta_3 \colon \mathbb{Z}^3 \longrightarrow \mathbb{Z}^3$ defined by $\theta_2(a, b) = (b, a)$ and $\theta_3(a, b, c) = \big(\max(c, b - a), a + c, \min(b - c, a)\big)$. The map θ_3 already appeared in Chapter 12, where it was denoted θ and defined in (12.3). It is perhaps better to write it this way:

$$\theta_3(a, b, c) = \big(b + c - \min(b, a + c), a + c, \min(b - c, a)\big).$$

Then we recognize it as the tropicalization of an algebraic map

$$(a, b, c) \mapsto \left(\frac{bc}{b + ac}, ac, \frac{b + ac}{c} \right).$$

Proposition 15.6. *Let $\mathbf{i}, \mathbf{j} \in \mathrm{Red}(w_0)$. Then there is a piecewise linear map $\widehat{\mathcal{R}}_{\mathbf{i},\mathbf{j}} \colon \mathbb{Z}^N \to \mathbb{Z}^N$ such that if $v \in \mathcal{B}_\infty$, then $\widehat{\mathcal{R}}_{\mathbf{i},\mathbf{j}}(\widehat{v}_{\mathbf{i}}) = \widehat{v}_{\mathbf{j}}$. These maps satisfy the relation*

$$\widehat{\mathcal{R}}_{\mathbf{j},\mathbf{k}} \circ \widehat{\mathcal{R}}_{\mathbf{i},\mathbf{j}} = \widehat{\mathcal{R}}_{\mathbf{i},\mathbf{k}}.$$

In case (M1), $\widehat{\mathcal{R}}_{\mathbf{i},\mathbf{j}}$ is θ_2 applied to the (a_k, a_{k+1}) substring of $\widehat{v}_{\mathbf{i}} = (a_1, \ldots, a_N)$. In case (M2), $\widehat{\mathcal{R}}_{\mathbf{i},\mathbf{j}}$ is θ_3 applied to the (a_k, a_{k+1}, a_{k+2}) substring.

Proof. This follows immediately from Proposition 12.3, where it is shown that θ_2 and θ_3 describe the isomorphisms $\mathcal{B}_i \otimes \mathcal{B}_j \to \mathcal{B}_j \otimes \mathcal{B}_i$ (when s_i and s_j commute) and $\mathcal{B}_i \otimes \mathcal{B}_j \otimes \mathcal{B}_i \to \mathcal{B}_j \otimes \mathcal{B}_i \otimes \mathcal{B}_j$ (when $s_i s_j$ has order 3). $\square$

The Lusztig parametrization is similar. It relies on a family of piecewise linear maps that we denote $\mathcal{R}_{\mathbf{i},\mathbf{j}} \colon \mathbb{Z}^N \to \mathbb{Z}^N$ that satisfy a similar compatibility $\mathcal{R}_{\mathbf{j},\mathbf{k}} \circ \mathcal{R}_{\mathbf{i},\mathbf{j}} = \mathcal{R}_{\mathbf{i},\mathbf{k}}$. The piecewise linear maps are different from the $\widehat{\mathcal{R}}_{\mathbf{i},\mathbf{j}}$, but they may similarly be described by reduction to the cases (M1) and (M2).

Proposition 15.7. *There exists a family of piecewise linear maps* $\mathcal{R}_{\mathbf{i},\mathbf{j}}\colon \mathbb{Z}^N \to \mathbb{Z}^N$ *satisfying*

$$\mathcal{R}_{\mathbf{j},\mathbf{k}} \circ \mathcal{R}_{\mathbf{i},\mathbf{j}} = \mathcal{R}_{\mathbf{i},\mathbf{k}} \qquad (15.5)$$

such that if $\mathbf{j}$ *is obtained from* $\mathbf{i}$ *as in (M1) then* $\mathcal{R}_{\mathbf{i},\mathbf{j}}$ *applied to* $(a_1,\ldots,a_N) \in \mathbb{Z}^N$ *consists of interchanging* a_k *and* a_{k+1}, *while if* $\mathbf{j}$ *is obtained from* $\mathbf{i}$ *as in (M2), then* $\mathcal{R}_{\mathbf{i},\mathbf{j}}$ *consists of applying* ϑ, *defined in (15.2), to the substring* (a_k, a_{k+1}, a_{k+2}).

Proof. We may lift this problem to a question of change of coordinates on affine N-space $\mathbb{A}^N$ as follows. We prove the existence of birational maps $\mathfrak{R}_{\mathbf{i},\mathbf{j}}\colon \mathbb{A}^N \to \mathbb{A}^N$ such that

$$\mathfrak{R}_{\mathbf{j},\mathbf{k}} \circ \mathfrak{R}_{\mathbf{i},\mathbf{j}} = \mathfrak{R}_{\mathbf{i},\mathbf{k}}$$

and such that in cases (M1) the map $\mathfrak{R}_{\mathbf{i},\mathbf{j}}$, when applied to $(a_1,\ldots,a_N)$ interchanges a_k and a_{k+1}, while in the case (M2), it applies the map $\vartheta_{\mathrm{alg}}\colon \mathbb{A}^3 \to \mathbb{A}^3$ defined by (15.1) to (a_k, a_{k+1}, a_{k+2}). We define $\mathfrak{R}_{\mathbf{i},\mathbf{j}}$ to be the birational map $\phi_{\mathbf{j}}^{-1}\phi_{\mathbf{i}}$ in terms of the map $\phi_{\mathbf{i}}$ introduced in the previous section. In the cases (M1) and (M2), the transition functions $\mathfrak{R}_{\mathbf{i},\mathbf{j}}$ may be computed by branching down to $A_1 \times A_1$ or A_2 and apply the results of the previous section.

We may simply define $\mathcal{R}_{\mathbf{i},\mathbf{j}}$ to be the tropicalization of $\mathfrak{R}_{\mathbf{i},\mathbf{j}}$, and it clearly has the required properties. □

We may now define a crystal $\mathcal{L}$ as follows. An element v of $\mathcal{L}$ consists of a family of "views" of v, one for each $\mathbf{i} \in \mathrm{Red}(w_0)$. The view $v_{\mathbf{i}}$ for $\mathbf{i} \in \mathrm{Red}(w_0)$ is an element of $\mathbb{N}^N$. These must be related by the rule $v_{\mathbf{j}} = \mathcal{R}_{\mathbf{i},\mathbf{j}}(v_{\mathbf{i}})$. It follows from (15.5) that given $v_{\mathbf{i}}$ for any $\mathbf{i}$, we may complete it uniquely to an element of $\mathcal{L}$ (since (15.5) ensures that the chosen reduced word does not matter). We define a weight function $\mathrm{wt}\colon \mathcal{L} \to \Lambda$ by

$$\mathrm{wt}(v_{\mathbf{i}}) = -\sum_{j=1}^{N} a_j \gamma_j \qquad (15.6)$$

where $v_{\mathbf{i}} = (a_1,\ldots,a_N)$ and $\gamma_i = \gamma_i(\mathbf{i})$ are as in (15.4). By Proposition 15.3, the γ_j are the positive roots in some order.

Lemma 15.8. *The weight function defined by (15.6) is independent of the choice of* $\mathbf{i}$.

Proof. It is sufficient to show that (15.6) gives the same result for $\mathbf{i}$ and $\mathbf{j}$ that are related by (M1) or (M2). Let $w = s_{i_1} \cdots s_{i_{k-1}}$. In the (M1) case, $(\gamma_k, \gamma_{k+1}) = (w(\alpha_i), w s_i(\alpha_j))$. Since s_i and s_j commute, $s_i(\alpha_j) = \alpha_j$, so $(\gamma_k, \gamma_{k+1}) = (w\alpha_i, w\alpha_j)$. Thus for $\mathbf{j}$, γ_k and γ_{k+1} are interchanged, as are a_k and a_{k+1}, so the sum (15.6) is unchanged. In the (M2) case, we have $(\gamma_k, \gamma_{k+1}, \gamma_{k+2}) = (w\alpha_i, w s_i(\alpha_j), w s_i s_j(\alpha_i))$. Since α_i and α_j may be identified with the simple roots in an A_2 root system, we have $s_i(\alpha_j) = \alpha_i + \alpha_j$ and

$s_i s_j(\alpha_i) = \alpha_j$. Thus $(\gamma_k, \gamma_{k+1}, \gamma_{k+2}) = (w\alpha_i, w(\alpha_i + \alpha_j), w\alpha_j)$, and for $\mathbf{j}$ these are replaced by $(w\alpha_j, w(\alpha_i + \alpha_j), w\alpha_i)$. On the other hand (a_k, a_{k+1}, a_{k+2}) is replaced by $(a'_k, a'_{k+1}, a'_{k+2}) = \vartheta(a_k, a_{k+1}, a_{k+2})$, which satisfy $a'_{k+1} + a'_{k+2} = a_{k+1} + a_k$, $a'_{k+1} + a'_k = a_{k+1} + a_{k+2}$. So the contribution of these three terms, which equals $(a_k + a_{k+1})w(\alpha_i) + (a_{k+1} + a_{k+2})w(\alpha_j)$, is unchanged. ☐

We may now define the crystal operations on $\mathcal{L}$. To define e_i, f_i, ε_i and φ_i we choose a word $\mathbf{i} \in \mathrm{Red}(w_0)$ such that $i_1 = i$. Then we define $\varepsilon_i(v) = a_1$ where $v_\mathbf{i} = (a_1, \ldots, a_N)$. We define $e_i(v) = 0$ if $a_1 = 0$. Otherwise, e_i decrements a_1 without affecting any of the other a_k. We define $\varphi_i(v)$ by requiring that $\varphi_i(v) - \varepsilon_i(v) = \langle \mathrm{wt}(v), \alpha_i^\vee \rangle$, and $f_i(v)$ increments a_1 without affecting any of the other a_k.

Lemma 15.9. *This definition is independent of the choice of $\mathbf{i}$, provided $i_1 = i$.*

Proof. If $\mathbf{j} \in \mathrm{Red}(w_0)$ also has $j_1 = i$, then applying Matsumoto's theorem to the word $s_i^{-1} w_0$ we obtain a sequence of words interpolating $(i_2, \ldots, i_k)$ and $(j_2, \ldots, j_k)$ in which consecutive words are related by (M1) or (M2). We prepend i to each of these, and so we see that there is a sequence $\mathbf{i}_1, \ldots, \mathbf{i}_r$ beginning with $\mathbf{i}$ and ending with $\mathbf{j}$ in which consecutive words differ by (M1) or (M2), and each word begins with i. Thus $\mathcal{R}_{\mathbf{i},\mathbf{j}} = \mathcal{R}_{\mathbf{i}_{r-1},\mathbf{i}_r} \circ \cdots \circ \mathcal{R}_{\mathbf{i}_1,\mathbf{i}_2}$ does not affect the first coordinate. The statement follows. ☐

Example 15.10. *In the A_2 case, let us compare this definition with the one we gave in Section 15.1. There are two reduced words $\mathbf{i} = (1, 2, 1)$ and $\mathbf{j} = (2, 1, 2)$. So if $v = (a, b, c) \in \mathbb{N}^3$, which we called $\mathcal{L}$ in Section 15.1, then we define the two views $v_\mathbf{i} = (a, b, c)$ and $v_\mathbf{j} = \vartheta(a, b, c)$. It is clear that this is the same crystal in slightly different notation.*

Theorem 15.11. *The crystal $\mathcal{L}$ is isomorphic to $\mathcal{B}_\infty$.*

Proof. We begin by showing that $\mathcal{L}$ is weakly Stembridge. It is not hard to see that it is upper seminormal, and $f_i(v)$ is never zero. Therefore Axioms S0 and S0′ are satisfied. To prove the remaining axioms, we will check S2 and S3 when α_i and α_j are not orthogonal, leaving aside the case where they are orthogonal to the reader. (We also leave the dual axioms to the reader.) In this case $s_i s_j s_i = s_j s_i s_j$ has length 3 and we find a reduced word $(i_4, \ldots, i_N)$ for $(s_i s_j s_i)^{-1} w_0$, to which we prepend (i, j, i). We thus obtain a reduced word $\mathbf{i}$ that begins with (i, j, i). Working with this reduced word, the verification of Axioms S2 and S3 is identical to Proposition 15.2.

Now there exists a crystal embedding τ of $\mathcal{L}$ into $\mathcal{B}_\infty$ by Theorem 12.9. We will show that τ is surjective. If not, let $x \in \mathcal{B}_\infty$ be an element that is maximal with respect to the property that it is not in the image of τ. Then $x = f_i(y)$ for some y, and by assumption $y = \tau(y')$ for some $y' \in \mathcal{L}$. But $f_i(y')$ is nonzero because by construction, the crystal maps f_i on $\mathcal{L}$ never take any element to 0. Therefore, $\tau(f_i(y')) = f_i(\tau(y')) = x$, which is a contradiction. ☐

The paper [Salisbury, Schultze and Tingley (2016a)] describes an efficient bracketing procedure for computing the crystal operations for particular **i**.

Remark 15.12. As we explained in the introduction to Chapter 12, a *root partition* of μ is a tuple (k_α) indexed by positive roots α in which $k_\alpha \in \mathbb{N}$ and $\sum_{\alpha \in \Phi^+} k_\alpha \alpha = \mu$. Also in Chapter 12, the Kostant partition function $P(\mu)$ was defined to be the number of root partitions. It was proved in Exercise 13.4 that the generating function (12.2) for $P(\mu)$ is the character of $\mathcal{B}_\infty$. But deducing this from Kashiwara's definition of $\mathcal{B}_\infty$ is not very straightforward. By contrast, this fact is built into the Lusztig parametrization. Indeed, this follows from the definition (15.6) of the weight function.

15.4 Weyl group action

This section and the next may be skipped without loss of continuity.

About the Weyl group action, [Berenstein and Kazhdan (2000)] write:

> Surprisingly many interesting rational actions of W come from geometric crystals. For example, the action of W on Grothendieck's simultaneous resolution $\widetilde{B} \to G$ comes from the structure of a geometric crystal on $\widetilde{G}$. Also all of the examples in [Braverman and Kazhdan (2000)] come from geometric crystals.

In Chapter 11, we discussed an action of the Weyl group on the finite, highest weight crystal $\mathcal{B}_\lambda$. For each index i, we defined

$$\sigma_i(x) = \begin{cases} f_i^k(x) & \text{if } k \geqslant 0, \\ e_i^{-k}(x) & \text{if } k < 0, \end{cases} \tag{15.7}$$

where $k = \langle \mathrm{wt}(x), \alpha_i^\vee \rangle$. We may try applying this definition to other crystals. For this definition to be valid in a crystal $\mathcal{B}$, it is necessary that $\sigma_i(x)$ defines an element of $\mathcal{B}$, never 0. This is true for any seminormal crystal $\mathcal{B}$, since if $k \geqslant 0$, then $0 \leqslant k \leqslant \varphi_i(x)$, while if $k < 0$ then $0 < -k \leqslant \varepsilon_i(x)$, and therefore $\sigma_i(x)$ defines an element of $\mathcal{B}$. This definition does not work for $\mathcal{L}$, however, because (15.7) may annilate x.

However we can use a variant $\widehat{\mathcal{L}}$ of $\mathcal{L}$ for which $e_i(x)$ and $f_i(x)$ are never 0. To construct this, recall that we defined $\mathcal{L}$ to be the set of families $\{v_{\mathbf{i}} \mid \mathbf{i} \in \mathrm{Red}(w_0)\}$ of elements $v_{\mathbf{i}} \in \mathbb{N}^N$ that satisfy $v_{\mathbf{j}} = \mathcal{R}_{\mathbf{i},\mathbf{j}}(v_{\mathbf{i}})$. We define $\widehat{\mathcal{L}}$ similarly, except that we take $v_{\mathbf{i}} \in \mathbb{Z}^N$. Whereas in $\mathcal{L}$ we defined $e_i(v) = 0$ if $\varepsilon_i(v) = 0$, in $\widehat{\mathcal{L}}$ we define the crystal operations so that both $e_i(v)$ and $f_i(v)$ are never zero. To define $e_i(v)$, we require that if $\mathbf{i}$ is chosen so that its first component is i, then $e_i(v) = v'$ where if $v_{\mathbf{i}} = (a_1, \ldots, a_N)$, then $v'_{\mathbf{i}} = (a_1 - 1, a_2, \ldots, a_N)$. The proofs go through without modification, so $\widehat{\mathcal{L}}$ is a crystal and $\mathcal{L}$ is a subcrystal.

Now the definition (15.7) makes sense for $\widehat{\mathcal{L}}$, and the operators σ_i have order two. Actually we need to generalize them slightly. If ν is any weight, define

$$\sigma_i^\nu(v) = \begin{cases} f_i^k(v) & \text{if } k \geqslant 0, \\ e_i^{-k}(v) & \text{if } k < 0, \end{cases}$$

where now $k = \langle \mathrm{wt}(v) + \nu, \alpha_i^\vee \rangle$. Let $\mathcal{T}_\nu = \{t_\nu\}$ be the singleton crystal defined in Example 2.28. Then clearly $\sigma_i(t_\nu \otimes v) = t_\nu \otimes \sigma_i^\nu(v)$.

Theorem 15.13. *The maps σ_i of $\mathcal{T}_\nu \otimes \widehat{\mathcal{L}}$ satisfy the braid relations. Thus there is an action of the Weyl group W on this crystal in which the simple reflection s_i acts via σ_i.*

Proof. The second statement follows from the first since W is a Coxeter group. It is clearly sufficient to show that the σ_i^ν acting on $\widehat{\mathcal{L}}$ satisfy the braid relations.

We will prove that $\sigma_i^\nu \sigma_j^\nu \sigma_i^\nu = \sigma_j^\nu \sigma_i^\nu \sigma_j^\nu$ if $\langle \alpha_i, \alpha_j^\vee \rangle = -1$, leaving the case $\langle \alpha_i, \alpha_j^\vee \rangle = 0$ to the reader. We reduce to the A_2 crystal, so we discuss this case first. Let F be a field and let M be a nonzero element. Define a rational map s_{alg}^M of $\mathbb{A}^3$ defined over F by

$$s_{\mathrm{alg}}^M(a,b,c) = \left(\frac{Mc}{ab}, b, c \right).$$

We also recall the operator ϑ_{alg} from Section 15.1. Suppressing alg from the notation,

$$\vartheta s_{M_2} \vartheta s_{M_1} \vartheta s_{M_2} = s_{M_1} \vartheta s_{M_2} \vartheta s_{M_1} \vartheta. \tag{15.8}$$

Indeed both maps, applied to (a,b,c), give (a',b',c') where

$$a' = \frac{bcM_2 + M_1 M_2}{a^2 b + abc + aM_2 + cM_2},$$

$$b' = \frac{a^2 bM_1 + abcM_1 + aM_1 M_2 + cM_1 M_2}{ab^2 c + abM_1},$$

$$c' = \frac{abcM_2 + aM_1 M_2}{a^2 bc + abc^2 + acM_2 + c^2 M_2}.$$

Tropicalizing, this yields

$$s_M(a,b,c) = (M + c - a - b, b, c),$$
$$\vartheta(a,b,c) = (b + c - \min(a,c), \min(a,c), a + b - \min(a,c)).$$

It follows from (15.8) that these maps satisfy the corresponding relation which we may write as

$$(\vartheta s_{M_2} \vartheta) s_{M_1} (\vartheta s_{M_2} \vartheta) = s_{M_1} (\vartheta s_{M_2} \vartheta) s_{M_1}.$$

Choose $M_1 = \langle \nu, \alpha_1^\vee \rangle$ and $M_2 = \langle \nu, \alpha_2^\vee \rangle$. Taking $\mathbf{i} = (1,2,1)$ and $\mathbf{j} = (2,1,2)$, we have

$$(\sigma_1 v)_{\mathbf{i}} = s_{M_1}(v_{\mathbf{i}}), \qquad (\sigma_2 v)_{\mathbf{j}} = s_{M_2}(v_{\mathbf{j}}), \qquad v_{\mathbf{j}} = \vartheta(v_{\mathbf{i}}).$$

Thus $(\sigma_2 v)_{\mathbf{i}} = \vartheta s_{M_2} \vartheta(v_{\mathbf{i}})$. We see that $\sigma_i^\nu \sigma_j^\nu \sigma_i^\nu = \sigma_j^\nu \sigma_i^\nu \sigma_j^\nu$ in this case.

Now we turn to the general case. We embed the A_2 root system into the given simply-laced root system Φ by taking the rank two root system generated by i and j with the same weight lattice Λ. We complete the word (i, j, i) to a reduced word $\mathbf{i} = (i, j, i, i_4, \ldots, i_N)$ of w_0, and let $\mathbf{i}' = (j, i, j, i_4, \ldots, i_N)$. Let $v \in \widehat{\mathcal{L}}$ and let $v_{\mathbf{i}} = (a_1, \ldots, a_N)$. Let $\mathcal{L}_3$ be the A_2 crystal that we have just discussed. Let $\gamma_m(\mathbf{i})$ be the sequence of roots as in (15.4). We note that $\gamma_m(\mathbf{i}) = \gamma_m(\mathbf{i}')$ if $m \geqslant 4$ since $s_i s_j s_i = s_j s_i s_j$. Let

$$\mu = \sum_{m=4}^{N} a_m \gamma_m(\mathbf{i}) = \sum_{m=4}^{N} a_m \gamma_m(\mathbf{i}').$$

We now exhibit a morphism $\alpha: t_{-\mu} \otimes \widehat{\mathcal{L}}_3 \longrightarrow \widehat{\mathcal{L}}$ by mapping $t_{-\mu} \otimes v$ with $v \in \widehat{\mathcal{L}}_3$ and $v_{(1,2,1)} = (x, y, z)$ to $\alpha(v)$ with

$$\alpha(t_{-\mu} \otimes v)_{\mathbf{i}} = (x, y, z, a_4, \ldots, a_N).$$

The image of α is preserved by the maps σ_i^ν and indeed we have $\alpha \circ \sigma_i^{\nu - \mu} = \sigma_i^\nu \circ \alpha$ and $\alpha \circ \sigma_j^{\nu - \mu} = \sigma_j^\nu \circ \alpha$. So the braid relation is reduced to the A_2 case. $\qquad\square$

15.5 The geometric weight map

This section may be skipped without loss of continuity.

In this section, G will be a reductive algebraic group scheme defined and split over $\mathbb{Z}$. Thus, if R is any commutative ring, we have the group $G(R)$ of points of G with coefficients in R. For example, such split reductive algebraic groups include $GL(r)$, or a Chevalley group ([Borel (1970); Steinberg (1968)]). As in the rest of this chapter, we will limit ourselves to the simply-laced case to save ourselves some extra work.

We have seen in this chapter how some aspects of crystal bases can be regarded as the tropicalizations of similar constructions for algebraic varieties. Particularly, the change of basis formulas for the Lusztig parametrization are tropicalizations of birational maps of the flag variety $X = G/B^-$, or of its dense open subset N^+. Taking the process of finding geometric liftings of piecewise linear constructions that appear in the theory of crystal bases one step further, [Berenstein and Kazhdan (2000, 2007)] showed how the crystal itself can have a geometric lifting. In this section we will take one more step towards the theory of geometric crystals by considering the geometric lifting of the weight map from a crystal to the weight lattice.

We have already mentioned that geometric crystals are not actually crystals of G, but of another group, the (connected) Langlands dual group $\widehat{G}$. The relationship between the two is that the weight lattices of G and $\widehat{G}$ are in duality, so that roots of G are actually coroots of $\widehat{G}$. However, since we are assuming that G is simply-laced, the bijection $\alpha \mapsto \alpha^\vee$ between the two root systems is an isometry, and so we will not worry too much about distinguishing Φ and $\Phi^\vee$. But there is another aspect of this distinction, that we will be more careful about. This is that weight lattice Λ is not the group of characters $X^*(T) = \text{Hom}(T, \mathbb{G}_m)$, where $\mathbb{G}_m$ is the multiplicative affine algebraic group such that $\mathbb{G}_m(R) \cong R^\times$ for any commutative ring R. Rather, Λ is the group of *cocharacters* $X_*(T) = \text{Hom}(\mathbb{G}_m, T)$.

For $\mathcal{B}_\infty$, the geometric crystal is X or N^+. (These are really the same in the category of varieties with rational maps, because the inclusion $N^+ \hookrightarrow X$ is a birational equivalence, that is, an isomorphism in this category.) The first step is to find a geometric lifting of wt: $\mathcal{B}_\infty \to \Lambda$. The other crystal operations must also be lifted, of course, but let us begin with wt.

The constructions here are related to those in Section 15.2. However, we need a bit more detail. For every index i, we assume that there exists a homomorphism $\iota_i \colon \text{SL}(2) \to G$ of algebraic groups with certain properties that we will explain. (See Corollary 2 to Theorem 7 in Chapter 5 of [Steinberg (1968)].) Let us denote

$$x_i(a) = \iota_i \begin{pmatrix} 1 & a \\ & 1 \end{pmatrix}, \quad y_i(a) = \iota_i \begin{pmatrix} 1 & \\ a & 1 \end{pmatrix}, \quad \overline{s}_i = \iota_i \begin{pmatrix} 0 & 1 \\ -1 & 0 \end{pmatrix}, \quad h_i(a) = \iota_i \begin{pmatrix} a & \\ & a^{-1} \end{pmatrix}.$$

This embedding may be chosen so that if $\mathfrak{g}$ is the Lie algebra of G and $d\iota_i \colon \mathfrak{sl}(2) \to \mathfrak{g}$ is the induced map of Lie algebras, then

$$d\iota_i \begin{pmatrix} 0 & 1 \\ 0 & 0 \end{pmatrix} = X_{\alpha_i}, \quad d\iota_i \begin{pmatrix} 0 & 0 \\ 1 & 0 \end{pmatrix} = X_{-\alpha_i}, \quad d\iota_i \begin{pmatrix} 1 & \\ & -1 \end{pmatrix} = H_i,$$

where X_α ($\alpha \in \Phi$) and H_i ($i \in I$) are a Chevalley basis of the Lie algebra of the derived Lie algebra $[\mathfrak{g}, \mathfrak{g}]$ relative to the Cartan subalgebra $\text{Lie}(T)$. (See Theorem 1 of [Steinberg (1968)] for the Chevalley basis.) We require that the image of h_i is contained in the maximal torus T, and that $\overline{s}_i$ is contained in the normalizer of T, so that it corresponds to an element s_i of the Weyl group. Finally we require that the $\overline{s}_i$ satisfy the braid relation, that is,

$$\overline{s}_i \overline{s}_j \overline{s}_i \cdots = \overline{s}_j \overline{s}_i \overline{s}_j \cdots \tag{15.9}$$

with $n(i, j)$ terms on both sides, where $n(i, j)$ is the order of $s_i s_j$. This may always be arranged. (See [Tits (1966)].)

For example, if $G = \text{GL}(r)$ and E_{ij} denotes the elementary matrix with a 1 in the i, j position and 0's elsewhere, then ι_i is just the embedding

$$g \mapsto \begin{pmatrix} I_{i-1} & & \\ & g & \\ & & I_{r-i-1} \end{pmatrix}, \quad 1 \leqslant i \leqslant r - 1.$$

It is not possible to arrange the map $s_i \mapsto \overline{s}_i$ to extend to a homomorphism $W \to G$, since this already fails when $G = \text{SL}(2)$. However, the following is a useful substitute.

Lemma 15.14. *It is possible to choose the representative $\overline{w}$ in the normalizer of T of the Weyl group element w for all w so that $\overline{ww'} = \overline{w} \cdot \overline{w'}$ if $\ell(w) + \ell(w') = \ell(ww')$.*

Proof. We want to define $\overline{w} = \overline{s}_{i_1} \cdots \overline{s}_{i_k}$ if $w = s_{i_1} \cdots s_{i_k}$ is a reduced expression. By Matsumoto's theorem, this is well defined in view of (15.9). □

Because the algebraic group G is defined over $\mathbb{Z}$, we may arrange that the Weyl group representatives $\overline{w}$ are in $G(\mathcal{O})$.

Let $\mathcal{K} = \mathbb{R}((\tau))$ be the field of formal power series in an indeterminate τ and let $\mathcal{O} = \mathbb{R}[[\tau]]$. Let $\mathcal{K}^+$ be the subset of the field $\mathcal{K} = \mathbb{R}((\tau))$ of formal power series in an indeterminate τ that are either zero, or whose leading coefficient is positive. Also let $\mathcal{O}^+ = \mathcal{O} \cap \mathcal{K}^+$.

The set $\mathcal{K}^+$ is closed under addition, multiplication and division, but not subtraction, so like $\mathbb{T}$ it is a semi-ring. The *valuation map* ord$: \mathcal{K}^+ \to \mathbb{Z} \cup \{\infty\}$ is defined by ord $\left(\sum a_k \tau^k \right) = n$, where n is the smallest integer such that $a_n \neq 0$; while ord$(0) = \infty$. Identify $\mathbb{Z} \cup \{\infty\}$ with its image in the tropical ring $\mathbb{T} = \mathbb{R} \cup \{\infty\}$. Thus, we make use of the tropical operations and relabel ∞ as $\mathbb{O}$ and 0 as $\mathbb{1}$.

Lemma 15.15. *The valuation map* ord$: \mathcal{K}^+ \to \mathbb{T}$ *is a homomorphism of semi-rings.*

Proof. The map ord satisfies

$$\text{ord}(f + g) = \min\big(\text{ord}(f), \text{ord}(g)\big) = \text{ord}(f) \oplus \text{ord}(g)$$

because when we add two nonzero elements of $\mathcal{K}^+$, the leading coefficients, both being positive, can never cancel. Also clearly ord$(fg) = \text{ord}(f) \,\text{ord}(g) = \text{ord}(f) \oplus \text{ord}(g)$. □

Lemma 15.16. *The group* $T(\mathcal{K})/T(\mathcal{O})$ *is isomorphic to the weight lattice* Λ.

Proof. Recall that $\Lambda \cong \text{Hom}(\mathbb{G}_m, T)$. We associate with $\mu \in \text{Hom}(\mathbb{G}_m, T)$ the coset $\mu(\tau)T(\mathcal{O})$. It is easy to see that this map $\text{Hom}(\mathbb{G}_m, T) \to T(\mathcal{K})/T(\mathcal{O})$ is an isomorphism. □

The valuation map gives another way of thinking about this isomorphism. The torus $T \cong \mathbb{G}_m^r$ for some r. Then the weight lattice is identified with $\mathbb{Z}^r = \text{Hom}(\mathbb{G}_m, \mathbb{G}_m)^r$, with the weight $\mu: \mathbb{G}_m \to T \cong \mathbb{G}_m^r$ being identical with the map $c \mapsto (c^{a_1}, \ldots, c^{a_r})$ for a unique r-tuple $(a_1, \ldots, a_r) \in \mathbb{Z}^r$. Now if $\mu(\tau)T(\mathcal{O})$ is the coset in $T(\mathcal{K})/T(\mathcal{O})$ associated with μ as in Lemma 15.16, then the integers $(a_1, \ldots, a_r)$ are the valuations of the components of $\mu(\tau)$.

We are now ready to define the rational map wt$: X \to T$, which will take the place of the weight map on the crystal. We must specify an open subset, which is to be the domain of wt. We will take this to be the image in $X = G/B^-$ of the intersection $B^- w_0 B^- \cap BB^-$. We define a map wt $: (B^- w_0 B^- \cap BB^-)/B^- \to T$ as follows. Recall that the Weyl group representatives, in particular $\overline{w}_0$, are in $G(\mathcal{O})$. Every coset in $(B^- w_0 B^- \cap BB^-)/B^-$ has a representative nB^-, where $n \in N^+$. Since $n \in B^- w_0 B^-$, we write $n\overline{w}_0^{-1} = n_- t n_+$, where $n_- \in N^-$ and $n_+ \in N^+$ and

$t \in T$. (Since w_0 has order two, both $\overline{w}_0$ and $\overline{w}_0^{-1}$ are representatives of w_0.) Then we define $\mathrm{wt}(nB^-) = t^{-1}$.

The notation wt for this rational map wt$\colon X \longrightarrow T$ will be justified by the fact that its tropicalization can be identified with the usual weight map wt$\colon \mathcal{L} \longrightarrow \Lambda$, where we are identifying Λ with $T(\mathcal{K})/T(\mathcal{O})$.

Let us see how this works for GL(3). Consider

$$
n = \begin{pmatrix} 1 & a & \\ & 1 & \\ & & 1 \end{pmatrix} \begin{pmatrix} 1 & & \\ & 1 & b \\ & & 1 \end{pmatrix} \begin{pmatrix} 1 & & c \\ & 1 & \\ & & 1 \end{pmatrix} \in B^- w_0 B^- \cap BB^-.
$$

We have $n\overline{w}_0^{-1} = n_- t n_+$ with

$$
\overline{w}_0 = \begin{pmatrix} & & 1 \\ & -1 & \\ 1 & & \end{pmatrix}, \qquad n_- = \begin{pmatrix} 1 & & \\ a^{-1} & 1 & \\ (ab)^{-1} & (a+c)/(bc) & 1 \end{pmatrix}, \qquad t = \begin{pmatrix} ab & & \\ & a^{-1}c & \\ & & (bc)^{-1} \end{pmatrix}
$$

and

$$
n_+ = \begin{pmatrix} 1 & -\frac{a+c}{ab} & (ab)^{-1} \\ & 1 & -c^{-1} \\ & & 1 \end{pmatrix}, \qquad \mathrm{wt}(n) = \begin{pmatrix} a^{-1}b^{-1} & & \\ & ac^{-1} & \\ & & bc \end{pmatrix}.
$$

Now tropicalizing the diagonal matrix wt(n) entry by entry gives

$$
(-a-b, a-c, b+c) = -(a+b)\alpha_1 - (b+c)\alpha_2,
$$

consistent with (15.3).

Theorem 15.17. *Let $\mathbf{i} = (i_1, \ldots, i_N)$ be a reduced word for w_0 and let $\mathfrak{N}$ be the subset of $N^+(\mathcal{O}^+)$ consisting of elements of the form $x_{i_1}(a_1) \cdots x_{i_N}(a_N)$ with a_k nonzero. Then $\mathfrak{N}$ does not depend on the choice of $\mathbf{i}$. There exists a surjective map $q\colon \mathfrak{N} \to \mathcal{L}$ such that if $n \in \mathfrak{N}$ and $v = q(n)$, then for any $\mathbf{i}$ we have*

$$
v_{\mathbf{i}} = \big(\mathrm{ord}(a_1), \ldots, \mathrm{ord}(a_N)\big), \qquad n = x_{i_1}(a_1) \cdots x_{i_N}(a_N). \tag{15.10}
$$

Identifying $T(\mathcal{K})/T(\mathcal{O})$ with Λ, the following diagram commutes:

$$
\begin{array}{ccc}
\mathfrak{N} & \xrightarrow{\ \mathrm{wt}\ } & T(\mathcal{K}) \\
{\scriptstyle q}\Big\downarrow & & \Big\downarrow{\scriptstyle \mathrm{proj}} \\
\mathcal{L} & \xrightarrow{\ \mathrm{wt}\ } & \Lambda.
\end{array} \tag{15.11}
$$

Proof. We may take (15.10) to be the definition of q, but we must check that v is independent of the choice of $\mathbf{i}$. By Matsumoto's theorem, it is sufficient to check that if $\mathbf{i}$ and $\mathbf{j}$ are related by (M1) or (M2) as in Section 15.3, then $v_{\mathbf{j}} = R_{\mathbf{i},\mathbf{j}}(v_{\mathbf{i}})$. We will discuss the case (M2). Thus $\mathbf{j} = (j_1, \ldots, j_N)$ differs from $\mathbf{i}$ by replacing a segment $(i_k, i_{k+1}, i_{k+2}) = (i, j, i)$ by $(j_k, j_{k+1}, j_{k+2}) = (j, i, j)$ and $j_m = i_m$ if $m \notin \{k, k+1, k+2\}$. Then

$$
n = x_{j_1}(b_1) \cdots x_{j_N}(b_N),
$$

where $b_m = a_m$ if $m \notin \{k, k+1, k+2\}$ and $(b_k, b_{k+1}, b_{k+2}) = \vartheta(a_k, a_{k+1}, a_{k+2})$ where ϑ is the map (15.1). Indeed, this follows from the identity $x_i(a_k)x_j(a_{k+1})x_i(a_{k+2}) = x_j(b_k)x_i(b_{k+1})x_j(b_{k+2})$ in the subgroup of N^+ generated by the root groups x_i and x_j, since that subgroup is isomorphic to the maximal unipotent subgroup of $SL(3)$, so we may use the calculation in Section 15.1. Let $\mathcal{R}_{i,j} \colon \mathbb{A}^N \to \mathbb{A}^N$ be the map in Proposition 15.7. We have proved that $(b_1, \ldots, b_N)$ is obtained from $(a_1, \ldots, a_N)$ by applying $\mathcal{R}_{i,j}$, and it follows from Lemma 15.15 that v_j is obtained from v_i by applying the corresponding tropical map $\mathcal{R}_{i,j}$.

Now we may explain why $\mathfrak{N}$ is independent of the choice of $\mathbf{i}$. Given $n \in N^+(\mathcal{O}^+)$ with a representation (15.10) with a_i nonzero elements of $\mathcal{O}^+$, we may write n in the form (15.5), where *a priori* the b_k are known to be elements of $\mathcal{K}$. The maps $\mathcal{R}_{i,j}$ preserve $\mathcal{K}^+$, and so the b_k are in $\mathcal{K}^+$. Moreover, $(\mathrm{ord}(b_1), \ldots, \mathrm{ord}(b_N))$ are obtained from $(\mathrm{ord}(a_1), \ldots, \mathrm{ord}(a_N))$ by applying $\mathcal{R}_{i,j}$, and since this piecewise-linear map preserves $\mathbb{N}^N$, we have $\mathrm{ord}(b_i) \geqslant 0$, so the b_i are in $\mathcal{O}^+$.

The map q is thus well-defined and it is clear that it is surjective. We need to prove the commutativity of (15.11). We fix $\mathbf{i}$. Let $\gamma_k = \gamma_k(\mathbf{i})$ be defined by (15.4). Let us denote $y_k = s_{i_1} \cdots s_{i_{k-1}}$, so that $\gamma_i = y_k(\alpha_{i_k})$. For $g \in SL(2)$ define

$$\iota_{\gamma_k}(g) = \overline{y}_k \iota_k(g) \overline{y}_k^{-1}.$$

Since $y_k(\alpha_{i_k}) = \gamma_k$, this is the embedding of $SL(2)$ into G generated by the root subgroups corresponding to the roots γ_k and $-\gamma_k$. We observe that

$$n\overline{w}_0^{-1} = \prod_{k=1}^{N} \overline{y}_k x_{i_k}(a_k)\overline{y}_{k+1}^{-1}$$

where

$$\overline{y}_k x_{i_k}(a_k)\overline{y}_{k+1}^{-1} = \overline{y}_k x_{i_k}(a_k)\overline{s}_k^{-1}\overline{y}_k^{-1} = \overline{y}_k \iota_{i_k}\left(\begin{pmatrix} a_k & -1 \\ 1 & \end{pmatrix}\right)\overline{y}_k^{-1}.$$

That is,

$$\overline{y}_k x_{i_k}(a_k)\overline{y}_{k+1}^{-1} = \iota_{\gamma_k}\begin{pmatrix} a_k & -1 \\ 1 & \end{pmatrix} = \iota_{\gamma_k}\begin{pmatrix} 1 & \\ a_k^{-1} & 1 \end{pmatrix}\iota_{\gamma_k}\begin{pmatrix} a_k & \\ & a_k^{-1} \end{pmatrix}\iota_{\gamma_k}\begin{pmatrix} 1 & -a_k^{-1} \\ & 1 \end{pmatrix}.$$

Multiplying these together, we may move the upper triangular matrices to the left, the upper triangular matrices to the right and find that $n\overline{w}_0^{-1} = n_- t n_+$, where

$$t = \prod_k \iota_{\gamma_k}\begin{pmatrix} a_k & \\ & a_k^{-1} \end{pmatrix}.$$

Thus $\mathrm{wt}(nB_-)$ is the inverse of this matrix. Now under the isomorphism $T(\mathcal{K})/T(\mathcal{O}) \cong \Lambda$, the element $\iota_{\gamma_k}\begin{pmatrix} t & \\ & t^{-1} \end{pmatrix}$ is mapped to $\mathrm{ord}(t)\gamma_k$. Hence $\mathrm{wt}(nB_-)$ projects onto $-\sum_k a_k\gamma_k$. The commutativity of (15.11) now follows from (15.6). $\square$

15.6   MV polytopes: The A_2 case

We will show in Section 15.8 that the Lusztig basis is encoded in certain polytopes in the ambient vector space of the weight lattice. These are called *MV polytopes* after Mirković and Vilonen, though they were actually introduced in [Anderson (2000)]. Every MV polytope P has a lowest weight vector λ_{low} and a highest weight vector λ_{high}. Any translate of P has the same meaning as P, so λ_{low} and λ_{high} may be chosen at our convenience. In this section we will (after some preliminaries) choose $\lambda_{\text{low}} = 0$; later we will make a different choice.

Let $v \in \mathcal{B}_\infty$, and let $w \in W$ have length r. We choose a reduced word $(i_1, \ldots, i_r)$ for $w = s_{i_1} \cdots s_{i_r}$. We complete this to a reduced word $\mathbf{i} = (i_1, \ldots, i_N) \in \text{Red}(w_0)$ for w_0. Let $w_k = s_{i_1} \cdots s_{i_k}$ so that in (15.4) we have $\gamma_k = w_{k-1}(\alpha_{i_k})$. Generalizing (15.6), let us define

$$\text{wt}(v, w) = \lambda_{\text{low}} + \sum_{j=1}^{r} a_j \gamma_j. \tag{15.12}$$

This is independent of the choice of $\mathbf{i}$ by the same argument as in Lemma 15.8. Clearly, $\text{wt}(v, 1) = \lambda_{\text{low}}$ while $\text{wt}(v, w_0) = \lambda_{\text{low}} - \text{wt}(v)$.

We will prove in Section 15.8 that these weights are the vertices of a convex polytope with remarkable properties, called the MV polytope, which we will denote $\text{MV}(v)$. In this section, we confirm this for type A_2.

Let $v_{\mathbf{i}} = (a, b, c)$ with $\mathbf{i} = (1, 2, 1)$ and $v_{\mathbf{i}'} = (a', b', c')$ with $\mathbf{i}' = (2, 1, 2)$. These are the only two reduced words for w_0 and we have

$$\gamma(\mathbf{i}) = (\alpha_1, \alpha_1 + \alpha_2, \alpha_2), \qquad \gamma(\mathbf{i}') = (\alpha_2, \alpha_1 + \alpha_2, \alpha_1).$$

Thus taking $\lambda_{\text{low}} = 0$, we obtain the weights in Table 15.1.

Table 15.1   The vertices of $\text{MV}(v)$

w	$\text{wt}(v, w) - \lambda_{\text{low}}$ (from $\mathbf{i}$)	w	$\text{wt}(v, w) - \lambda_{\text{low}}$ (from $\mathbf{i}'$)
1	0	1	0
s_1	$a\alpha_1$	s_2	$a'\alpha_2$
$s_1 s_2$	$(a + b)\alpha_1 + b\alpha_2$	$s_2 s_1$	$(a' + b')\alpha_2 + b\alpha_1$
w_0	$(a + b)\alpha_1 + (b + c)\alpha_2$	w_0	$(a' + b')\alpha_2 + (b' + c')\alpha_1$

We have $(a', b', c') = \vartheta(a, b, c)$ with ϑ as in (15.2), which ensures the necessary consistency $(a + b)\alpha_1 + (b + c)\alpha_2 = (a' + b')\alpha_2 + (b' + c')\alpha_1$.

To obtain a closer look at the polytope, we distinguish two cases: $a \geqslant c$ and $a \leqslant c$. In the first case, $(a', b', c') = (b, c, a + b - c)$, while in the second case, $(a', b', c') = (b + c - a, a, b)$. Thus we have the two cases in Table 15.2.

The MV polytopes in the two cases are illustrated in Figure 15.1. There are degenerate situations where one or more of the six edges (each parallel to a positive root) collapses to length zero. More generally, a polytope in the ambient vector space of the weight lattice that is a translate of one of these by a weight is called an *MV polytope* for Cartan type A_2. See Definition 15.23 below for the general definition.

Table 15.2   Left: the case $a \geqslant c$.

w	$\mathrm{wt}(v, w) - \lambda_{\mathrm{low}}$
1	0
s_1	$a\alpha_1$
s_2	$b\alpha_2$
$s_1 s_2$	$(a + b)\alpha_1 + b\alpha_2$
$s_2 s_1$	$c\alpha_1 + (b + c)\alpha_2$
w_0	$(a + b)\alpha_1 + (b + c)\alpha_2$

Right: the case $a \leqslant c$.

w	$\mathrm{wt}'(v, w) - \lambda_{\mathrm{low}}$
1	0
s_1	$a\alpha_1$
s_2	$(b + c - a)\alpha_2$
$s_1 s_2$	$(a + b)\alpha_1 + b\alpha_2$
$s_2 s_1$	$a\alpha_1 + (b + c)\alpha_2$
w_0	$(a + b)\alpha_1 + (b + c)\alpha_2$

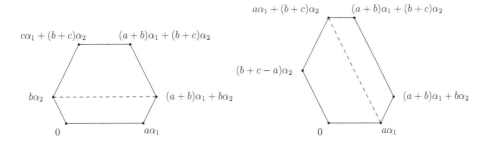

Fig. 15.1   MV Polytopes for A_2. Left: $a \geqslant c$; Right $a \leqslant c$. The vertex v is labeled by $\mathrm{wt}(v) - \lambda_{\mathrm{low}}$.

Remark 15.18. We may now observe (in the A_2 case) an important property of the MV polytopes. The polytope corresponding to $v \in \mathcal{B}_\infty$ has a lowest weight λ_{low} and a highest weight λ_{high}. Now given any reduced word $\mathbf{i} = (i_1, \dots, i_N)$ let $v_{\mathbf{i}} = (a_1, \dots, a_N)$ be the corresponding view. Then the path connecting the sequence of weights

$$\lambda_{\mathrm{low}}, \quad \lambda_{\mathrm{low}} + a_1 \gamma_1(\mathbf{i}), \quad \lambda_{\mathrm{low}} + a_1 \gamma_1(\mathbf{i}) + a_2 \gamma_2(\mathbf{i}), \quad \dots$$

is a path from the lowest weight to the highest weight, around the edge of the polytope. The lengths of the segments of this path are $a_1, a_2, \dots$. Thus the polytope encodes in its dimensions every view in the Lusztig parametrization. In Figure 15.1, the path corresponding to $\mathbf{i} = (1, 2, 1)$ proceeds around the right edge of the polytope, and the lengths of the three sides (giving each positive root length 1) are a, b, c. The other path corresponding to $\mathbf{i} = (2, 1, 2)$ proceeds around the other side, and the lengths of the three sides are a', b', c'. Since

$$a\alpha_1 + b(\alpha_1 + \alpha_2) + c\alpha_2 = a'\alpha_2 + b'(\alpha_1 + \alpha_2) + c'\alpha_1,$$

both paths arrive at the same terminus. See also Figure 15.2.

15.7   Tropical Plücker relations

In this section and the next, we will impose the simplifying assumption that the weight lattice is of semisimple type. Before we consider MV polytopes in this setting, let us define a more general class of polytopes, the *generalized Weyl polytope*.

Let $\Lambda^\vee$ be the lattice of *coweights*, which is the dual lattice to Λ. In $\mathbb{Q} \otimes \Lambda^\vee$ we choose vectors $\varpi_i^\vee$ for $i \in I = \{1, \dots, r\}$ such that $\langle \alpha_i, \varpi_j^\vee \rangle = \delta_{ij}$, called *fundamental coweights*. A coweight of the form $w(\varpi_i^\vee)$ for w in the Weyl group W is called a *chamber coweight*. Let

$$\mathrm{CW} = \{ w(\varpi_i^\vee) \mid w \in W, i \in I \}$$

be the set of chamber coweights.

There exists a permutation $i \mapsto i'$ of the index set I such that $-w_0 \alpha_i = \alpha_{i'}$.

Lemma 15.19. *If $\nu^\vee$ is a chamber coweight, then so is $-\nu^\vee$.*

Proof. Indeed, if $\nu^\vee = w \varpi_i^\vee$, then we claim that $w w_0 \varpi_{i'}^\vee = -\nu^\vee$. It is enough to show that $w_0 \varpi_{i'}^\vee = -\varpi_i$. If j is another index, then $\langle \alpha_j, w_0 \varpi_{i'}^\vee \rangle = \langle w_0 \alpha_j, \varpi_{i'}^\vee \rangle = \langle -\alpha_{j'}, \varpi_{i'}^\vee \rangle = -\delta_{i,j}$. Since we are assuming that the weight lattice is semisimple, the statement follows. □

Now let us fix a collection $M_\bullet$ of integers $M_{\nu^\vee}$ for $\nu^\vee \in \mathrm{CW}$. Let

$$P(M_\bullet) = \{ x \in \mathbb{R} \otimes \Lambda \mid \langle x, \nu^\vee \rangle \geqslant M_{\nu^\vee} \text{ for } \nu^\vee \in \mathrm{CW} \}.$$

For $w \in W$, there is a unique $\mu_w \in \mathbb{R} \otimes \Lambda$ such that

$$\langle \mu_w, w \varpi_i^\vee \rangle = M_{w \varpi_i^\vee}. \tag{15.13}$$

Indeed, these equations put μ_w on the intersection of $|I|$ independent hyperplanes in a real vector space of the same dimension, and this intersection consists of a single point. Define μ_w to be the unique solution to (15.13). Later in Theorem 15.25 we will show that $\mu_w = \mathrm{wt}(v, w)$, but until we prove this we will need two separate notations.

We make one assumption about the $M_{\nu^\vee}$.

Assumption 15.1. We assume that μ_w be vertices of the polytope $P(M_\bullet)$.

We do not assume that the μ_w are distinct.

Let us consider how Assumption 15.1 could fail. If one of the $M_{\nu^\vee}$ is very small (negative), then the hyperplane $M_{\nu^\vee} = \langle x, \nu^\vee \rangle$ could completely miss the polytope (as determined by the inequalities for the other chamber coweights). This will cause those μ_w that lie on this hyperplane not to lie on the polytope. The meaning of Assumption 15.1 is that this does not occur.

On Assumption 15.1, the data $\{\mu_w\}$ (including the parametrization of this set by the Weyl group) is called a *GGMS datum* (for Gelfand, Goresky, MacPherson and Serganova). Then $P(M_\bullet)$ is called a *generalized Weyl polytope*. As an example, if λ is dominant, then the convex hull of the Weyl group orbit $W\lambda$ is a generalized Weyl polytope. A polytope $P(M_\bullet)$ of this form is called a *Weyl polytope*.

Proposition 15.20. *If Assumption 15.1 is satisfied, then*

$$M_{w \varpi_i^\vee} + M_{w s_i \varpi_i^\vee} \leqslant \sum_{j \neq i} -\langle \alpha_j, \alpha_i^\vee \rangle M_{w \varpi_j^\vee}. \tag{15.14}$$

The inequalities (15.14) are called the *edge inequalities*. Observe that $-\langle \alpha_j, \alpha_i^\vee \rangle$ is a nonnegative integer.

Proof. We observe that

$$w\varpi_i^\vee + ws_i\varpi_i^\vee = \sum_{j \neq i} -\langle \alpha_j, \alpha_i^\vee \rangle w\varpi_j^\vee.$$

To prove this, we may assume $w = 1$. Then, since Λ is assumed to be of semisimple type, it is sufficient to prove that both sides have the same inner product with each α_j. If $j = i$, then the inner product of either side of this identity is zero. If $j \neq i$, the inner product of either side is $-\langle \alpha_j, \alpha_i^\vee \rangle$.

Now we take the inner product with μ_w. Using (15.13) and the fact that since $\mu_w \in P(M_\bullet)$ we have $M_{ws_i\varpi_i^\vee} \leqslant \langle \mu_w, ws_i\varpi_i^\vee \rangle$, we obtain

$$M_{w\varpi_i^\vee} + M_{ws_i\varpi_i^\vee} \leqslant \langle \mu_w, w\varpi_i^\vee \rangle + \langle \mu_w, ws_i\varpi_i^\vee \rangle$$
$$= \sum_{j \neq i} -\langle \alpha_j, \alpha_i^\vee \rangle \langle \mu_w, w\varpi_j^\vee \rangle = \sum_{j \neq i} -\langle \alpha_j, \alpha_i^\vee \rangle M_{w\varpi_j^\vee},$$

proving (15.14). $\square$

An MV polytope is a generalized Weyl polytope, but not every generalized Weyl polytope is an MV polytope. Let us illustrate this in the A_2 case, where we have already computed the MV polytopes. We see in Figure 15.1 that there are two types of MV polytopes, both characterized by the fact that a diagonal (indicated with a dashed line) is parallel to one of the roots α_1 or α_2. We would like to abstract this property in a way that will generalize.

This is accomplished by the *tropical Plücker relations* due to [Berenstein and Zelevinsky (1996)]. We say that the *tropical Plücker relations* are satisfied if whenever $ws_i > w$ and $ws_j > w$, where $\langle \alpha_i, \alpha_j^\vee \rangle = -1$, we have

$$M_{ws_i\varpi_i^\vee} + M_{ws_j\varpi_j^\vee} = \min(M_{w\varpi_i^\vee} + M_{ws_is_j\varpi_j^\vee}, M_{w\varpi_j^\vee} + M_{ws_js_i\varpi_i^\vee}). \tag{15.15}$$

If $\langle \alpha_i, \alpha_j^\vee \rangle = 0$, there is no condition. We are assuming that the root system is simply-laced, so all the roots have the same length. In the general situation, α_i and α_j may have different lengths, in which case they generate a rank two root system of type C_2 or G_2. Then the tropical Plücker relation must be modified. See [Kamnitzer (2010)] for more information.

Lemma 15.21. *If $P(M_\bullet)$ satisfies the tropical Plücker relations, so does the translate $P(M_\bullet) + \lambda$ for any $\lambda \in \Lambda$.*

Proof. We see that $P(M_\bullet) + \lambda = P(M'_\bullet)$, where

$$M'_{\nu^\vee} = M_{\nu^\vee} + \langle \lambda, \nu^\vee \rangle.$$

Thus it is sufficient to show that

$$\langle \lambda, ws_i\varpi_i^\vee \rangle + \langle \lambda, ws_j\varpi_j^\vee \rangle = \langle \lambda, w\varpi_i^\vee \rangle + \langle \lambda, ws_is_j\varpi_j^\vee \rangle = \langle \lambda, w\varpi_j^\vee \rangle + \langle \lambda, ws_js_i\varpi_i^\vee \rangle,$$

and after replacing λ by $w\lambda$, we may assume that $w = 1$. We will actually show that

$$s_i \varpi_i^\vee + s_j \varpi_j^\vee = \varpi_i^\vee + s_i s_j \varpi_j^\vee = \varpi_j^\vee + s_j s_i \varpi_i^\vee.$$

We first note that $\varpi_i^\vee + s_i s_j \varpi_j^\vee$ is orthogonal to α_i and α_j. Indeed,

$$\langle \alpha_i, \varpi_i^\vee + s_i s_j \varpi_j^\vee \rangle = \langle \alpha_i, \varpi_i^\vee \rangle + \langle s_j s_i \alpha_i, \varpi_j^\vee \rangle = \langle \alpha_i, \varpi_i^\vee \rangle + \langle -\alpha_i - \alpha_j, \varpi_j^\vee \rangle = 1 - 1 = 0,$$

and similarly for α_j. Since applying any element of the group of order 6 generated by s_i and s_j to a vector that is orthogonal to α_i and α_j leaves it unmoved, the statement follows.   □

In Proposition 15.22 below, we will confirm the tropical Plücker relation for type A_2. Since s_1 and s_2 are to be right ascents of w, the only case to check is $w = 1$. Thus we need to show that

$$M_{s_1 \varpi_1^\vee} + M_{s_2 \varpi_2^\vee} = \min(M_{\varpi_1^\vee} + M_{s_1 s_2 \varpi_2^\vee}, M_{\varpi_2^\vee} + M_{w s_2 s_1 \varpi_1^\vee}). \tag{15.16}$$

We embed the A_2 root system in Example 2.5. The polytopes that we consider are the polytopes from Figure 15.1, optionally translated by an element λ of Λ.

Proposition 15.22. *A translate of one of the polytopes in Figure 15.1 satisfies the tropical Plücker relation (15.16).*

Proof. Although we are asserting the tropical Plücker relations for a translate of one of the polytopes in Figure 15.1, by Lemma 15.21 we may omit the translation and work with the polytope in Figure 15.1.

Define the data $M_\bullet$ as in Table 15.3. With this data, it is easy to check that (15.16) is satisfied.

The reader will easily check in both cases $a \geqslant c$ or $a \leqslant c$ that the six lines $\langle x, \nu^\vee \rangle = M_{\nu^\vee}$ are the edges of the first hexagon in Figure 15.1 and that the intersection of the half-planes $\langle x, \nu^\vee \rangle \geqslant M_{\nu^\vee}$ is its interior. Hence this hexagon is $P(M_\bullet)$.   □

Definition 15.23. An MV polytope in $\mathbb{R} \otimes \Lambda$ is a generalized Weyl polytope $P(M_\bullet)$ that satisfies the tropical Plücker relations (15.15).

It follows from Proposition 15.22 that any translate of one of the polytopes in Figure 15.1 is an MV polytope.

We digress to explain the term *tropical Plücker relation*. For the $GL(r)$ flag variety, the Plücker coordinates are in bijection with the chamber weights. Indeed, choose a matrix $g \in GL(r)$ that may be projected into the flag variety $X = G/B$. Let S be a proper nonempty subset of $\{1, \ldots, r\}$, and let p_S be the minors formed with g_{ij} where $i \in I$ and $r - |S| < j \leqslant r$. When $|S| = |S'|$ the ratio $p_S/p_{S'}$ is constant on the coset gB, so these ratios are functions on X, and the Plücker coordinates p_S are homogenous coordinates on X. On the other hand, S corresponds to the weight $\sum_{i \in S} \mathbf{e}_i$, and these are exactly the chamber weights.

Table 15.3 The values of $M_\bullet$ for A_2.

$\nu^\vee$	$M_{\nu^\vee}$
$\varpi_1^\vee$	0
$\varpi_2^\vee$	0
$s_2\varpi_2^\vee$	$-b - c + \min(a, c)$
$s_1\varpi_1^\vee$	$-a$
$s_1 s_2\varpi_2^\vee$	$-a - b$
$s_2 s_1\varpi_1^\vee$	$-b - c$

The Plücker relations between the homogenous coordinates include three term relations such as $p_{\{1\}}p_{\{2,3\}} - p_{\{2\}}p_{\{1,3\}} + p_{\{3\}}p_{\{1,2\}} = 0$ on GL(3). The relations (15.15) are obtained by tropicalizing the three-term Plücker relations.

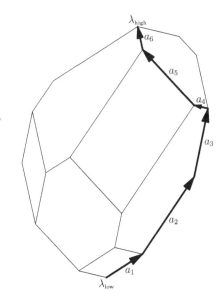

Fig. 15.2 An A_3 MV polytope. Corresponding to the 14 reduced words for w_0, there are 14 paths from the lowest weight λ_{low} to the highest weight λ_{high} in the polytope; the one shown corresponds to the reduced word $\mathbf{i} = (1, 2, 3, 2, 1, 2)$. The lengths of the edges are the components of the view $v_{\mathbf{i}} = (a_1, \ldots, a_6)$, and the directions are the positive roots $\alpha_1, s_1(\alpha_2) = \alpha_1 + \alpha_2, \ldots$ in the order given by (15.4). **Caution**: in contrast with most figures in this book, the arrows are not f_i operators. They are simply drawn here to indicate a path from λ_{low} to λ_{high}.

15.8 The crystal structure on MV polytopes

We will consider MV polytopes to be elements of crystals. For this purpose, any translate of a polytope may be considered to be equivalent to any other. It will be convenient, however, to arrange so that if λ is the highest weight element of the crystal, then every MV polytope has $\lambda_{\text{high}} = \lambda$. Let $\mathcal{L}$ be the crystal described in

Section 15.3, and let λ_{low} be a fixed element of Λ. We will associate with every $v \in \mathcal{L}$ and λ_{low} an MV polytope $\mathrm{MV}(v) = \mathrm{MV}(v, \lambda_{\text{low}})$ such that, for every reduced word $\mathbf{i} = (i_1, \ldots, i_N)$ for w_0, the components in the view $v_{\mathbf{i}} = (a_1, \ldots, a_N)$ are the lengths of the edges of a path from 0 to $\mathrm{wt}(v)$ in the weight lattice. This is illustrated in Figure 15.2.

We characterize $\mathrm{MV}(v)$ by specifying its vertices, which were already defined in (15.12). We thus define $\mathrm{MV}(v)$ to be the convex hull of the weights $\mathrm{wt}(v, w)$.

Lemma 15.24. *Let $w, w' \in W$ and $v \in \mathcal{L}$. Then*

$$w^{-1} \mathrm{wt}(v, w) \preccurlyeq w^{-1} \mathrm{wt}(v, w'), \tag{15.17}$$

in terms of the partial order $\succcurlyeq$ defined in Section 2.1.

Proof. First we show that for any $w \in W$ and any simple reflection s_i, we have

$$\mathrm{wt}(v, ws_i) - \mathrm{wt}(v, w) = cw(\alpha_i), \qquad c \geqslant 0. \tag{15.18}$$

We may assume $ws_i > w$, since if $ws_i < w$ then interchanging w and ws_i, and remembering that $s_i(\alpha_i) = -\alpha_i$, we reduce to the other case. Now we pick a reduced word $(i_1, \ldots, i_k)$ for w. Then $(i_1, \ldots, i_k, i)$ is a reduced word for ws_i, which we complete to a reduced word $\mathbf{i}$ for w_0. Let $(a_1, \ldots, a_N) = v_{\mathbf{i}}$. Applying the definition (15.12), we see that $\mathrm{wt}(v, ws_i) - \mathrm{wt}(v, w) = a_{k+1}\gamma_{k+1}$, where $\gamma_{k+1} = w(\alpha_i)$ by (15.4). This proves (15.18).

We may now prove (15.17). Let $w^{-1}w' = s_{j_1} \cdots s_{j_\ell}$ be a reduced decomposition. Then applying (15.18) repeatedly,

$$w^{-1}\big(\mathrm{wt}(v, w') - \mathrm{wt}(v, w)\big) = \sum_{j=1}^{\ell} w^{-1}\big(\mathrm{wt}(v, ws_{i_1} \cdots s_{i_j}) - \mathrm{wt}(v, ws_{i_1} \cdots s_{i_{j-1}})\big)$$

is a linear combination with nonnegative integer coefficients of $\alpha_{i_1}, s_{i_1}(\alpha_{i_2}), \ldots$ and by Proposition 15.3 these are positive roots. The statement follows. $\square$

Theorem 15.25. *The polytope $\mathrm{MV}(v)$ is an MV polytope. If $w \in W$, then $\mu_w = \mathrm{wt}(v, \lambda)$.*

Proof. First let us show that $\mathrm{MV}(v)$ is a generalized Weyl polytope. We will define integers $M_{\nu^\vee}$ for chamber coweights $\nu^\vee$ such that $\mathrm{MV}(v) = P(M_\bullet)$.

Let $\nu^\vee = w\varpi_i^\vee$ be a chamber coweight. We have $\langle \mathrm{wt}(v, w'), \nu^\vee \rangle = \langle w^{-1}\mathrm{wt}(v, w'), \varpi_i^\vee \rangle$. Since $\varpi_i^\vee$ is nonnegative on the positive roots, it follows from Lemma 15.24 that $\langle \mathrm{wt}(v, w'), \nu^\vee \rangle$ is minimized on the set $S_{\nu^\vee}$ of w' such that $\nu^\vee = w'\varpi_i^\vee$. This is a coset in W of the stabilizer of $\varpi_i^\vee$. The convex hull $F_{\nu^\vee}$ of the set of weights $\mathrm{wt}(v, w')$ with $w' \in S_{\nu^\vee}$ is contained in a hyperplane $H_{\nu^\vee}$ orthogonal to $\nu^\vee$. If $M_{\nu^\vee}$ is the constant value of $\langle \mathrm{wt}(x), \nu' \rangle$ on this hyperplane, then we see that $\langle x, \nu^\vee \rangle \geqslant M_{\nu^\vee}$ for $x \in \mathrm{MV}(v)$, with equality on $H_{\nu^\vee}$. Thus $F_{\nu^\vee}$ is a face (possibly degenerate) of the convex hull of the $\mathrm{wt}(v, w)$. Thus $\mathrm{MV}(v)$ is a generalized Weyl polytope.

Next we prove the tropical Plücker relations. Let $\langle \alpha_i, \alpha_j^\vee \rangle = -1$, and let w be such that $ws_i > w$ and $ws_j > w$. This assumption implies that $\ell(ws_is_js_i) = \ell(w) + 3$ and that $s_is_js_i = s_js_is_j$. We choose a reduced word $(i_1, \ldots, i_k)$ for w. Then $(i_1, \ldots, i_k, i, j, i)$ and $(i_1, \ldots, i_k, j, i, j)$ are two reduced words for $ws_is_js_i$. We complete these to reduced words $\mathbf{i}$ and $\mathbf{j}$ of w_0. Let $(a_1, \ldots, a_N) = v_{\mathbf{i}}$ and $(b_1, \ldots, b_N) = v_{\mathbf{j}}$. Then by definition of the map $\mathcal{R}_{\mathbf{i},\mathbf{j}}$ in Section 15.3 we have $(b_{k+1}, b_{k+2}, b_{k+3}) = \vartheta(a_{k+1}, a_{k+2}, a_{k+3})$, while $a_i = b_i$ if $i \notin \{k+1, k+2, k+3\}$. We denote $(a, b, c) = (a_{k+1}, a_{k+2}, a_{k+3})$ and $(a', b', c') = (b_{k+1}, b_{k+2}, b_{k+3}) = \vartheta(a, b, c)$.

Now the root system Φ contains an A_2 root system spanned by α_i and α_j. In the notation of Section 15.1, (a, b, c) is an element of the $\mathcal{B}_\infty$ crystal with the Lusztig parametrization for this A_2 root system. Since $\vartheta(a, b, c) = (a', b', c')$, the weights

$$
\begin{array}{ccc}
0, & a\alpha_1, & a\alpha_1 + b(\alpha_1 + \alpha_2), \\
a\alpha_1 + b(\alpha_1 + \alpha_2) + c\alpha_2 & a'\alpha_2 & a'\alpha_2 + b'(\alpha_1 + \alpha_2)
\end{array}
\tag{15.19}
$$

are the vertices of an MV polytope $P(M'_\bullet)$ for the A_2 root system, where $M'_\bullet$ has values given in Table 15.3. The affine transformation $x \mapsto wx + \mathrm{wt}(v, w)$ takes the six weights in (15.19) to the six weights $\mathrm{wt}(v, wy)$, where y runs through the A_2 Weyl group generated by s_i and s_j. Thus by Proposition 15.22 the restrictions of the hyperplanes pulling back the hyperplanes $\langle x, \nu^\vee \rangle = M_{\nu^\vee}$ under this map bound an A_2 MV polytope. Hence the tropical Plücker relation can be reduced to the A_2 case.

Now let us prove that $\mathrm{wt}(w, v) = \mu_w$. We will prove this at first for one particular MV polytope, which is the Weyl polytope whose vertices are at the W-orbit of the Weyl vector ρ. Then $\lambda_{\mathrm{low}} = -\rho$ while $\lambda_{\mathrm{high}} = \rho$. This Weyl polytope is realized as an MV polytope by taking $M_{\nu^\vee} = -1$ for every chamber coweight $\nu^\vee$. For this particular MV polytope, we will prove that $\mathrm{wt}(w, v)$ and μ_w both equal $w(-\rho)$. Given $w \in W$, choose a reduced word $\mathbf{i} = (i_1, \ldots, i_k)$ of w and complete it to a reduced word $(i_1, \ldots, i_N)$ of w_0. It is easy to see that for the element v of $\mathcal{L}$ that produces this Weyl polytope as its MV polytope, we have $v_{\mathbf{i}} = (1, \ldots, 1)$ for every $\mathbf{i}$. Hence

$$
\mathrm{wt}(w, v) = \lambda_{\mathrm{low}} + \gamma_1(\mathbf{i}) + \cdots + \gamma_k(\mathbf{i}) = -\rho + \sum_{j=1}^{k} \gamma_j(\mathbf{i}).
$$

Now $\gamma_1(\mathbf{i}), \ldots, \gamma_k(\mathbf{i})$ are simply the positive roots α such that $w^{-1}(\alpha)$ is a negative root. Therefore

$$
\mathrm{wt}(w, v) = -\rho + \sum_{\substack{\alpha \in \Phi^+ \\ w^{-1}(\alpha) \in \Phi^-}} \alpha = w(-\rho).
$$

On the other hand, this Weyl polytope is invariant under the action of W on Λ, and the action must be compatible with the action on chamber coweights. From this we deduce that $\mu_w = w(\mu_1) = w(-\rho)$.

We have verified that for one particular $P(M_\bullet)$ having distinct vertices, we have $\mathrm{wt}(w, v) = \mu_w$ for all w. Now we wish to deform this polytope through the space of data $M_\bullet$ that satisfy the edge and tropical Plücker relations. For this variation, we do not require that the $M_\bullet$ be integers, so the vertices μ_w and $\mathrm{wt}(w,v)$ of the polytope will be in $\mathbb{R} \otimes \Lambda$ but not necessarily Λ. The vectors $v_i = (a_1, \ldots, a_N)$ are not required to be integers, but we still have $v_j = \mathcal{R}_{i,j}(v_i)$. The vertices of the polytopes are parametrized as both μ_w and $\mathrm{wt}(w,v)$, and both vary continously with the data $M_\bullet$. Thus the identity $\mathrm{wt}(w,v) = \mu_w$ remains valid as we deform the polytope. We may then specialize it to the one we want.   □

Now let $\mathcal{B}$ be either $\mathcal{B}_\infty$ or $\mathcal{B}_\lambda$, where λ is a dominant weight. If $\mathcal{B} = \mathcal{B}_\infty$, we take $\lambda = 0$. In either case, $\mathcal{B}$ is a highest weight crystal with highest weight λ, and we may embed $\mathcal{B} \hookrightarrow \mathcal{T}_\lambda \otimes \mathcal{L}$. Suppose that $v \in \mathcal{B}$ and $t_\lambda \otimes u$ is the corresponding element of $\mathcal{T}_\lambda \otimes \mathcal{L}$, then for $\mathbf{i} \in \mathrm{Red}(w_0)$. We will denote $v_{\mathbf{i}} = u_{\mathbf{i}}$. We obtain an MV polytope $\mathrm{MV}(v)$ with vertices as in (15.12). We choose $\lambda_{\mathrm{low}} = \mathrm{wt}(v)$. This results in an MV polytope whose lowest weight is $\mathrm{wt}(v)$ and whose highest weight is λ.

Due to the crystal structure on $\mathcal{B}$ (or on $\mathcal{T}_\lambda \otimes \mathcal{L}$) we obtain a crystal structure on MV polytopes. Assume $f_i(v) \neq 0$. We consider the effect of f_i on the polytope $\mathrm{MV}(v)$. Choose the reduced word $\mathbf{i}$ so that $i_1 = i$. Then if $v_{\mathbf{i}} = (a_1, a_2, \ldots, a_N)$, by the definition of the crystal structure in $\mathcal{L}$ we have $(f_i v)_{\mathbf{i}} = (a_1 + 1, a_2, \ldots, a_N)$. This means that the face of $\mathrm{MV}(v)$ adjacent to the lowest weight vector whose bounding hyperplane is

$$\langle x, \varpi_i^\vee \rangle = M_{\varpi_i^\vee}, \qquad M_{\varpi_i^\vee} = \langle \mathrm{wt}(v), \varpi_i^\vee \rangle$$

is pushed out, increasing $M_{\varpi_i^\vee}$ by 1.

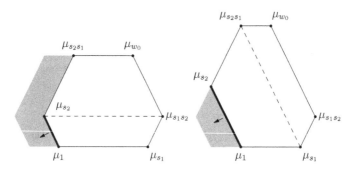

Fig. 15.3   Effect of f_1 on the MV polytope $\mathrm{MV}(v)$ with $v_{(1,2,1)} = (a,b,c)$. The shaded area is added to the polytope when f_1 is applied. The face that is pushed out in the indicated direction is marked. In the first case, this causes growth along another face in order to preserve the property of being an MV polytope. Left: The first case $a \geqslant c$. Right: the second case $a < c$.

It is worth considering how this works in the A_2 case. We may determine the effect of f_i from the data in Figure 15.1. Applying f_1 replaces (a, b, c) by $(a+1, b, c)$, and we follow this change by a translation by $-\alpha_1$ to keep the highest weight vertex at 0. We see that there are two cases, illustrated in Figure 15.3. Suppose that we start with a polytope $\mathrm{MV}(v)$ with $v_{(1,2,1)} = (a, b, c)$. If it is in the first case in

Figure 15.3, then we see that $\mathrm{MV}(f_1 v)$ is also, so applying f_1 repeatedly always gives back a polytope of the first case. On the other hand if we apply f_1 repeatedly to a polytope of the second case, it will transition to a polytope of the first case after $c - a$ applications of f_1.

15.9   The $\star$-involution

Let G be a reductive Lie group with maximal torus T and maximal unipotent N^+. The group G has an involution $\star: G \to G$ such that $(g_1 g_2)^\star = g_2^\star g_1^\star$ that preserves T and N^+, and that induces the bijection $\alpha \mapsto -w_0 \alpha$ of the positive roots. For example, if $G = \mathrm{GL}(r)$, the involution has the form

$$g^\star = \overline{w}_0 \cdot {}^t g \cdot \overline{w}_0^{-1},$$

where $\overline{w}_0$ is a representative of the long Weyl group element and ${}^t g$ is the matrix transpose.

As we will explain, the $\star$-involution on $\mathcal{B}_\infty$ introduced in Chapter 14 is the tropicalization of the antiautomorphism of N^+ induced by this antiautomorphism of G. Let us introduce a bijection $\star$ of $\mathcal{L}$ that is the tropicalization of the geometric map $\star$.

If $\mathbf{i} = (i_1, \ldots, i_N)$ is a reduced word for w_0, let $\mathbf{i}' = (i_N', \ldots, i_1')$, where $i \mapsto i'$ is the bijection of the index set defined by $\alpha_{i'} = -w_0 \alpha_i$.

Proposition 15.26. *There is weight-preserving bijection* $\star: \mathcal{L} \longrightarrow \mathcal{L}$ *such that if* $v_\mathbf{i} = (a_1, \ldots, a_N)$, *then*

$$v_{\mathbf{i}'}^\star = (a_N, \ldots, a_1). \tag{15.20}$$

Proof. We must show that with $v_\mathbf{i}^\star$ defined as in (15.20), then $\mathcal{R}_{\mathbf{i}', \mathbf{j}'}(v_{\mathbf{i}'}) = v_{\mathbf{j}'}$. It is sufficient to check this when $\mathbf{i}'$ and $\mathbf{j}'$ are related by (M1) or (M2). In this case, so are $\mathbf{i}$ and $\mathbf{j}$, so $\mathcal{R}_{\mathbf{i}, \mathbf{j}}(v_\mathbf{i}) = v_\mathbf{j}$. Thus we reduce to the fact that the map ϑ defined by (15.2) commutes with the reversal map $(a, b, c) \longmapsto (c, b, a)$.

We also have to check that $\mathrm{wt}(v^\star) = \mathrm{wt}(v)$. Indeed, this is easy to deduce from the fact that, with $\gamma_k(\mathbf{i})$ defined as in (15.4), we have

$$\gamma(\mathbf{i}') = \big(\gamma_N(\mathbf{i}), \ldots, \gamma_1(\mathbf{i})\big). \tag{15.21}$$

See Exercise 15.9. $\qquad\qquad\qquad\qquad\qquad\qquad\qquad\qquad\qquad\qquad\qquad\qquad\square$

Observe that this means that the $\star$-involution on $\mathcal{L}$ is the tropicalization of the geometric map $\star$. Indeed, the geometric map $\star$ has the effect on N^+ of sending

$$x_{i_1}(a_1) \cdots x_{i_N}(a_N) \longmapsto x_{i_N'}(a_N) \cdots x_{i_1'}(a_1)$$

and tropicalizing gives precisely the description in Proposition 15.26.

We have now defined $\star$ twice: once for $\mathcal{B}_\infty$ in Chapter 14 and once for $\mathcal{L}$ in this section.

Theorem 15.27. *If we identify $\mathcal{L}$ and $\mathcal{B}_\infty$, the two definitions of the involution are the same.*

We will not prove this, but see Exercise 15.8 for a proof in the A_2 case. See [Berenstein and Zelevinsky (2001), Proposition 3.3] for the general case, making use of the theory of quantum groups. And see Remark 15.31 below for an outline of how this can probably be proved combinatorially.

As in Chapter 14, we may use $\star$ to define a modified crystal structure on $\mathcal{L}$. Thus we define $\varepsilon_i^\star(v) = \varepsilon_i(v^\star)$, $\varphi_i^\star(v) = \varphi_i(v^\star)$, and $e_i^\star$, $f_i^\star$ are e_i and f_i conjugated by $\star$. Since the $\star$ map is weight-preserving, these give alternative crystal structures on $\mathcal{L}$.

We next reprove Proposition 14.9 using the definition of $\star$ from this chapter.

Lemma 15.28. *Suppose i and j are indices such that $j \neq i$. Then we may choose $\mathbf{i} \in \mathrm{Red}(w_0)$ such that $i_1 = i$ and $i_N = j'$.*

Proof. We claim that the length $\ell(s_i w_0 s_{j'}) = N - 2$. If not, then $\ell(s_i w_0 s_{j'}) = N$, so $s_i w_0 s_{j'} = w_0$, which implies that $s_{j'} = w_0 s_i w_0 = s_{i'}$. Since we are assuming that $i \neq j$, this is a contradiction. $\qquad\square$

Proposition 15.29. *If $i \neq j$, then the operators $e_i^\star$ and $f_i^\star$ on $\mathcal{L} \cup \{0\}$ commute with e_j and f_j. For $x \in \mathcal{L}$, we have $\varepsilon_j(f_i^\star(x)) = \varepsilon_j(x)$.*

Proof. We choose the reduced word as in Lemma 15.28. Then writing $v_{\mathbf{i}} = (a_1, \ldots, a_N)$, the effect of e_i and f_i is to decrease or increase a_1, while the effect of $e_i^\star$ and $f_i^\star$ is to decrease or increase a_N. These operations clearly commute. $\qquad\square$

The star involution has a simple effect on MV polytopes: up to a translation, $\mathrm{MV}(v^\star)$ is the negative of $\mathrm{MV}(v)$, that is $\mathrm{MV}(v^\star) = -\mathrm{MV}(v) = \{-x \mid x \in \mathrm{MV}(v)\}$, or (at our convenience) a translate of this polytope. See Exercise 15.6.

We can now describe the effect of $f_i^\star$ and $e_i^\star$ on $\mathrm{MV}(v)$. By Exercise 15.6, these are the conjugates of f_i and e_i by the map $x \mapsto -x$, followed at our convenience by a translation to keep the highest weight of the polytope at 0. In other words, where as in Figure 15.3 f_i pushes out a bottom face of $\mathrm{MV}(v)$, $f_i^\star$ pushes out a top face.

Proposition 15.30. *Let $v \in \mathcal{B}_\lambda$, and let $\mathrm{MV}(v)$ be the corresponding MV polytope with highest weight $\lambda_{\mathrm{high}} = \lambda$ and lowest weight $\lambda_{\mathrm{low}} = \mathrm{wt}(v)$. Then $\varepsilon_i(v)$ is the length of the edge coming out of λ_{low} in the α_i direction, and $\varepsilon_i^\star(v)$ is the length of the edge coming out of λ in the $-\alpha_i$ direction. In other words,*

$$\varepsilon_i(v) = \max\{k \mid \lambda_{\mathrm{low}} + k\alpha_i \in \mathrm{MV}(v)\}, \qquad \varepsilon_i^\star(v) = \max\{k \mid \lambda - k\alpha_i \in \mathrm{MV}(v)\}.$$

Proof. Let $\mathbf{i} = (i_1, \ldots, i_N)$ be a reduced word for w_0 chosen so that $i_1 = i$. Let $\gamma_i(\mathbf{i})$ be as in (15.4). The vertex adjacent to the lowest vertex λ_{low} in the α_i direction is

$$\mathrm{wt}(v, s_i) = \lambda_{\mathrm{low}} + a_1 \gamma_1(\mathbf{i}), \qquad v_{\mathbf{i}} = (a_1, \ldots, a_N).$$

Since $\gamma_1(\mathbf{i}) = \alpha_i$, the interpretation of $\varepsilon_i(v)$ as the length of the edge from λ_{low} to $\text{wt}(v, s_i)$ is clear. The corresponding statement for $\varepsilon_i^\star(v) = \varepsilon_i(v^\star)$ follows from Exercise 15.6. $\qquad\square$

Remark 15.31. Although we do not prove Theorem 15.27, let us make some remarks about how this can probably be proved combinatorially. Let $\mathcal{B}_i$ be the elementary crystal defined in Section 12.2. We may define a map $\psi_i \colon \mathcal{L} \to \mathcal{B}_i \otimes \mathcal{L}$ by imitating the definition in Chapter 14. That is, let

$$\psi_i(x) = u_i(-t) \otimes y,$$

where $t = \varepsilon_i^\star(x)$ and $y = e_i^{\star t}(x)$, using the definitions of $e_i^\star$ and $\varepsilon_i^\star$ in this chapter. What must be checked is that ψ_i is a crystal morphism. Indeed, since the crystal $\mathcal{L}$ is connected, it has a *unique* embedding into $\mathcal{B}_i \otimes \mathcal{L}$, and so if ψ_i is a morphism it must be the same as the morphism ψ_i in Chapter 14. This implies that $e_i^\star$ has the same meaning as in Chapter 14. Moreover, it follows from Proposition 15.29 that this map ψ_i commutes with e_j if $j \neq i$. We thus need to show that $\psi_i \circ e_i = e_i \circ \psi_i$. Using the definition of the crystal tensor product, this means that we need to check the following dichotomy. Let $x' = e_i(x)$, $t' = \varepsilon_i^\star(x')$, $y' = e_i^{\star t'}(x')$. Then we need to check that if $\varepsilon_i^\star(x) > \varphi_i(y)$ then $t' = t - 1$ and $y' = y$, while if $\varepsilon_i^\star(x) \leqslant \varphi_i(y)$, then $t' = t$ and $y' = e_i(y)$. Peter Tingley has suggested a way of doing this. In [Salisbury, Schultze and Tingley (2016a)] it is shown that particular reduced words, called *simply-braided* have properties that make them suitable for checking this fact. One can find a reduced expression for $\mathbf{i} = (i_1, \ldots, i_N)$ for w_0 which is simply-braided for i, and such that $i_N = i'$, so that $e_i^\star$ is easy to compute. The simply-braided condition means that e_i, ε_i can also be computed in the Lusztig parametrization by a bracketing procedure similar to that described in Section 2.4. This gives a method of checking the dichotomy.

15.10 MV polytopes and the finite crystals $\mathcal{B}_\lambda$

As an application of the $\star$ involution, we determine the MV polytopes of the finite crystal $\mathcal{B}_\lambda$.

Proposition 15.32 (Kamnitzer). *Let P be an MV polytope whose highest weight is λ. The following are equivalent:*

(i) $M_{w_0 s_i \cdot \varpi_i^\vee} \geqslant \langle w_0 \lambda, \varpi_i^\vee \rangle$ *for all $i \in I$,*
(ii) P *is contained in the Weyl polytope that is the convex hull of $W \cdot \lambda$.*

Proof. See [Kamnitzer (2007), Section 6] and [Kamnitzer (2010), Section 8], where this is proved using the affine Grassmannian and a result from [Anderson (2003)]. It seems desirable to give a more direct proof of this fact. (Exercise 15.12.) $\qquad\square$

Theorem 15.33 ([Kamnitzer (2007)]). *Let λ be a dominant weight. Suppose that P is an MV polytope with highest weight λ. Then $P = \mathrm{MV}(v)$ for some $v \in \mathcal{B}_\lambda$ if and only if P is contained in the Weyl polytope that is the convex hull of the W-orbit of λ.*

Proof. We have to reinterpret the quantity $M_{w_0 s_i \cdot \varpi_i^\vee}$ that appears in Proposition 15.32. We will prove that

$$M_{w_0 s_{i'} \varpi_{i'}^\vee} - \langle w_0 \lambda, \varpi_{i'}^\vee \rangle = \langle \lambda, \alpha_i^\vee \rangle - \varepsilon_i^\star(v). \tag{15.22}$$

As in Lemma 15.19, we have $w_0 \varpi_{i'}^\vee = -\varpi_i^\vee$ and $w_0 s_{i'} \varpi_{i'}^\vee = -s_i \varpi_i^\vee$. So the left-hand side equals $M_{-s_i \varpi_i^\vee} - \langle \lambda, w_0 \varpi_{i'}^\vee \rangle = M_{-s_i \varpi_i^\vee} + \langle \lambda, \varpi_i^\vee \rangle$. The point $\lambda - \varepsilon_i^\star(v) \alpha_i$ is the lower endpoint of the edge of $\mathrm{MV}(v)$ coming out of λ in the $-\alpha_i$ direction, so it is on the hyperplane $\langle x, -s_i \varpi_i^\vee \rangle = M_{-s_i \varpi_i^\vee}$, which is a face of the polytope through this vertex. Therefore

$$M_{-s_i \varpi_i^\vee} = \langle \lambda - \varepsilon_i^\star(v) \alpha_i, -s_i \varpi_i^\vee \rangle = -\langle s_i(\lambda - \varepsilon_i^\star(v) \alpha_i), \varpi_i^\vee \rangle.$$

Since $s_i(\lambda - \varepsilon_i^\star(v) \alpha_i) = \lambda - \langle \lambda, \alpha_i^\vee \rangle \alpha_i + \varepsilon_i^\star(v) \alpha_i$, Equation (15.22) follows.

We see that the criterion in Proposition 15.32 for $\mathrm{MV}(v)$ to be contained in the convex hull of $W \cdot \lambda$ is equivalent to the assumption that $\varepsilon_i^\star(v) \leqslant \langle \lambda, \alpha_i^\vee \rangle$. By Theorem 14.19, this is equivalent to $v \in \mathcal{B}_\lambda$. $\square$

[Jiang and Sheng (2016)] used virtual crystals to define a crystal structure on MV polytopes for non-simply-laced cases, extending work of [Naito and Sagaki (2008)] for types B and C.

Exercises

Exercise 15.1. Take the root system for $\mathfrak{sl}_4$ and the reduced word $\mathbf{i} = (1, 2, 3, 1, 2, 1)$ for w_0. Compute the action of the crystal operators f_3 and f_2 on $v_{\mathbf{i}} = (2, 3, 1, 2, 4, 2) \in \mathcal{L}$ and express the answer in the same "view".

Hint: You can compare your answers with [Tingley (2016), Section 7].

Exercise 15.2. The goal of this exercise is to compute the piecewise-linear transition maps $\mathcal{R}_{\mathbf{i}, \mathbf{j}}$ for the Lusztig parametrization for type C_2, by doing the analogous computations in C_2 to the ones done for A_2 in Section 15.1.

(i) Consider the matrix generators of the maximal unipotent subgroup of the symplectic group $\mathrm{Sp}(4)$.

$$x_1(a) = \begin{pmatrix} 1 & a & & \\ & 1 & & \\ & & 1 & \\ & & -a & 1 \end{pmatrix}, \qquad x_2(a) = \begin{pmatrix} 1 & & & \\ & 1 & a & \\ & & 1 & \\ & & & 1 \end{pmatrix},$$

where the unspecified entries are zero. Equate

$$x_1(a) x_2(b) x_1(c) x_2(d) = x_2(a') x_1(b') x_2(c') x_1(d')$$

and express the parameters a', b', c', d' in terms of a, b, c, d. Show that

$$a' = \frac{bc^2 d}{\pi_2}, \quad b' = \frac{\pi_2}{\pi_1}, \quad c' = \frac{\pi_1^2}{\pi_2}, \quad d' = \frac{abc}{\pi_1},$$

where

$$\pi_1 = ab + (a + c)d, \quad \pi_2 = a^2 b + (a + c)^2 d$$

provides a solution.

Hint: See [Berenstein and Zelevinsky (2001), Proposition 7.1] and [Berenstein and Zelevinsky (1997), Theorem 3.1].

(ii) Tropicalize this to find the transition maps $\mathcal{R}_{\mathbf{i},\mathbf{j}}$ for type C_2.

(iii) Discuss how this gives a Lustig parametrization for all Cartan types in which the Dynkin diagram has only single or double bonds, that is, all the finite Cartan types except G_2. (See the references in the hint to (a) for G_2.)

The following exercise is based on [Kamnitzer (2007), Lemma 6.8].

Exercise 15.3. Let $M_\bullet$ be data that satisfies the tropical Plücker relations (15.15). Let $\mathbf{i} = (i_1, \ldots, i_N)$ be a reduced word of w_0 and define $y_k^{\mathbf{i}} = s_{i_1} \cdots s_{i_k}$ for all $1 \leqslant k \leqslant N$. Set $p^{\mathbf{i}} = (p_1^{\mathbf{i}}, \ldots, p_N^{\mathbf{i}})$, where

$$p_k^{\mathbf{i}} = M_{y_{k-1}^{\mathbf{i}} \varpi_{i_k}^\vee} - M_{y_k^{\mathbf{i}} \varpi_{i_k}^\vee}. \tag{15.23}$$

Let $\widehat{\mathcal{R}}_{\mathbf{i},\mathbf{j}}$ be as in Proposition 15.6. Prove that $p^{\mathbf{j}} = \widehat{\mathcal{R}}_{\mathbf{i},\mathbf{j}}(p^{\mathbf{i}})$.

Hint: It is sufficient to assume that $\mathbf{i}$ and $\mathbf{j}$ are related as in (M1) or (M2) in Section 15.3. For (M2), use the tropical Plücker relation and remember that $s_i \varpi_j^\vee = \varpi_j^\vee$ and $s_j \varpi_i^\vee = \varpi_i^\vee$.

Let us define "downward" string data that is analogous to $\mathrm{string}_{\mathbf{i}}(v)$, but using the f_i instead of the e_i. Let $\mathcal{B}$ be a seminormal crystal and let $v \in \mathcal{B}$. Consider $\mathbf{i} = (i_1, \ldots, i_N)$ a reduced word for w_0. Define $\mathrm{string}_{\mathbf{i}}^f(v) = (a_1, \ldots, a_N)$ as follows. First, $a_1 = \varphi_{i_1}(v)$. This is the number of times f_{i_1} may be applied to v, so that $f_{i_1}^{a_1} v \neq 0$. Then let $a_2 = \varphi_{i_2}(f_{i_1}^{a_1}(v))$, etc. We may consider the stations of v in analogy with (11.1) to be the sequence

$$v_0 = v, \quad v_1 = f_{i_1}^{a_1} v, \quad v_2 = f_{i_2}^{a_2} f_{i_1}^{a_1} v, \quad \ldots, \quad v_N = f_{i_N}^{a_N} \cdots f_{i_1}^{a_1} v. \tag{15.24}$$

Exercise 15.4. Let $v \in \mathcal{B}_\lambda$. Prove that the final station v_N in the sequence (15.24) is the lowest weight element of $\mathcal{B}_\lambda$.

The following exercise is based on [Kamnitzer (2007), Lemma 6.6].

Exercise 15.5. Let $v \in \mathcal{B}_\lambda$. Let $\mathrm{MV}(v)$ be the corresponding MV polytope, oriented so that its highest weight vector $\lambda_{\mathrm{high}} = \lambda$. The goal of this exercise is to prove that

$$\mathrm{string}_{\mathbf{i}}^f(v) = (p_1^{\mathbf{i}}, \ldots, p_N^{\mathbf{i}}) \tag{15.25}$$

with $p_k^{\mathbf{i}}$ as in (15.23).

(i) If v is the lowest weight element of $\mathcal{B}_\lambda$, show that $P(M_\bullet)$ is the Weyl polytope, which is the convex hull of $W \cdot \lambda$. Show that $M_{w \cdot \varpi_i^\vee}$ is independent of $w \in W$, and that each $p_k^i = 0$. Deduce that (15.25) is satisfied in this case.

(ii) With v fixed, show that if (15.25) is satisfied for one **i** then it is true for all **i**. (Use Exercise 15.3.)

(iii) Suppose that $v \in \mathcal{B}_\lambda$ is such that (15.25) is true for v and $e_i(v) \neq 0$. Choose **i** so that $i_1 = i$. Show that (15.25) is also true for $e_i(v)$.

(iv) Prove (15.25) is true for all $v \in \mathcal{B}_\lambda$.

Exercise 15.6. Let $v \in \mathcal{B}_\infty$ and $P = \mathrm{MV}(v)$ be the corresponding MV polytope. Let $-P = \{-x \mid x \in P\}$ be its negative.

(i) If $P = P(M_\bullet)$, define $M'_{\nu^\vee} = M_{-\nu^\vee}$. Prove that $-P = P(M'_\bullet)$.

(ii) Show that the $M'_\bullet$ satisfy the tropical Plücker relations and deduce that $-P$ is an MV polytope.

(iii) Show that $\mathrm{MV}(v^\star)$ is a translate of $-P$.

Exercise 15.7. Let $\mathcal{B}_\lambda$ be a $\mathrm{GL}(n)$ crystal of tableaux. Let T be a $\mathrm{GL}(n)$ tableau of shape λ and let v be the corresponding element of $\mathcal{L}$ in type A_{n-1}, so that $T \mapsto t_\lambda \otimes v$ in the crystal embedding $\mathcal{B}_\lambda \to \mathcal{T}_\lambda \otimes \mathcal{L}$. Fix the reduced word

$$\mathbf{i} = (1, 2, \ldots, n-1, 1, 2, \ldots, n-2, \ldots, 1, 2, 1)$$

for w_0. Let a_{ij} be the number of j in the i-th row of T. Prove that

$$v_\mathbf{i} = (a_{12}, a_{13}, \ldots, a_{1n}, a_{23}, \ldots, a_{2n}, \ldots, a_{n-1,n}).$$

Remark: This exercise is the analogue of Proposition 11.2 for the Lusztig data instead of the string data.

Hint: Explicitly compute the crystal operators for the fixed view determined by **i** and show that the described map is indeed a crystal embedding. (For type D, this was carried out in [Salisbury, Schultze and Tingley (2016b)] using marginally large tableaux; in type A this can be done similarly using the marginally large tableaux of Exercise 12.1.) For another approach see [Claxton and Tingley (2014)].

Exercise 15.8. The goal of this exercise is to verify by direct computation that the map $\star$ on $\mathcal{L}$ defined in this chapter agrees with that in the last chapter in the A_2 case.

(i) Let $\mathbf{i} = (1, 2, 1)$, so $\mathbf{i}' = (2, 1, 2)$. Show that if $v_\mathbf{i} = (a, b, c)$, then $\mathrm{string}_\mathbf{i}(v) = (a, b+c, b)$.

(ii) Make use of the maps θ and τ that were introduced in Section 14.1. Recall that

$$\tau\big(\mathrm{string}_\mathbf{i}(v)\big) = \mathrm{string}_\mathbf{i}(v^\star), \qquad \theta\big(\mathrm{string}_\mathbf{i}(v)\big) = \mathrm{string}_{\mathbf{i}'}(v).$$

Check that $\theta\tau(a, b+c, b) = (c, a+b, b)$ and deduce that $v^\star_{(2,1,2)} = (c, b, a)$.

(iii) Show that Theorem 15.27 is true for A_2.

Exercise 15.9. Prove (15.21) and deduce that $\mathrm{wt}(v) = \mathrm{wt}(v^\star)$.

Exercise 15.10. Let $c \otimes u_\lambda$ be a highest weight element of $\mathcal{B}_\mu \otimes \mathcal{B}_\lambda$. Let $u_\mu^{\mathrm{low}} \otimes b$ be the lowest weight element of the component containing $c \otimes u_\lambda$. Let $(p_1, \ldots, p_N)$ be the downward string data $\mathrm{string}_{\mathbf{i}}^{f}(c)$ for c with respect to $\mathbf{i}$, and $(q_1, \ldots, q_N)$ the usual (upward) string data $\mathrm{string}_{\mathbf{i}^{\mathrm{rev}}}(b)$ for b with respect to $\mathbf{i}^{\mathrm{rev}} := (i_N, \ldots, i_1)$. Prove that, for all k,

$$p_k + q_{N-k+1} = \langle \nu, w_{k-1}^{\mathbf{i}} \cdot \alpha_{i_k}^\vee \rangle,$$

where $\nu = \mathrm{wt}(c \otimes u_\lambda)$.

Exercise 15.11. Use Exercises 15.5 and 15.10 to prove Theorem 14.21.
Hint: See [Kamnitzer and Tingley (2009b)].

Exercise 15.12. (Open) Give a combinatorial proof of Proposition 15.32 using the properties of MV polytopes.

Chapter 16

Further Topics

There are many other developments within the theory of crystal bases, that we have not addressed in this book. In this chapter, we will briefly mention a few further topics and point the reader to the literature for additional details. This is not intended to be an exhaustive survey and within each topic we have undoubtedly not cited even all of the most important papers.

16.1 Kirillov–Reshetikhin crystals

In this book we have restricted ourselves to classical roots systems, which, as we have seen in Chapter 2, are intimately related to finite-dimensional Lie algebras. Another very important class of root systems are the affine root systems [Macdonald (1972); Kac (1990)] that are associated to affine Kac–Moody Lie algebras [Kac (1990)]. For affine Lie algebras there are two weight lattices, the weight lattice of the full Kac–Moody Lie algebra, which includes a derivation, and the weight lattice for the derived Lie algebra, which discards the derivation. For the weight lattice without a derivation, there exist finite-dimensional representations that are no longer highest weight, unlike in the case of classical root systems. The irreducible, finite-dimensional modules of the quantum groups associated to affine root systems were classified by [Chari and Pressley (1995, 1998)] in terms of Drinfeld polynomials.

A special class of such finite-dimensional affine modules were studied by [Kirillov and Reshetikhin (1987)], called *Kirillov–Reshetikin (KR) modules*. It was conjectured by [Hatayama, Kuniba, Okado, Takagi and Yamada (1999); Hatayama, Kuniba, Okado, Takagi and Tsuboi (2002)] and later proven for type D by [Okado (2007)] and all nonexceptional cases by [Okado and Schilling (2008)], that the KR modules admit crystal bases. These *Kirillov–Reshetikin (KR) crystals* $B^{k,s}$ are labeled by two positive integers, where k is in the index set of the underlying classical root system and s a positive integer. The KR modules/crystals and their tensor products have interesting characters. It was shown by [Nakajima (2003a); Hernandez (2006)] in the untwisted cases and [Hernandez (2010)] in general, that they solve the Q-system, which has deep connections to quiver varieties and exactly solvable lattice models in statistical mechanics. Q-systems are functional equations that the characters satisfy [Hatayama, Kuniba, Okado, Takagi and Yamada (1999); Hatayama, Kuniba, Okado, Takagi and Tsuboi (2002); Kuniba, Nakanishi and Tsuboi (2002)].

The index set of affine type $A_r^{(1)}$ is $I = \{0, 1, 2, \ldots, r\}$. By Levi branching, that is, by discarding the index 0, an $A_r^{(1)}$ crystal may be regarded as an A_r crystal. As a type A_r crystal, the KR crystal

$$B^{k,s} \cong \mathcal{B}_{(s^k)}$$

is isomorphic to the highest weight crystal of tableaux of highest weight a rectangle of width s and height k. The crystal operators f_0 and e_0 were constructed by [Shimozono (2002)] using the promotion operator [Schützenberger (1972); Haiman (1992)] $\mathfrak{pr}$ as follows

$$f_0 = \mathfrak{pr}^{-1} \circ f_1 \circ \mathfrak{pr}, \qquad e_0 = \mathfrak{pr}^{-1} \circ e_1 \circ \mathfrak{pr}.$$

This idea exploits the rotational symmetry of the affine Dynkin diagram of type $A_r^{(1)}$. [Kwon (2013)] provided another combinatorial model for the type A KR crystals in terms of nonnegative integral matrices, whereas [Kus (2013)] gave a model in terms of polytopes.

For other nonexceptional types, combinatorial models for KR crystals were constructed in [Fourier, Okado and Schilling (2009)], in many cases again exploiting certain automorphisms of the affine Dynkin diagrams. For exceptional cases, some of the KR crystals were constructed in [Yamane (1998); Kashiwara, Misra, Okado and Yamada (2007); Jones and Schilling (2010)].

As will be explained in Section 16.3, the property of perfectness of a KR crystal is essential in the construction of affine, infinite-dimensional highest weight crystal using the Kyoto path model. Perfectness of KR crystals for nonexceptional types was proven in [Fourier, Okado and Schilling (2010)].

Demazure crystals inside tensor products of KR crystals were studied in [Fourier, Schilling and Shimozono (2007); Schilling and Tingley (2012)]. This is closely related to the energy function on tensor products of KR crystals and has applications to the affine Demazure character formula and q-deformed Whittaker functions. Recently, relations to Khovanov–Lauda–Rouquier algebras were also pointed out in [Kvinge and Vazirani (2015)].

Uniform models in all untwisted affine types for tensor products of single-column KR crystals $B^{k,1}$ were given in [Lenart, Naito, Sagaki, Schilling and Shimozono (2015a, 2016, 2015b)]. These models involve generalizations of the Lakshmibai–Seshadri paths (in the theory of the Littelmann path model; see Section 16.2), which are based on the graph on parabolic cosets of a Weyl group known as the parabolic quantum Bruhat graph.

Recently, an $A_r^{(1)}$ geometric crystal structure on the Grassmannian $\mathrm{Gr}(r-k;r)$ was constructed by [Frieden (2016)], which tropicalizes to the corresponding KR crystal $B^{r,s}$ for any $s > 0$. It was previously conjectured by [Kashiwara, Nakashima and Okado (2008)] that the KR crystals for general types form a family of coherent crystals and that there is a geometric crystal whose tropicalization is the limit of the coherent family.

16.2 Littelmann path and alcove path models

In this book, we have discussed several models for crystals: the tableaux model for crystals (see Chapters 3 and 6), a model on decreasing factorizations (see Chapter 10), MV polytopes (see Chapter 15), and the string and Lusztig parametrization for $\mathcal{B}_\infty$ (see Chapters 12 and 15). Other very important classes of models are the Littelmann path model [Littelmann (1994, 1995b, 1997)], LS galleries [Gaussent and Littelmann (2005)], and the alcove path model [Lenart and Postnikov (2008)].

The *Littelmann path model* takes its origin in the standard monomial theory of [Lakshmibai and Seshadri (1991)]. It associates to each irreducible representation a rational vector space with basis given by paths from the origin to a weight as well as a pair of root operators acting on paths for each simple root. More precisely, let $[0,1]_\mathbb{Q} := [0,1] \cap \mathbb{Q}$. A *Littelmann path* is a piecewise-linear mapping

$$\pi\colon [0,1]_\mathbb{Q} \to \Lambda \otimes_\mathbb{Z} \mathbb{Q},$$

such that $\pi(0) = 0$ and $\pi(1) \in \Lambda$. Two paths π_1, π_2 are considered identical if there is a piecewise-linear, nondecreasing, surjective, continuous map $\phi\colon [0,1]_\mathbb{Q} \to [0,1]_\mathbb{Q}$ such that $\pi_1 = \pi_2 \circ \phi$.

To define the root operators, let $h_\alpha(t) = \langle \pi(t), \alpha^\vee \rangle$ for each $\alpha \in \Phi$ and set $m_\alpha = \min_{t \in [0,1]_\mathbb{Q}} \{h_\alpha(t)\}$. Let $\ell_\alpha(t)$ and $r_\alpha(t)$ be nondecreasing mappings on $[0,1]_\mathbb{Q}$ defined by

$$\ell_\alpha(t) = \min_{t \leqslant s \leqslant 1} \{1, h_\alpha(s) - m_\alpha\}, \qquad r_\alpha(t) = 1 - \min_{0 \leqslant s \leqslant t} \{1, h_\alpha(s) - m_\alpha\}.$$

Hence $\ell_\alpha(t) = 0$ until the last time that $h_\alpha(s) = m_\alpha$ and $r_\alpha(t) = 1$ after the first time that $h_\alpha(s) = m_\alpha$. Define new paths π_ℓ and π_r by

$$\pi_\ell(t) = \pi(t) - \ell_\alpha(t)\alpha, \qquad \pi_r(t) = \pi(t) + r_\alpha(t)\alpha.$$

The *root operators* are then defined as

$$f_\alpha \pi = \begin{cases} \pi_\ell & \text{if } \ell(1) = 1, \\ 0 & \text{otherwise,} \end{cases}$$

$$e_\alpha \pi = \begin{cases} \pi_r & \text{if } r(0) = 0, \\ 0 & \text{otherwise.} \end{cases}$$

It was shown by [Joseph (1995); Kashiwara (1996)] that the Littelmann path model provides a model for the Kashiwara crystals corresponding to highest weight representations of any Lie algebra or symmetrizable Kac–Moody Lie algebra. [Stembridge (2003)] showed that the Littelmann path model satisfies the Stembridge axioms. [Littelmann (1994)] deduced a character formula, a decomposition or Littlewood–Richardson rule, and branching rules from the Littelmann path model.

[Gaussent and Littelmann (2005)] subsequently introduced a variant of the Littelmann path model in terms of LS galleries and showed that these are closely related to Mirković–Vilonen cycles (see [Mirković and Vilonen (2000)]).

The *alcove path model* [Lenart and Postnikov (2008)] is a discrete counterpart to the Littelmann path model. Instead of allowing any piecewise-linear path from 0 to some weight λ, elements in the alcove path model are given by sequences of adjacent alcoves, which are cut out by affine hyperplanes. This can also be formulated in terms of so-called λ chains, which are chains of positive roots defined by certain interlacing conditions. The root operators are then defined by conditions on the λ chain.

16.3 Kyoto path model

Integrable representations of Kac–Moody Lie algebras are important because they play a role analogous to the finite-dimensional representations of finite-dimensional complex Lie algebras. The *Kyoto path model* is a model for integrable, highest weight, infinite-dimensional crystals associated to affine Kac–Moody Lie algebras [Kang, Kashiwara, Misra, Miwa, Nakashima and Nakayashiki (1992b,a)]. It is based on the notion of *perfect crystals*.

For an affine Kac–Moody Lie algebra, the index set is $I = \{0, 1, 2, \ldots, r\}$. The weight lattice is given by $\Lambda = \bigoplus_{i \in I} \mathbb{Z}\varpi_i \oplus \mathbb{Z}\delta$, where ϖ_i are the fundamental weights and δ is the null root, and let $\Lambda^+ = \bigoplus_{i \in I} \mathbb{Z}_{\geqslant 0}\varpi_i$ be the dominant weights. Each irreducible integrable representation (or crystal) of a Kac–Moody Lie algebra has a unique highest weight element whose weight is dominant. For a positive integer ℓ, the set of level-ℓ weights is

$$\Lambda_\ell^+ = \{\lambda \in \Lambda^+ \mid \langle c, \lambda \rangle = \ell\},$$

where c is the canonical central element of the affine Kac–Moody Lie algebra (see [Kac (1990)]). We denote by $\overline{\Lambda}$ the weight lattice of the underlying classical Lie algebra, which is obtained by dropping the 0-node in the affine Dynkin diagram.

Let $\mathcal{B}$ be a crystal associated to an affine Kac–Moody Lie algebra. We define $\varepsilon(b) = \sum_{i \in I} \varepsilon_i(b)\varpi_i$ and $\varphi(b) = \sum_{i \in I} \varphi_i(b)\varpi_i$. For a positive integer ℓ, we say that $\mathcal{B}$ is *perfect* of level ℓ if the following conditions are satisfied:

(1) $\mathcal{B} \otimes \mathcal{B}$ is connected.
(2) There exists a $\overline{\lambda} \in \overline{\Lambda}$, such that $\overline{\mathrm{wt}}(\mathcal{B}) \subset \overline{\lambda} + \sum_{i \in I \setminus \{0\}} \mathbb{Z}_{\leqslant 0}\alpha_i$ and there is a unique element in $\mathcal{B}$ of classical weight $\overline{\mathrm{wt}}$ given by $\overline{\lambda}$. Here classical weight refers to the weight function of the crystal if considered as an $I \setminus \{0\}$-crystal.
(3) For all $b \in \mathcal{B}$, $\langle c, \varepsilon(b) \rangle \geqslant \ell$.
(4) For all $\omega \in \Lambda_\ell^+$, there exist unique elements $b_\omega, b^\omega \in \mathcal{B}$, such that

$$\varepsilon(b_\omega) = \omega = \varphi(b^\omega).$$

The most important condition is Condition 4, which states that $\varepsilon, \varphi \colon \mathcal{B}_{\min} \to \Lambda_\ell^+$ are bijections, where

$$\mathcal{B}_{\min} = \{b \in \mathcal{B} \mid \langle c, \varepsilon(b) \rangle = \ell\}.$$

The Kyoto path model now provides a model for the integrable, highest weight crystal $\mathcal{B}(\lambda)$ with highest weight element u_λ by recursively making use of the following crystal isomorphism. Given a dominant weight λ_0 of level ℓ and a perfect crystal $\mathcal{B}^0$ of level ℓ, define the crystal isomorphism

$$\mathcal{B}(\lambda_0) \cong \mathcal{B}^0 \otimes \mathcal{B}(\lambda_1)$$

defined by $u_{\lambda_0} \mapsto b^0 \otimes u_{\lambda_1}$, where b^0 is the unique element in $\mathcal{B}^0$ such that $\varphi(b^0) = \lambda_0$ and $\varepsilon(b^0) = \lambda_1$. Iterating this isomorphism, one obtains

$$\mathcal{B}(\lambda_0) \cong \mathcal{B}^0 \otimes \mathcal{B}^1 \otimes \cdots \otimes \mathcal{B}^N \otimes \mathcal{B}(\lambda_{N+1}).$$

In the limit $N \to \infty$, this models $\mathcal{B}(\lambda)$ as a semi-infinite tensor product of perfect crystals of level ℓ, where $\langle c, \lambda \rangle = \ell$.

Recall from Section 16.1, that certain Kirillov–Reshetikhin crystals are perfect and can hence be used in this recursive fashion to construct the infinite-dimensional highest weight crystals $\mathcal{B}(\lambda)$ for $\lambda \in \Lambda^+$.

16.4   Nakajima monomial model

Let $\mathcal{M}$ be the set of monomials in the commuting variables $Y_{i,k}$, where $i \in I$ is the index of a simple root and $k \in \mathbb{Z}$. [Nakajima (2003b)] gave a crystal structure on $\mathcal{M}$. A variation of this monomial realization was given by [Kashiwara (2003)], who also proved that a connected component of a monomial with a dominant integral weight λ is isomorphic to the crystal $\mathcal{B}_\lambda$.

[Kang, Kim and Shin (2007)] provide a Nakajima monomial model for $\mathcal{B}_\infty$ in terms of *modified Nakajima monomials*, which we will now introduce. Let $\widehat{\mathcal{M}}$ be the set of monomials of the form

$$M = \prod_{(i,k) \in I \times \mathbb{Z}_{\geqslant 0}} Y_{i,k}^{y_i(k)} \mathbf{1},$$

where $\mathbf{1}$ is a variable commuting with all $Y_{i,k}$. For a given set of integers $C = (c_{ij})_{i \neq j}$ such that $c_{ij} + c_{ji} = 1$, the crystal operators on $M \in \widehat{\mathcal{M}}$ are defined as

$$e_i M = \begin{cases} A_{i,k_e} M & \text{if } \varepsilon_i(M) > 0, \\ 0 & \text{if } \varepsilon_i(M) = 0, \end{cases}$$

$$f_i M = A_{i,k_f}^{-1} M,$$

where

$$A_{i,k} = Y_{i,k} Y_{i,k+1} \prod_{j \neq i} Y_{j,k+c_{ji}}^{a_{ji}}$$

and (a_{ij}) is the Cartan matrix. Furthermore,

$$\mathrm{wt}(M) = \sum_{i \in I} \left(\sum_{k \geqslant 0} y_i(k) \right) \varpi_i,$$

$$\varphi_i(M) = \max\left\{ \sum_{0 \leqslant j \leqslant k} y_i(j) \mid k \geqslant 0 \right\},$$

$$\varepsilon_i(M) = \varphi_i(M) - \langle \mathrm{wt}(M), \alpha_i^\vee \rangle,$$

$$k_f = \min\left\{ k \geqslant 0 \mid \varphi_i(M) = \sum_{0 \leqslant j \leqslant k} y_i(j) \right\},$$

$$k_e = \max\left\{ k \geqslant 0 \mid \varphi_i(M) = \sum_{0 \leqslant j \leqslant k} y_i(j) \right\}.$$

Monomial realizations for level-0 fundamental representations were given in [Hernandez and Nakajima (2006)]. A certain quotient of these also gives a model for the single column KR crystals. One of the main motivations for the Nakajima monomial model is the relation to q-characters of finite-dimensional representations of quantum affine algebras introduced by [Frenkel and Reshetikhin (1999)].

16.5 Crystals on rigged configurations

Rigged configurations arose in work of [Kerov, Kirillov and Reshetikhin (1986); Kirillov and Reshetikhin (1986)] in the Bethe Ansatz study of exactly solvable lattice models. They label the solutions of the Bethe equations, which index the eigenvalues of the underlying Hamiltonian. As combinatorial objects, they are sequences $(\nu^{(1)}, \ldots, \nu^{(r)})$ of r partitions $\nu^{(i)}$ (where r is the rank of the root system) together with a set of *riggings* for each partition. Riggings are nonnegative integer labels attached to each part of the partition subject to the constraint that they are bounded by the so-called vacancy numbers, which depend on a shape (or more generally tensor product of crystals). In type A, the number of these configurations of a fixed shape λ and weight μ coincides with the corresponding Kostka number $K_{\lambda,\mu}$, which is the number of semistandard Young tableaux of shape λ and weight μ. This suggests a bijection between rigged configurations and semistandard Young tableaux, which was indeed established in [Kirillov and Reshetikhin (1986); Kirillov, Schilling and Shimozono (2002)].

Kostka numbers $K_{\lambda,\mu}$ count the number of irreducible representations V_λ in $V_{(\mu_1)} \otimes V_{(\mu_2)} \otimes \cdots$ in type A. In the language of crystals, this would be the number of highest weight elements of weight λ in $\mathcal{B}_{(\mu_1)} \otimes \mathcal{B}_{(\mu_2)} \otimes \cdots$. Hence it is natural to ask whether there is a generalization of the set of rigged configurations with a crystal structure, such that the above rigged configurations are highest weight elements. This was affirmatively answered for all classical simply-laced types by [Schilling (2006)] using Stembridge axioms and extended in [Schilling and Scrimshaw (2015)] to non-simply-laced cases using virtual crystals. [Salisbury and Scrimshaw (2015)] introduced a rigged configuration model for $\mathcal{B}_\infty$ in all Kac–

Moody types. The ⋆-involution on the rigged configurations for $\mathcal{B}_\infty$ was given in [Salisbury and Scrimshaw (2016)].

16.6 Modular branching rules of the symmetric group and crystal bases

Whereas the representation theory of the symmetric group S_n is well-understood over $\mathbb{C}$, many open questions remain regarding the representation theory over fields of positive characteristic, also known as modular representation theory. By work of [Dipper and James (1986)], modular representation theory of the symmetric group S_n at characteristic p resembles the representation theory of the corresponding Iwahori–Hecke algebra $H_n(\mathbb{C}, e^{2\pi i/p})$ over $\mathbb{C}$ at a p-th root of unity. As we will explain, this representation theory is connected with the theory of crystal bases through an analogy in which induction and restriction correspond to the crystal operations e_i and f_i in affine $A_{p-1}^{(1)}$ crystals. [Misra and Miwa (1990)] computed the crystal basis of the *basic representation* of the affine Kac–Moody Lie algebras of type A. This is the integrable representation of highest weight ϖ_0, which is important in mathematical physics. Observing the similarity between its crystal and the branching lattice for modular representations of the symmetric groups described by [Kleshchev (1995)], [Lascoux, Leclerc and Thibon (1996)] conjectured a very precise and surprising connection between crystal bases over affine Kac–Moody algebras and projective indecomposable modules over the Iwahori–Hecke algebras. [Lascoux, Leclerc and Thibon (1996)] also conjectured an efficient combinatorial algorithm for computing decomposition numbers, that is, the multiplicities of the irreducible $H_n(\mathbb{C}, e^{2\pi i/p})$-modules in the corresponding Specht modules. The conjectures were proven by [Ariki (1996)]. This proof was later generalized by [Varagnolo and Vasserot (1999)] using the quantized Schur algebra. See also [Grojnowski (1999)].

   This area of research has led to many exciting new directions, KLR algebras being one of them. Excellent accounts of this topic and recent developments were written by [Kleshchev (2005, 2010)]. See also [Vazirani (2002)].

16.7 Tokuyama's formula

[Tokuyama (1988)] gave a deformation of the Weyl character formula for $\mathrm{GL}(r)$. For a dominant weight λ of $\mathrm{GL}(r)$, let χ_λ be the irreducible character corresponding to λ, so that $\chi_\lambda(z) = s_\lambda(z_1, \ldots, z_r)$, where z_i are the eigenvalues of $z \in \mathrm{GL}(r)$. We take $z \in T$, the diagonal subgroup. The Weyl character formula implies that

$$\prod_{\alpha \in \Phi^+} (1 - z^{-\alpha}) \chi_\lambda(z) = \sum_{w \in W} (-1)^{\ell(w)} z^{w(\lambda+\rho)-\rho}, \tag{16.1}$$

where $\rho = (r-1, r-2, \ldots, 0)$. Tokuyama's formula replaces the "Weyl denominator" $\prod_{\alpha \in \Phi^+}(1 - z^{-\alpha})$ by a deformation; namely, it is a formula for

$$\prod_{\alpha \in \Phi^+} (1 - qz^{-\alpha})\chi_\lambda(z). \tag{16.2}$$

Tokuyama stated his formula in terms of Gelfand–Tsetlin patterns, and in view of the bijection between these and tableaux described in Section 11.2, it is straightforward to translate it into the crystal language. This was done in [Brubaker, Bump and Friedberg (2011b)], but there the string patterns were made with the f_i, and ended up at the lowest weight elements of the crystal. Our statement is therefore slightly different. We refer to [Brubaker, Bump and Friedberg (2011b)] for further explanations and proofs.

We replace the sum over the Weyl group in (16.1) with a sum over the crystal $\mathcal{B}_{\lambda+\rho}$. To define what is summed, let $v \in \mathcal{B}_{\lambda+\rho}$. Let the expression $G(v)$ be the product over $\text{string}_\Omega(v)$, where Ω is the particular reduced word $\Omega = (1, 2, 1, 3, 2, 1, \ldots, 1)$ for w_0 as in (11.3). We let

$$\text{string}_\Omega(v) = (a_1, a_2, \ldots, a_N) = (a_{11}, a_{21}, a_{22}, a_{31}, \ldots).$$

Both notations a_i and a_{ij} will be useful and $a_{11} = a_1$, $a_{21} = a_2$, etc.

We decorate the entries $\text{string}_\Omega(v)$ by drawing circles around some, and boxes around others. The set $\{\text{string}_\Omega(v) \mid v \in \mathcal{B}_{\lambda+\rho}\}$ may be described by inequalities on the a_{ij}. If a_{ij} satisfies its lower bound, we circle it. If it satisfies its upper bound, we box it.

To make this precise, let us describe the circling rule in concrete terms. It follows from Proposition 11.2 that:

$$\begin{array}{ccccc} \ddots & & & & \vdots \\ a_{31} \geqslant & a_{32} \geqslant & a_{33} \geqslant 0 & & \\ & a_{21} \geqslant & a_{22} \geqslant 0 & & \\ & & a_{11} \geqslant 0 & & \end{array}$$

We circle a_{ij} if $i > j$ and $a_{ij} = a_{i,j+1}$, or if $i = j$ and $a_{ij} = 0$. This is equivalent to saying that a_{ij} cannot be decreased and remain a string pattern of the crystal.

To describe the boxing rule, let $v_1 = v, v_2, \ldots, v_{N+1} = u_{\lambda+\rho}$ be the stations of v as defined in Section 11.2. Thus $v_{i+1} = e_{\Omega_i}^{a_i} v_i$. Then we box a_i if $f_{\Omega_i} v_i = 0$. This is equivalent to saying that v_i is the lowest weight element of the Ω_i-root string through v_i.

Now we may define

$$G(v) = \prod_{i=1}^N G_i(v), \quad \text{where} \quad G_i(v) = \begin{cases} 1 & \text{if } a_i \text{ is circled but not boxed,} \\ 1-q & \text{if } a_i \text{ is neither boxed nor circled,} \\ -q & \text{if } a_i \text{ is boxed but not circled,} \\ 0 & \text{if } a_i \text{ is both boxed and circled.} \end{cases}$$

Theorem 16.1. *With this definition,*

$$\prod_{\alpha \in \Phi^+} (1 - qz^{-\alpha})\chi_\lambda(z) = \sum_{v \in \mathcal{B}_{\lambda+\rho}} G(v)z^{\mathrm{wt}(v)-\rho}. \tag{16.3}$$

Analogs of this result exist for other Cartan types. See, for example [Chinta and Gunnells (2012); Beineke, Brubaker and Frechette (2012); Friedberg and Zhang (2014); Friedlander, Gaudet and Gunnells (2015)]. In these generalizations, it seems important to choose the reduced word representing w_0 carefully. For an arbitrary reduced word, there may exist a function $G(v)$ that makes (16.3) true, but this formula may be difficult to make explicit.

Tokuyama's formula closely resembles the important result of [Casselman and Shalika (1980)] in the representation theory of p-adic groups. A generalization of this result expresses the p-adic Whittaker functions on the n-fold metaplectic covers of GL(r) as a similar sum over $\mathcal{B}_{\lambda+\rho}$. See [Brubaker, Bump, Friedberg and Hoffstein (2007); McNamara (2011); Brubaker, Bump, Chinta, Friedberg and Gunnells (2012); Puskás (2016); Brubaker, Buciumas and Bump (2016)]. Another direction of generalization of the Casselman-Shalika formula seeks to represent Whittaker functions that are fixed by the Iwahori subgroup as Demazure characters or nonsymmetric Macdonald polynomials. See [Puskás (2016); Patnaik and Puskás (2015)].

As a variant, one may apply the same definition to the crystal $\mathcal{B}_\infty$. See [Bump and Nakasuji (2010)]. Boxing does not occur since $f_i(v)$ is never 0. This produces

$$\prod_{\alpha \in \Phi^+} \frac{1 - qz^{-\alpha}}{1 - z^{-\alpha}}.$$

Like (16.3) this expression also occurs in the representation theory of p-adic groups, as the Gindikin-Karpelevich formula.

Tokuyama's theorem has another, different interpretation. In this, (16.2) is expressed as the partition function of a statistical mechanical system in the six-vertex model, a well-known solvable lattice model. We will not state this precisely, but see [Hamel and King (2002, 2005); Brubaker, Bump and Friedberg (2011b,a); Brubaker and Schultz (2015)]. In this formulation, the left-hand side of (16.3) is the partition function, which is a sum over the states of the system, which are in bijection with $\{v \in \mathcal{B}_{\lambda+\rho} \mid G(v) \neq 0\}$.

16.8 Crystals of Lie superalgebras

Lie superalgebras (or *graded Lie algebras*) are generalizations of Lie algebras that have a $\mathbb{Z}/2\mathbb{Z}$ grading. They reflect a geometry in which both commuting and anti-commuting variables can interact with each other. A *super* vector space $V = V_0 \oplus V_1$ is one with a $\mathbb{Z}/2\mathbb{Z}$ grading. If a is a homogeneous element, that is, if $a \in V_i$ with $i \in \{0, 1\}$, then we denote $|a| = i$. We write $\dim(V) = (m|n)$, where $m = \dim(V_0)$

and $n = \dim(V_1)$. A *Lie superalgebra* is a super vector space with a bilinear operation $[\,,\,]$ that satisfies modified versions of the usual axioms:

$$[b, a] = (-1)^{|a||b|}, \qquad [a, [b, c]] = [[a, b], c] + (-1)^{|a||b|}[b, [a, c]].$$

An example is $\mathfrak{gl}(m|n)$, which is $\mathrm{End}(V)$ with $\dim(V) = (m|n)$. The $\mathbb{Z}/2\mathbb{Z}$ grading
on $\mathrm{End}(V)$ in which the homogeneous part $\mathrm{End}(V)_0$ is $\mathrm{Hom}(V_0, V_0) \oplus \mathrm{Hom}(V_1, V_1)$
and $\mathrm{End}(V)_1$ is $\mathrm{Hom}(V_0, V_1) \oplus \mathrm{Hom}(V_1, V_0)$. The bracket operation is defined by
$[a, b] = ab - (-1)^{|a||b|}ba$ for homogeneous a and $b \in \mathrm{End}(V)$, and extended to all of
$\mathrm{End}(V)$ by linearity. Finite-dimensional Lie superalgebras were classified by [Kac
(1977)].

Quantized enveloping algebras of Lie superalgebras were constructed by [Yamane
(1994)]. The theory of crystal bases for Lie superalgebras was initiated by [Benkart,
Kang and Kashiwara (2000)], who obtained crystal bases with a tableaux model for
the representations of $\mathfrak{gl}(m|n)$. The foundational paper of [Jeong (2001)] proves the
existence of crystal bases to the general case.

Like $\mathrm{GL}(n)$, the Lie superalgebra $\mathfrak{gl}(n|m)$ has characters indexed by partitions,
and these are given by polynomials called *supersymmetric Schur functions*. See
[Macdonald (1992)] and [Berele and Regev (1987)], where they are called *hook Schur
functions*. Let λ be a partition and let $\alpha = (\alpha_1, \ldots, \alpha_n)$ and $\beta = (\beta_1, \ldots, \beta_m)$.
Denote the Schur function s_λ in all $n + m$ variables α_i and β_j as $s_\lambda(\alpha, \beta)$. By the
last characterization of the Littlewood–Richardson coefficients in Theorem B.5,

$$s_\lambda(\alpha, \beta) = \sum_{\mu, \nu} c^\lambda_{\mu, \nu} s_\mu(\alpha)\, s_\nu(\beta).$$

The supersymmetric Schur function, denoted $s_\lambda(\alpha|\beta)$ is obtained by applying the
involution in the ring of symmetric functions to the set of variables. Since the
involution interchanges s_ν and $s_{\nu'}$, where ν' is the conjugate partition, this means

$$s_\lambda(\alpha|\beta) = \sum_{\mu, \nu} c^\lambda_{\mu, \nu} s_\mu(\alpha)\, s_{\nu'}(\beta).$$

The term is interpreted as zero if $\ell(\mu) > n$ or $\ell(\nu') > m$. Just as the Schur
polynomial $s_\lambda(\alpha)$ is a character of $\mathrm{GL}(n)$, the supersymmetric Schur polynomial
$s_\lambda(\alpha|\beta)$ is a character of $\mathfrak{gl}(n|m)$. This is by no means the end of the story of the
finite-dimensional modules of $\mathfrak{gl}(n|m)$. See [Serganova (1996); Brundan (2003)].

Dualities analogous to Schur–Weyl duality and the $\mathrm{GL}(n) \times \mathrm{GL}(m)$ duality
that we discuss in Appendices A and B were developed by [Sergeev (1984)]. See
also [Howe (1989)] for an influential early paper on dualities that used Lie superalgebras, and [Cheng and Wang (2012)] for a current exposition. Just as the
$\mathrm{GL}(n) \times \mathrm{GL}(m)$ duality, that we discuss in Appendix B, is related to the Cauchy
identity, its generalization to Lie superalgebras is related to the following combinatorial formula due to [Berele and Remmel (1985)]:

$$\sum_\lambda s_\lambda(\alpha|\beta) s_\lambda(\gamma|\delta) = \frac{\prod_{i,j}(1 + \alpha_i \delta_j) \prod_{k,\ell}(1 + \beta_k \gamma_\ell)}{\prod_{p,q}(1 - \alpha_p \gamma_q) \prod_{s,t}(1 - \beta_s \delta_t)}. \tag{16.4}$$

This identity generalizes both the Cauchy identity and the dual Cauchy identity. A crystal analog of Sergeev duality, which implies (16.4), may be found in [Kwon (2007)].

The second type of *Sergeev duality* is an analog of Schur–Weyl duality. It concerns the projective representations of the symmetric group, which were known already to Schur to be related to *Schur Q-polynomials*. These are representations of central extensions of the symmetric group S_k that do not factor through S_k itself. See [Macdonald (1995)] Section III.8 for the Schur Q-polynomials. In [Stembridge (1989)] the projective representations of S_k are related to the theory of shifted tableaux, whose theory was developed in [Sagan (1987); Worley (1984)].

[Sergeev (1984)] found an analog of Schur–Weyl duality in which the "queer" Lie superalgebra $\mathfrak{q}(n)$ is used in place of the general linear group in a version of Schur–Weyl duality that explains the projective representations of S_k. A quantum version was given by [Olshanski (1992)], similar to the quantum version of Schur–Weyl duality in [Jimbo (1986)]. Recently the theory of crystal bases for $\mathfrak{q}(n)$ has been developed in [Grantcharov, Jung, Kang and Kim (2010); Grantcharov, Jung, Kang, Kashiwara and Kim (2014); Jung and Kang (2012)], with this duality in mind.

Appendix A

Schur–Weyl Duality

In the two appendices, we review facts about the representation theory of reductive complex Lie groups, such as $\mathrm{GL}(n, \mathbb{C})$, which have analogs in the main part of the book. We include an exposition of these facts because of the importance of these analogies in motivating questions in the main part. Roughly, the representation theoretic material in this Appendix A corresponds to the combinatorics in Chapter 8, and the material in the next Appendix B corresponds to the combinatorics in Chapter 9. We assume without proof one main fact about representation theory: the Weyl character formula for $G = \mathrm{GL}(n, \mathbb{C})$. We give proofs for most other facts.

A.1   Generalities

If X is an affine algebraic variety over $\mathbb{C}$, then we identify X with its set of complex points. This means that elements of the *affine algebra* or *coordinate ring* $\mathcal{O}(X)$ of X may be considered to be functions on X. These will be called *regular functions*. If Y is another affine variety and $\phi\colon X \longrightarrow Y$ is a map, then we say that ϕ is *regular* if $f \circ \phi \in \mathcal{O}(X)$ for all $f \in \mathcal{O}(Y)$. It follows from this definition and the Nullstellensatz that regular maps $X \longrightarrow Y$ correspond bijectively to algebra homomorphisms $\mathcal{O}(Y) \longrightarrow \mathcal{O}(X)$. If X is a vector space, elements of the dual space X^* are regular, and the ring of regular functions may be identified with the symmetric algebra $\bigvee(X^*)$. It is a polynomial ring $\mathbb{C}[\lambda_1, \ldots, \lambda_n]$, where λ_i form a basis of X^*. If f is a nonzero regular function on an irreducible affine variety X, then $X_f = \{x \in X \mid f(x) \neq 0\}$ may also be regarded as an affine variety. The affine algebra of X_f is obtained from $\mathcal{O}(X)$ by adjoining f^{-1}. Since we are assuming X is irreducible, $\mathcal{O}(X)$ is an integral domain and f^{-1} is an element of its field of fractions.

A *finite-dimensional representation* of $G = \mathrm{GL}(n, \mathbb{C})$, is a pair (π, V), where V is a finite-dimensional complex vector space and π is homomorphism $\pi\colon G \longrightarrow \mathrm{GL}(V)$ is a *regular* map. By this we mean the following. Let $g = (g_{ij}) \in G$. If we identify $\mathrm{GL}(V) = \mathrm{GL}(n, \mathbb{C})$, then $\pi(g) = (\pi(g)_{kl})$ is a matrix. Then *regularity* of π means that the matrix coefficients $\pi(g)_{kl}$ are polynomials in the g_{ij} and in $\det(g)^{-1}$. If $\det(g)^{-1}$ does not appear, we say that the representation π is *polynomial*.

Thus in any case $\det^N \otimes \pi$ is polynomial for sufficiently large N.

We may refer to V in a finite-dimensional representation (π, V) as a *G-module*. As usual, the representation π or G-module V is *irreducible* if V has no invariant subspaces.

Let (π, V) be a finite-dimensional representation. Let $(\hat{\pi}, V^*)$ be the *dual* or *contragredient representation* in which $\hat{\pi}(g)$ is the adjoint of $\pi(g^{-1})$. If π is isomorphic to its contragredient, then we say π is *self-dual*.

The Lie algebra $\mathfrak{gl}(n, \mathbb{C})$ of $\mathrm{GL}(n, \mathbb{C})$ is the associative algebra $\mathrm{Mat}_n(\mathbb{C})$ with the Lie bracket $[X, Y] = XY - YX$. If (π, V) is a representation, then the Lie algebra acts on V by

$$Xv = \frac{d}{dh}\pi(e^{hX})v\Big|_{h=0}. \tag{A.1}$$

Let $U(n)$ be the maximal compact subgroup of G consisting of unitary matrices. The Lie algebra $\mathfrak{u}(n)$ of $U(n)$ is the Lie subalgebra of $\mathfrak{gl}(n, \mathbb{C})$ consisting of skew-hermitian matrices. Its elements are characterized by the property that $e^{tX} \in U(n)$ for all t if and only if $X \in \mathfrak{u}(n)$.

Let T be the subgroup of G consisting of diagonal matrices. A regular character of T is called a *weight*. The group Λ of weights is called the *weight lattice*. It may be identified with $\mathbb{Z}^n$ as follows. If $\mu = (\mu_1, \ldots, \mu_n) \in \mathbb{Z}^n$, we use the notation

$$t \mapsto t^\mu, \quad t = \begin{pmatrix} t_1 & & \\ & \ddots & \\ & & t_n \end{pmatrix}, \quad t^\mu = \prod_{i=1}^n t_i^{\mu_i} \tag{A.2}$$

for the corresponding weight. We regard Λ as an additive group, so $t^{\lambda+\mu} = t^\lambda t^\mu$.

In Section 2.6, we regarded t^μ as a formal symbol. Here, t^μ is a character μ of $T(n, \mathbb{C})$ applied to an element t. Despite the different interpretations, the formalism is parallel, and many important formulas (such as the formulas for Schur functions) are identical in both settings.

Let (π, V) be a representation of $\mathrm{GL}(n)$. If μ is a weight, let

$$V_\mu = \{v \in V \mid \pi(t)v = t^\mu v, \ t \in T\} \tag{A.3}$$

be the *weight space* of μ. If $V_\mu \neq 0$, then μ is called a *weight* of the representation π.

The *adjoint representation* of G is the action of G on the Lie algebra induced by conjugation. As above, we identify the Lie algebra of $\mathrm{GL}(n, \mathbb{C})$ with $\mathrm{Mat}_n(\mathbb{C})$, and the adjoint representation is thus interpreted as the action of $\mathrm{GL}(n, \mathbb{C})$ on the vector space $\mathrm{Mat}_n(\mathbb{C})$ by conjugation. The *roots* are the nonzero weights in the adjoint representation, and the *root system* Φ is the set of roots. The roots are the weights $\mathbf{e}_i - \mathbf{e}_j$ where $1 \leqslant i, j \leqslant n$ and $i \neq j$, where $\mathbf{e}_i$ are the standard basis vectors in $\mathbb{Z}^n$. The roots are partitioned into *positive roots* and *negative roots*, where $\mathbf{e}_i - \mathbf{e}_j$ is called *positive* if $i < j$. The *simple positive roots* $\alpha_1, \ldots, \alpha_{n-1}$ are those that are not expressible as a sum of other positive roots. They are

$$\alpha_1 = (1, -1, 0, \ldots), \ \alpha_2 = (0, 1, -1, 0, \ldots), \quad \ldots, \quad \alpha_{n-1} = (\ldots, 0, 1, -1)$$

as in Chapter 2. The set of positive (resp. negative) roots is denoted Φ^+ (resp. Φ^-). Let Λ_{root} be the sublattice of Λ spanned by the roots. See also Example 2.5.

If α is a root, let X_α be an element of the Lie algebra $\mathfrak{g}$ such that $\text{Ad}(t)X_\alpha = t^\alpha X_\alpha$, where $\text{Ad}\colon G \longrightarrow \text{GL}(\mathfrak{g})$ is the adjoint representation. Because we identify $\mathfrak{g}$ with $\text{Mat}_n(\mathbb{C})$, $\text{Ad}(t)X_\alpha$ is just $tX_\alpha t^{-1}$. For example, if α is the simple root α_i, then identifying $\mathfrak{g}$ with $\text{Mat}_n(\mathbb{C})$, we may take X_α to be the matrix that has 1 in the $i, i+1$ position, and zeros elsewhere. The *root operator* X_α acts on the Lie algebra by (A.1).

Proposition A.1. *The operator* X_α *maps* V_μ *into* $V_{\mu+\alpha}$.

Proof. We write gv instead of $\pi(g)v$ for $g \in G$, $v \in V$. Making use of the action (A.1), for $t \in T$ and $v \in V_\mu$ we have

$$tXv = \frac{d}{dh}te^{hX}v\Big|_{h=0} = \frac{d}{dh}e^{h\,\text{Ad}(t)X}tv\Big|_{h=0} = \frac{d}{dh}e^{ht^\alpha X}t^\mu v\Big|_{h=0}.$$

Now t^α may not be real, but by regularity this is an analytic function of X, so we are justified in using the chain rule and obtaining $t^{\mu+\alpha}Xv$. $\square$

Let $E_i = X_{\alpha_i}$ and $F_i = X_{-\alpha_i}$. These operators map V_μ into $V_{\mu+\alpha_i}$ and $V_{\mu-\alpha_i}$. They are the analogs of the Kashiwara raising and lowering operators e_i and f_i in the theory of crystals.

Proposition A.2 (Weyl's unitarian trick). *Let* (π, V) *be a finite-dimensional representation of* G. *If* $W \subseteq V$ *is a* $U(n)$-*invariant subspace, then it is* $\text{GL}(n, \mathbb{C})$-*invariant. Thus* V *is irreducible as a* G-*module if and only if it is irreducible as a* $U(n)$-*module.*

Proof. If W is invariant under $U(n)$, then it is invariant under the action of $\mathfrak{u}(n)$, which consists of skew-hermitian matrices. Observe that since $\pi\colon G \longrightarrow \text{GL}(V)$ is regular, for fixed $v \in V$ and $X \in \mathfrak{gl}(n, \mathbb{C})$ the map $t \mapsto e^{tX}v$ is analytic. Hence Xv defined by (A.1) is $\mathbb{C}$-linear in X as well as in v. Now the complex Lie algebra $\mathfrak{g} = \mathfrak{gl}(n, \mathbb{C})$ is the complexification of the real Lie algebra $\mathfrak{u}(n)$, i.e.

$$\mathfrak{g} = \mathfrak{u}(n) \oplus i\mathfrak{u}(n).$$

Therefore W is invariant under $\mathfrak{g}$ and by exponentiating under G. $\square$

Proposition A.3 (Weyl). *Every finite-dimensional representation decomposes into a direct sum of irreducibles.*

Proof. Complete reducibility of any finite-dimensional module V of $U(n)$ follows from the fact that $U(n)$ is compact. Indeed, we may find a $U(n)$-invariant inner product on V. A nonzero invariant subspace of minimal dimension is irreducible. Its orthogonal complement is invariant and decomposes into irreducible spaces by induction on dimension. These subspaces are irreducible for $\text{GL}(n, \mathbb{C})$ by the unitarian trick. $\square$

If (π, V) is a polynomial representation, we say that π is *homogeneous of degree* k if the coefficients $\pi(g)_{ij}$ are homogeneous polynomials of degree k. It is not hard to see that a polynomial irreducible representation is homogeneous, so if (π, V) is a polynomial representation, we may write $V = \bigoplus_{k \in \mathbb{N}} V_k$ where V_k is a homogeneous polynomial representation of degree k.

The *character* χ_π of a representation is the trace $\chi_\pi(g) = \text{tr } \pi(g)$. Because of our regularity assumption, it is a polynomial function of the g_{ij} and $\det^{-1}(g)$; if π is polynomial, $\det^{-1}(g)$ is not needed.

We will occasionally encounter representations that are not finite-dimensional. However, we will require that such a representation Π be a direct sum of finite-dimensional representations, each occurring with finite multiplicity. If only polynomial representations occur, then we may define the *graded character*. Indeed, we may write

$$\Pi = \bigoplus_{k=0}^{\infty} \Pi_k$$

where Π_k is the homogenous part of degree k; that is, each matrix coefficient $\pi(g)_{ij}$ is a homogeneous polynomial of degree k in the g_{ij}. Now the *graded character* is the power series

$$\chi_\Pi(g; q) = \chi_\Pi(q) = \sum_{k=0}^{\infty} \chi_{\Pi_k} q^k.$$

For example, consider the symmetric algebra $\Pi = \bigvee(\mathbb{C}^n)$ over $\mathbb{C}^n$, so $\Pi_k = \bigvee^k(\mathbb{C}^n)$ is the k-th symmetric power. If $t_1, \ldots, t_n$ are the eigenvalues of g, then

$$\chi_{\Pi_k}(g) = h_k(t_1, \ldots, t_n),$$

where

$$h_k(t_1, \ldots, t_n) = \sum_{i_1 \leqslant i_2 \leqslant \cdots \leqslant i_k} t_{i_1} \cdots t_{i_k} \tag{A.4}$$

is the k-th *complete symmetric polynomial*. It is the sum of all monomials of degree k in the t_i. The graded character is

$$H(q) = \sum_{k=0}^{\infty} h_k(t_1, \ldots, t_n) q^k = \prod_{i=1}^{n} (1 - qt_i)^{-1}. \tag{A.5}$$

Similarly, consider the exterior algebra $\Pi = \bigwedge(\mathbb{C}^n)$ over $\mathbb{C}^n$. The k-th exterior power $\bigwedge^k(\mathbb{C}^n)$ is zero unless $k \leqslant n$, so unlike the symmetric algebra, $\bigwedge(\mathbb{C}^n)$ is finite-dimensional. The graded character is

$$E(q) = \sum_{k=0}^{n} e_k(t_1, \ldots, t_n) q^k = \prod_{i=1}^{n} (1 + qt_i), \tag{A.6}$$

where

$$e_k(t_1, \ldots, t_n) = \sum_{i_1 < \cdots < i_k} t_{i_1} \cdots t_{i_k} \tag{A.7}$$

is the k-th *elementary symmetric polynomial.*

If $\pi\colon G \longrightarrow \mathrm{GL}(V)$ is a representation, we may restrict π to T, and then V decomposes into the direct sum of *weight spaces* V_μ defined by (A.3). The dimension $\dim(\mu) = \dim V_\mu$ is called the *weight multiplicity*. With t as in (A.2),

$$\chi_\pi(t) = \sum_{\mu \in \Lambda} \dim(\mu)\, t^\mu. \tag{A.8}$$

Let $N(T)$ be the normalizer of T, the group of monomial matrices. The quotient $W = N(T)/T \cong S_n$ is the *Weyl group*. The Weyl group W acts on Λ by conjugation. If we identify $\Lambda = \mathbb{Z}^n$, this is just the permutation action of S_n on the coordinates of a weight as element of $\mathbb{Z}^n$. Since the character is invariant under conjugation, its restriction to T is invariant under conjugation by $N(T)$ and so the function $\dim(\mu)$ in (A.8) is invariant under this action of W on Λ. The μ such that $\dim(\mu) \neq 0$ are called the *weights of* π.

The Weyl group W is generated by the *simple reflections* s_i, $i = 1, \ldots, n-1$. In the action on $\lambda \in \Lambda$, s_i interchanges λ_i and λ_{i+1}. Geometrically, it is the reflection in the hyperplane orthogonal to α_i.

A weight $\lambda = (\lambda_1, \ldots, \lambda_n)$ is called *dominant* if

$$\lambda_1 \geqslant \lambda_2 \geqslant \cdots \geqslant \lambda_n. \tag{A.9}$$

Under the action of the Weyl group, every weight is equivalent to a unique dominant weight. The dominant weights may be characterized by the assumption that $\langle \lambda, \alpha_i \rangle \geqslant 0$ with respect to the W invariant standard inner product on Λ, which coincides with the usual Euclidean inner product on $\mathbb{Z}^n$.

We recall that a partition λ of length $\leqslant n$ is a weakly decreasing sequence $(\lambda_1, \ldots, \lambda_k)$ with $k \leqslant n$ of nonnegative integers. We recall that we identify two such sequences if they differ by some trailing zeros, so λ is identified with the sequence $(\lambda_1, \ldots, \lambda_n)$ where $\lambda_i = 0$ for $i > k$. Thus a partition of length $\leqslant n$ is a dominant weight for $\mathrm{GL}(n, \mathbb{C})$. Not every dominant weight is a partition, since some of its entries could be negative. If $\lambda_n \geqslant 0$, then a dominant weight λ is a partition.

Let $\mathcal{C}$ be the cone of dominant weights, called the *positive Weyl chamber*. The *dual cone* $\mathcal{C}'$ may be defined as the set of all λ such that $\langle \lambda, \nu \rangle \geqslant 0$ for all $\nu \in \mathcal{C}$. Thus the condition for $\lambda - \mu$ to be in $\mathcal{C}'$ is that

$$\lambda_1 \geqslant \mu_1, \quad \lambda_1 + \lambda_2 \geqslant \mu_1 + \mu_2, \quad \lambda_1 + \lambda_2 + \lambda_3 \geqslant \mu_1 + \mu_2 + \mu_3, \quad \ldots.$$

We will write $\lambda \succcurlyeq \mu$ if $\lambda - \mu \in \mathcal{C}'$. (This definition is consistent with Definition 2.4.) This condition implies that $\sum_i \lambda_i = \sum_i \mu_i$ since both $(1, \ldots, 1)$ and its negative are dominant. It is not hard to see that $\lambda \succcurlyeq \mu$ if and only if $\lambda - \mu$ is a linear combination of $\alpha_1, \ldots, \alpha_{n-1}$ with nonnegative integer coefficients. If λ and μ are partitions of the same integer k, this partial order is called the *dominance order* but we warn the reader that this terminology is misleading, since the dominance order is really dual to the partial order that has the dominant weights as its positive cone. That is, $\lambda \succcurlyeq \mu$ means that $\langle \lambda - \mu, \nu \rangle \geqslant 0$ for all dominant weights ν.

If π is a finite-dimensional representation of $\mathrm{GL}(n,\mathbb{C})$, not necessarily irreducible, a *maximal weight* for π is a weight that is maximal with respect to the order $\succeq$. That is, λ is a *maximal weight* if the weight multiplicity $m(\lambda) \neq 0$, but $m(\mu) = 0$ whenever $\mu \succ \lambda$. If π is irreducible, we use the term *highest weight* for a maximal weight. If π is not irreducible, then by a *highest weight* we mean a weight that is the highest weight of some irreducible constitutent of V. With this usage, a highest weight may not be maximal.

It is clear that every finite-dimensional representation has maximal weights, but without assuming that π is irreducible, we cannot be sure that it does not have more than one.

Lemma A.4. *A maximal weight is dominant.*

Proof. Given a maximal weight λ, if λ is not dominant, then $\lambda_i < \lambda_{i+1}$ for some i. Let $\mu = s_i(\lambda)$. Then the weight multiplicity $m(\mu)$ equals $m(\lambda)$ and is hence nonzero. But $\mu \succ \lambda$ contradicting the assumption that λ is a maximal weight. $\square$

Theorem A.5 (The Weyl character formula). *Let π be an irreducible finite-dimensional representation of $\mathrm{GL}(n,\mathbb{C})$. Then π has a unique highest weight λ. The multiplicity $\dim(\lambda)$ of this highest weight equals 1. Moreover, π is the unique irreducible representation with this highest weight. In addition, every dominant weight is the highest weight of a unique irreducible representation, and so $\pi \leftrightarrow \lambda$ is a bijection between the isomorphism classes of irreducible representations and dominant weights. Let $t \in T$. Then*

$$\chi_\pi(t) = \frac{\displaystyle\sum_{w \in W} \mathrm{sgn}(w) t^{w(\lambda+\rho)}}{\displaystyle\sum_{w \in W} \mathrm{sgn}(w) t^{w(\rho)}}, \tag{A.10}$$

where $\rho = (n-1, n-2, \ldots, 0)$.

Here sgn is the sign character of S_n. For a proof, see [Bump (2013), Theorem 36.2 or Theorem 22.3]. A caveat here: in (A.10), both the numerator and denominator vanish if t has a repeated eigenvalue. However the ratio can be interpreted as a limiting value.

We denote by $\pi_\lambda^{\mathrm{GL}(n)}$ the irreducible representation with highest weight λ. If λ is a partition of k, then λ and indeed every weight of $\pi = \pi_\lambda^{\mathrm{GL}(n)}$ is a monomial of degree k. So the character χ_π is a homogeneous polynomial $s_\lambda = s_\lambda^{(n)}$ of degree k in the eigenvalues. The polynomials s_λ are called *Schur polynomials*. They are the same as the Schur polynomials defined in (3.3).

Both the numerator and the denominator in (A.10) may be written as determinants, so

$$s_\lambda(t_1, \ldots, t_n) = \frac{\det(t_i^{\lambda_j + n - 1 - j})}{\det(t_i^{n-1-j})}. \tag{A.11}$$

The denominator in (A.10) may be rewritten

$$\sum_{w \in W} \text{sgn}(w) t^{w(\rho)} = t^{\rho} \prod_{\alpha \in \Phi^+} (1 - t^{-\alpha}).$$

This is the *Weyl denominator formula* for $\text{GL}(n)$. The Weyl denominator formula for $\text{GL}(n)$ is equivalent to the well-known Vandermonde determinant formula

$$\det(t_i^{n-1-j}) = \prod_{i<j}(t_i - t_j).$$

Proposition A.6. *Let λ and μ be dominant weights. Then $\pi_{\lambda}^{\text{GL}(n)} \otimes \pi_{\mu}^{\text{GL}(n)}$ contains $\pi_{\lambda+\mu}^{\text{GL}(n)}$ with multiplicity one.*

Proof. We observe that, if ν_1 is any weight of $\pi_{\lambda} = \pi_{\lambda}^{\text{GL}(n)}$, then $\lambda \succcurlyeq \nu_1$. (This is a consequence of the fact that an irreducible representation has a unique maximal weight.)

Let mult_{λ} and mult_{μ} be the weight multiplicity functions for π_{λ} and π_{μ}, and let mult be the weight multiplicity function for their tensor product. Then

$$\text{mult}(\nu) = \sum_{\nu=\nu_1+\nu_2} \text{mult}_{\lambda}(\nu_1) \, \text{mult}_{\mu}(\nu_2). \tag{A.12}$$

The first step in the proof is to show that $\lambda + \mu$ is the unique maximal weight in $\pi_{\lambda} \otimes \pi_{\mu}$, and moreover that $\text{mult}(\lambda + \mu) = 1$. If ν is a weight of $\pi_{\lambda} \otimes \pi_{\mu}$, then $\nu = \nu_1 + \nu_2$, where $\lambda \succcurlyeq \nu_1$ and $\mu \succcurlyeq \nu_2$, so $\lambda + \mu \succcurlyeq \nu_1 + \nu_2$. This proves that $\lambda + \nu$ is the unique maximal weight of $\pi_{\lambda} \otimes \pi_{\mu}$. The only nonzero contribution to $\text{mult}(\lambda + \mu)$ in (A.12) is from $\nu_1 = \lambda$, $\nu_2 = \mu$, and since $\text{mult}_{\lambda}(\lambda) = \text{mult}_{\mu}(\mu) = 1$, we see that $\text{mult}(\lambda + \mu) = 1$.

Now decomposing $\pi_{\lambda} \otimes \pi_{\nu}$ into irreducibles, the weight $\lambda + \mu$ must appear in some summand π_{θ}. Thus $\theta \succcurlyeq \lambda + \mu$. Since $\lambda + \mu$ is a maximal weight vector, it follows that $\lambda + \mu = \theta$ and so $\pi_{\lambda+\mu}$ appears (with multiplicity exactly one) in $\pi_{\lambda} \otimes \pi_{\mu}$. $\qquad\square$

Proposition A.7. *Let λ be any dominant weight for $\text{GL}(n, \mathbb{C})$ that is a partition of k. Then $\pi_{\lambda}^{\text{GL}(n)}$ appears in the $\text{GL}(n, \mathbb{C})$-module $(\mathbb{C}^n)^{\otimes k}$.*

Proof. We observe that this is true if $\lambda = (1^k)$ so that $s_{\lambda} = e_k$, which was defined in (A.7). Indeed, in this case π_{λ} is $\bigwedge^k \mathbb{C}^n$ which of course can be realized as a summand in $(\mathbb{C}^n)^{\oplus k}$. To emphasize that (1^k) is a dominant weight, we denote it by ϖ_k. Any dominant weight λ that is a partition can be written as $\sum_{i=1}^{n} c_i \varpi_i$ with nonnegative integers c_i. So applying Proposition A.6 (or its immediate generalization to a tensor product of several irreducibles), we see that π_{λ} is a direct summand in

$$\left(\bigwedge^i \mathbb{C}^n\right)^{\otimes c_i} \subset \left((\mathbb{C}^n)^{\otimes i}\right)^{\otimes c_i} = (\mathbb{C}^n)^{\otimes k}$$

since $\sum c_i \cdot i = k$. $\qquad\square$

A.2 The Schur–Weyl duality correspondence

Schur–Weyl duality (also sometimes called *Frobenius–Schur duality* or *Schur duality*) is a relationship between the representation theory of S_k and the representation theory of $G = \mathrm{GL}(n, \mathbb{C})$. Consider the standard module $\mathbb{C}^n$ of G. Both G and S_k act on $(\mathbb{C}^n)^{\otimes k}$. The group G acts diagonally by

$$g(v_1 \otimes \cdots \otimes v_k) = gv_1 \otimes \cdots \otimes gv_k. \tag{A.13}$$

On the other hand, S_k acts by permuting the entries. Writing this as a right action, $w \in S_k$ acts as follows:

$$(v_1 \otimes \cdots \otimes v_k)w = v_{w(1)} \otimes \cdots \otimes v_{w(k)}. \tag{A.14}$$

So we may regard $(\mathbb{C}^n)^{\otimes k}$ as a $(G \times S_k)$-module. Suppose that Ω is a vector space and A is a ring of endomorphisms of Ω. Then the *commuting ring* of A is the ring of endomorphisms that commute with every element of A.

Proposition A.8. *Let A be the ring of endomorphisms of $(\mathbb{C}^n)^{\otimes k}$ generated the transformations (A.14), and let B be the ring of endomorphisms of $(\mathbb{C}^n)^{\otimes k}$ generated by the transformations (A.13). Then B is the commuting ring of A and A is the commuting ring of B.*

Proof. In this proof we denote $V = \mathbb{C}^n$ and $\Omega = V^{\otimes k}$.

Since B is a homomorphic image of the group algebra $\mathbb{C}[S_k]$, which is a semisimple ring, B is semisimple. It follows from the Jacobson Density Theorem (see [Lang (2002), p. 647]) that if A' is the commuting ring of B, then B is the commuting ring of A'. So it is enough to show that A is the commuting ring of B.

If X and Y are finite-dimensional vector spaces, then we have a bilinear map $X^* \times Y \longrightarrow \mathrm{Hom}(X, Y)$ in which $x^* \otimes y \in X^* \otimes Y$ is mapped to the linear transformation $x \mapsto \langle x, x^* \rangle y$. This bilinear map induces a natural isomorphism $X^* \otimes Y \cong \mathrm{Hom}(X, Y)$. In particular, we have a natural isomorphism

$$\mathrm{Hom}(X_1 \otimes \cdots \otimes X_k, Y_1 \otimes \cdots \otimes Y_k) \cong \bigotimes_{i=1}^{k} \mathrm{Hom}(X_i, Y_i) \tag{A.15}$$

since both are isomorphic to

$$\left(\bigotimes_{i=1}^{k} X_i^* \right) \otimes \left(\bigotimes_{i=1}^{k} Y_i \right).$$

We apply this with all X_i, Y_i equal to V. Define

$$\beta \colon \mathrm{End}(V) \times \cdots \times \mathrm{End}(V) \longrightarrow \mathrm{End}(\Omega)$$

by letting $\beta(f_1, \ldots, f_k)$ be the image of $f_1 \otimes \cdots \otimes f_k$ ($f_i \in \mathrm{End}(V)$) under the isomorphism (A.15). Let $Q(f) = \beta(f, \ldots, f)$ (k factors). Conjugating $\beta(f_1, \ldots, f_k)$ by an element of S_k acting on Ω via the action (A.14) on $V^{\otimes k}$ permutes the f_i.

Therefore, an element of the commuting ring of A is in the linear span of the image of the symmetrization

$$\beta'(f_1,\ldots,f_k) = \frac{1}{k!}\sum_{w\in S_k}\beta(f_{w(1)},\ldots,f_{w(k)}).$$

Let $Q(f) = \beta'(f,\ldots,f) = \beta(f,\ldots,f)$. Fix $f_1,\ldots,f_k$. If I is a subset of $[k] := \{1,\ldots,k\}$ let $f_I = \sum_{i\in I}f_i$. By inclusion-exclusion,

$$\beta'(f_1,\ldots,f_k) = \frac{1}{k!}\sum_{I\subseteq[k]}(-1)^{k-|I|}Q(f_I).$$

It follows that every element of the commuting ring of A is a linear combination of elements of $\mathrm{End}(\Omega)$ of the form $Q(f)$, $f\in\mathrm{End}(V)$. If f is invertible, then $Q(f)$ is a transformation (A.13), and in any case $Q(f)$ may be approximated by such transformations. It is thus clear that the commuting ring of A is B. $\qquad\square$

Proposition A.9. *Let Ω be a finite-dimensional vector space and let A and B be subalgebras of $\mathrm{End}(\Omega)$. Assume that A is the commuting ring of B and B is the commuting ring of A. Then $\mathrm{End}(\Omega)$ is a homomorphic image of $A\otimes B$, so Ω is a module for $A\times B$. Assume that Ω is isomorphic to $\bigoplus_i U_i\otimes W_i$ where U_i are A-modules and W_i are B-modules. If $i\neq j$, then U_i is not isomorphic to U_j and W_i is not isomorphic to W_j.*

Proof. If $U_i\cong U_j$ as A-modules let $\phi\colon U_i\longrightarrow U_j$ be an A-module isomorphism. Let $\psi\colon W_i\longrightarrow W_j$ be an arbitrary map. We consider the endomorphism of Ω that is $\phi\otimes\psi$ on the summand isomorphic to $U_i\otimes W_i$ and zero on the other summands. This endomorphism commutes with A but is not in B since it does not map $U_i\otimes W_i$ into itself. This is a contradiction since B is the commuting ring of A. $\qquad\square$

Theorem A.10. *The $\mathrm{GL}(n,\mathbb{C})\times S_k$ module $(\mathbb{C}^n)^{\otimes k}$ decomposes as follows:*

$$(\mathbb{C}^n)^{\otimes k} = \bigoplus_\lambda \pi_\lambda^{\mathrm{GL}(n)}\otimes\pi_\lambda^{S_k}. \tag{A.16}$$

Here λ runs through the partitions of k of length $\leqslant n$; $\pi_\lambda^{\mathrm{GL}(n)}$ is the irreducible representation of $\mathrm{GL}(n,\mathbb{C})$ with highest weight λ. The representation $\pi_\lambda^{S_k}$ is an irreducible representation of S_k. Every irreducible representation of $\mathrm{GL}(n,\mathbb{C})$ whose highest weight is a partition of k appears in this representation. Moreover if $n\geqslant k$, then every irreducible representation of S_k occurs exactly once in this decomposition.

Proof. In this proof, let $V = \mathbb{C}^n$ and $\Omega = V^{\otimes k}$. The weights of Ω are all homogeneous monomials of degree k, since if t is as in (A.2), t multiplies the basis vector $\mathbf{e}_{i_1}\otimes\cdots\otimes\mathbf{e}_{i_k}$ by $t_{i_1}\cdots t_{i_k}$. Therefore the irreducible representations of $\mathrm{GL}(n)$ that can occur are the $\pi_\lambda^{\mathrm{GL}(n)}$, where the highest weight λ is a partition of k. Thus we have a decomposition (A.16) where $\pi_\lambda^{S_k}$ are some representations of S_k. By Proposition A.9 and Proposition A.8, there are no repetitions among the $\pi_\lambda^{S_k}$ and they are irreducible.

The fact that all λ that are partitions of k of length $\leqslant n$ appear in this decomposition follows from Proposition A.7. These are exactly the highest weights of irreducible representations of $\mathrm{GL}(n, \mathbb{C})$ that are polynomial and homogeneous of degree k.

If $n \geqslant k$, the number of such λ equals $p(k)$, the number of partitions of k, so we have constructed this many distinct irreducible representations of S_k. However, S_k has this number of conjugacy classes, so we have constructed all of the irreducibles.

$\square$

The concept of a *correspondence* has been emphasized by Howe. (See for example [Howe (1989, 1995)].) Let Ω be a representation of a direct product $G_1 \times G_2$. For our purposes, we say that Ω is a *correspondence* if, when decomposed into irreducibles:

$$\Omega = \bigoplus_i \pi_i^{G_1} \otimes \pi_i^{G_2} \tag{A.17}$$

and there are no repetitions among the $\pi_i^{G_1}$ or the $\pi_i^{G_2}$. (This definition must be modified if Ω is not completely reducible.)

Given a correspondence (A.17), there is a bijection between a set of representations of G_1, namely the $\pi_i^{G_1}$, and a set of representations of G_2, namely the $\pi_i^{G_2}$. We use the following notation to indicate this:

$$\pi_i^{G_1} \xleftarrow{\ \Omega\ } \pi_i^{G_2}. \tag{A.18}$$

We see that if $\Omega = V^{\otimes k}$, then Ω is a correspondence for S_k and $\mathrm{GL}(n, \mathbb{C})$, called the *Schur* correspondence. This fact is known as *Schur–Weyl duality*, or *Schur duality*, or *Frobenius–Schur duality*.

A.3   Symmetric functions

We recall some facts about symmetric function theory. See [Macdonald (1995), Chapter I] as well as [Bump (2013), Part II], [Stanley (1999), Chapter 7], and [Sagan (2001)] for further details.

Let $x_1, \ldots, x_n$ be indeterminates and let $\mathrm{Sym}^{(n)}$ be the ring of symmetric polynomials in the x_i with integer coefficients. We have a homomorphism $\mathrm{Sym}^{(n)} \longrightarrow \mathrm{Sym}^{(n-1)}$ which sends $x_n \mapsto 0$, and the ring Sym is the inverse limit. We call elements of Sym *symmetric functions* (even though by this definition they are not really functions).

Lemma A.11. *Assume that $\ell(\lambda) \leqslant n - 1$. Then $s_\lambda^{(n)} \mapsto s_\lambda^{(n-1)}$ under the homomorphism* $\mathrm{Sym}^{(n)} \longrightarrow \mathrm{Sym}^{(n-1)}$ *that sends $x_n \mapsto 0$.*

Proof. Since $\ell(\lambda) \leqslant n-1$ we have $\lambda_n = 0$. Therefore the last columns of the matrices (t_i^{n-1-j}) and $(t_i^{\lambda_j + n - 1 - j})$ are the same, and turn into the column vector ${}^t(0, \ldots, 0, 1)$ when t_n is replaced by 0. Making minor expansions in the last columns of the numerator and denominator in (A.11) gives the numerator and denominator for $s_\lambda^{(n-1)}$. □

It follows from this compatibility that there is a unique element $s_\lambda \in \mathrm{Sym}$ whose image in $\mathrm{Sym}^{(n)}$ is $s_\lambda^{(n)}$ for all n. The ring Sym is graded by degree, and the s_λ with λ a partition of k form a basis of the homogenous part Sym_k, which is a free abelian group of rank $p(k)$, the number of partitions of k. We call the s_λ *Schur functions*.

If $\lambda = (k)$, then $s_{(k)} = h_k$, where h_k is as in (A.4). This is because $h_k(t_1, \ldots, t_n)$ is the character of an irreducible module, namely $\bigvee^k(\mathbb{C}^n)$, and its highest weight is the monomial $t^\lambda = t_1^k$. Similarly $s_{(1^k)} = e_k$, where (1^k) denotes the partition $(1, \ldots, 1)$ (k times), because it is the character of an irreducible representation $\bigwedge^k(\mathbb{C}^n)$ and its highest weight is (1^k). As with the general Schur polynomials, we have $e_k^{(n)} \mapsto e_k^{(n-1)}$ under the homomorphisms $\mathrm{Sym}^{(n)} \longrightarrow \mathrm{Sym}^{(n-1)}$ that sends x_n to zero, so there is a unique element e_k of Sym whose image in $\mathrm{Sym}^{(n)}$ is $e_k^{(n)}$. An element $h_k \in \mathrm{Sym}$ is defined similarly. These are the *elementary* and *complete* symmetric functions, respectively.

Proposition A.12. *The ring* Sym *is a polynomial ring in either the e_k or h_k:*

$$\mathrm{Sym} = \mathbb{Z}[e_1, e_2, e_3, \ldots] = \mathbb{Z}[h_1, h_2, h_3, \ldots].$$

The ring Sym *has an involution* $\iota \colon \mathrm{Sym} \longrightarrow \mathrm{Sym}$ *that interchanges the e_k and the h_k.*

Now let us define a graded ring $\mathcal{R}$ from the representations of the symmetric group. The homogeneous part $\mathcal{R}_k$ consists of the Grothendieck group of virtual representations of S_k. We describe the multiplicative structure of this ring. Let ϕ and ψ be representations of S_k and S_ℓ, and let $[\phi] \in \mathcal{R}_k$ and $[\psi] \in \mathcal{R}_\ell$ be the corresponding classes. Noting that $S_k \times S_\ell$ is embedded as a subgroup of $S_{k+\ell}$, we may induce its module $\phi \otimes \psi$ to $S_{k+\ell}$, and the product $[\phi] \cdot [\psi]$ is defined to be the class of this induced representation. We will sometimes call the subgroup $S_k \times S_\ell$ of $S_{k+\ell}$ a *Levi subgroup*.

Proof. It is well-known (and due to Newton) that the polynomials $e_1^{(n)}, \ldots, e_k^{(n)}$ are algebraically independent and generate $\mathrm{Sym}^{(n)}$ as a ring. See [Lang (2002), pg. 192]. So the e_i are algebraically independent in Sym and generate it as a polynomial ring.

Hence there is a ring homomorphism $\iota \colon \mathrm{Sym} \longrightarrow \mathrm{Sym}$ that maps e_i to h_i. We observe the identity $E(q)H(-q) = 1$, with $E(q)$ and $H(q)$ as in (A.6) and (A.5). This follows from the product expressions in those equations, and it implies $H(q)E(-q) = 1$. Thus the relations that express the h_i in terms of the e_i also express the e_i in terms of the h_i. Therefore ι is an involution. As a consequence the h_i are algebraically independent and generate Sym as a polynomial ring. □

We will show that the ring $\mathcal{R}$ is isomorphic to Sym. This is a reflection of Schur duality, with class $[\pi_\lambda^{S_k}]$ in $\mathcal{R}$ corresponding to the Schur function s_λ in Sym. It follows from Theorem A.10 that this description gives a linear map $\mathrm{ch} \colon \mathcal{R} \longrightarrow \mathrm{Sym}$ that is a bijection. This is Frobenius' *characteristic map*. We will prove in the next section that it is a ring homomorphism.

A.4   See-saws

Before we prove that ch is a ring homomorphism, we introduce another tool, called a *see-saw*. The terminology is from [Kudla (1984)]. A useful paper with applications of various see-saws to branching rules is [Howe, Tan and Willenbring (2005)], and a recent paper in the same spirit (using crystals) is [Kwon (2015)].

Suppose that G_1 and G_2 are groups and $H_i \subset G_i$ are subgroups. We assume that Ω is a vector space with actions of G_1 and G_2. We do not assume that the action of G_1 commutes with the action of G_2, but instead we assume that the action of G_1 commutes with the action of H_2 and that the action of G_2 commutes with the action of H_1. Thus Ω becomes either a $(G_1 \times H_2)$-module or a $(G_2 \times H_1)$-module. We assume that Ω is a correspondence for either of these actions. Thus we have a diagram:

$$
\begin{matrix}
G_1 & & G_2 \\
\uparrow & \times & \uparrow \\
H_1 & & H_2
\end{matrix}
\tag{A.19}
$$

The vertical arrows are inclusion of subgroups and the diagonal lines are correspondences.

Here is an example of a see-saw. Let G_1 be $S_{k+\ell}$ acting on $V^{\otimes(k+\ell)}$ where $V = \mathbb{C}^n$, and let H_2 be $\mathrm{GL}(V)$ acting as in (A.13). Now let H_1 be the *parabolic subgroup* $S_k \times S_\ell$ of $S_{k+\ell}$, in which the S_k acts on $\{1, \ldots, k\}$ and the S_ℓ acts on $\{k+1, \ldots, k+\ell\}$. Let G_2 be the group $\mathrm{GL}(V) \times \mathrm{GL}(V)$ acting via the following modification of (A.13):

$$
(g, h)(v_1 \otimes \cdots \otimes v_k) = g v_1 \otimes \cdots \otimes g v_k \otimes h v_{k+1} \otimes \cdots \otimes h v_{k+\ell}.
$$

We embed $H_2 = \mathrm{GL}(V)$ into $G_2 = \mathrm{GL}(V) \times \mathrm{GL}(V)$ diagonally via $g \mapsto (g, g)$. Then Ω is a correspondence for $G_1 \times H_2$ by Propositions A.8 and A.9. It is a correspondence for $G_2 \times H_1$ for the same reason. Thus we obtain the following see-saw:

$$
\begin{matrix}
S_{k+\ell} & & \mathrm{GL}(n) \times \mathrm{GL}(n) \\
\uparrow & \times & \uparrow \\
S_k \times S_\ell & & \mathrm{GL}(n)
\end{matrix}
\tag{A.20}
$$

Proposition A.13. *Let (A.19) be a see-saw diagram, let π^{G_i} be representations of G_i (i = 1, 2), and let σ^{H_i} be representations of the H_i. Suppose, in the notation (A.18), that*

$$\pi^{G_1} \xleftarrow{\ \Omega\ } \sigma^{H_2}, \qquad \pi^{G_2} \xleftarrow{\ \Omega\ } \sigma^{H_1}.$$

Then the multiplicity of σ^{H_1} in the restriction of π^{G_1} to H_1 equals the multiplicity of σ^{H_2} in the restriction of π^{G_2} to H_2.

Proof. We will show that both multiplicities equal the dimension of

$$\mathrm{Hom}_{H_1 \times H_2}(\sigma^{H_1} \otimes \sigma^{H_2}, \Omega). \tag{A.21}$$

Indeed, $\mathrm{Hom}_{H_2}(\sigma^{H_2}, \Omega)$ is naturally a G_1-module isomorphic to π^{G_1} and (A.21) may be identified with

$$\mathrm{Hom}_{H_1}(\sigma^{H_1}, \mathrm{Hom}_{H_2}(\sigma^{H_2}, \Omega)) \cong \mathrm{Hom}_{H_1}(\sigma^{H_1}, \pi^{G_1}).$$

Its dimension is the multiplicity of σ^{H_1} in the restriction of π^{G_1} restricted to H_1. By a symmetrical argument, it is also the multiplicity of σ^{H_2} in the restriction of π^{G_2} restricted to H_2. □

Theorem A.14. *The characteristic map* ch: $\mathcal{R} \longrightarrow$ Sym *is a ring homomorphism. If 1_{S_k} and sgn_{S_k} denote the trivial and sign characters of S_k, then $\mathrm{ch}[1_{S_k}] = h_k$ and $\mathrm{ch}[\mathrm{sgn}_{S_k}] = e_k$.*

Proof. The main task here is to show that ch is multiplicative. Let $\lambda \vdash k$, $\mu \vdash \ell$ and $\nu \vdash k + \ell$. Let $[\pi_\lambda^{S_k}]$, $[\pi_\mu^{S_\ell}]$ and $[\pi_\nu^{S_{k+\ell}}]$ be their classes in $\mathcal{R}$. By definition of the multiplication in the ring $\mathcal{R}$ the product $[\pi_\lambda^{S_k}] \cdot [\pi_\mu^{S_\ell}]$ is the class of the induced representation in $S_{k+\ell}$, and by Frobenius reciprocity the multiplicity of π_ν in this product equals the multiplicity of the representation $\pi_\lambda^{S_k} \otimes \pi_\mu^{S_\ell}$ in the restriction of $\pi_\nu^{S_{k+\ell}}$ to this subgroup. Applying Proposition A.13 to the see-saw in (A.20), this also equals the multiplicity of $\pi_\nu^{\mathrm{GL}(n)}$ in $\pi_\lambda^{\mathrm{GL}(n)} \otimes \pi_\mu^{\mathrm{GL}(n)}$. Taking the character, this equals the multiplicity of s_ν in $s_\lambda s_\mu$ when it is expanded in Schur functions. This tells us that ch is multiplicative and therefore is a ring homomorphism.

To identify $\mathrm{ch}[1_{S_k}]$ and $\mathrm{ch}[\mathrm{sgn}_{S_k}]$ we observe that the k-th symmetric and exterior powers of $\mathbb{C}^n$ appear as submodules of $(\mathbb{C}^n)^{\otimes k}$, showing that in the Schur correspondence

$$\bigvee^{\mathrm{GL}(n)} \xleftarrow{\ \mathrm{Schur}\ } 1_{S_k}, \qquad \bigwedge^{\mathrm{GL}(n)} \xleftarrow{\ \mathrm{Schur}\ } \mathrm{sgn}_{S_k}.$$

The corresponding symmetric polynomials are h_k and e_k. □

Appendix B

The Cauchy Correspondence

If G is a group and H a subgroup, a *branching rule* is an explicit description of how irreducible representations of G decompose into irreducibles of H when they are restricted. In this chapter we will discuss, among other things, branching rules for the general linear groups and symmetric groups, and relationships between different branching rules. The principal tool, $\mathrm{GL}(n) \times \mathrm{GL}(m)$ duality, might well be called the *Cauchy correspondence* because of its close relationship with the Cauchy identity. The results in this Appendix are analogous to the combinatorics in Chapter 9 on $\mathrm{GL}(n) \times \mathrm{GL}(m)$ bicrystals.

B.1   The Cauchy identity

An *affine algebraic group* Γ is an affine algebraic variety that is also a group such that the group multiplication map $\Gamma \times \Gamma \longrightarrow \Gamma$ and the inverse map $\Gamma \longrightarrow \Gamma$ are regular. Let $A = \mathcal{O}(\Gamma)$ denote the coordinate ring or affine algebra of regular functions on Γ. It is a finitely generated $\mathbb{C}$-algebra that is a reduced Noetherian ring. Since $\mathcal{O}(\Gamma \times \Gamma) = A \otimes A$, the multiplication $\Gamma \times \Gamma \longrightarrow \Gamma$ corresponds to an algebra homomorphism $A \longrightarrow A \otimes A$. This *comultiplication* $A \longrightarrow A \otimes A$ makes A into a commutative Hopf algebra, with the antipode corresponding to the inverse map $\Gamma \longrightarrow \Gamma$.

By a *finite-dimensional representation* (π, V) of Γ, we mean a homomorphism $\pi \colon \Gamma \longrightarrow \mathrm{GL}(V)$, where V is a complex vector space, that is a regular map. If $\Gamma = \mathrm{GL}(n, \mathbb{C})$ this agrees with our previous definition. We assume that it is known that every finite-dimensional representation decomposes into a direct sum of irreducibles; for example if $\Gamma = \mathrm{GL}(n, \mathbb{C})$, this follows from the unitarian trick (see Propositions A.2 and A.3). Let $\Gamma \times \Gamma$ act on A by right and left multiplication. That is, $(g_1, g_2) f(x) = f(g_2^{-1} x g_1)$.

Proposition B.1. *With this notation, A decomposes as a $\Gamma \times \Gamma$-module as*

$$A \cong \bigoplus \pi \otimes \hat{\pi},$$

where the summation runs over all finite-dimensional representations π of Γ, and $\hat{\pi}$ denotes the contragredient representation.

Proof. Let (π, V) be a finite-dimensional representation. We may embed $V \otimes V^*$ into A by mapping $v \otimes L$, where $v \in V$ and $L \in V^*$, to the function $g \mapsto L(\pi(g)v)$. It may easily be checked that this embedding is $\Gamma \times \Gamma$-equivariant. It must be shown that every element f of A is a linear combination of such functions. Indeed, let $\Delta f = \sum_i \phi_i \otimes \psi_i$, where $\phi_i, \psi_i \in A$. Thus, if $x, y \in \Gamma$, $f(xy) = \sum_i \phi_i(x)\psi_i(y)$. It follows that the space of right translates V of f is finite-dimensional, spanned by the ϕ_i. Let $\rho \colon \Gamma \longrightarrow \text{End}(V)$ denote the action by right translation, so $\rho(g)\phi(x) = \phi(xg)$, and let $L \colon V \longrightarrow \mathbb{C}$ be the map $L(\phi) = \phi(1)$. Then $f(g) = L(\rho(g)f)$. Decomposing V into irreducibles, we obtain the stated decomposition. $\square$

Proposition B.2. *If π is a representation of $\text{GL}(n, \mathbb{C})$, then the contragredient representation $\hat{\pi}$ is equivalent to the representation $\pi'(g) = \pi({}^t g^{-1})$.*

Here ${}^t g$ is the transpose of the matrix g.

Proof. Indeed, if $t_1, \ldots, t_n$ are the eigenvalues of g, then $\chi_{\hat{\pi}}(g) = \sum_{i=1}^n t_i^{-1}$. Now, because g is conjugate to its transpose, we see that $\hat{\pi}$ and π' have the same character, so they are equivalent. $\square$

Theorem B.3. *Let $G_1 = \text{GL}(n, \mathbb{C})$ and $G_2 = \text{GL}(m, \mathbb{C})$. Let $G_1 \times G_2$ act on $\mathbb{C}^n \otimes \mathbb{C}^m$. Then $G_1 \times G_2$ acts on the symmetric algebra $\bigvee(\mathbb{C}^n \otimes \mathbb{C}^m)$. We have the decomposition*

$$\bigvee(\mathbb{C}^n \otimes \mathbb{C}^m) = \bigoplus_\lambda \pi_\lambda^{\text{GL}(n)} \otimes \pi_\lambda^{\text{GL}(m)}, \tag{B.1}$$

where λ runs through all partitions of length $\leqslant \min\{m, n\}$.

If $t_1, \ldots, t_n$ are the eigenvalues of $g_1 \in \text{GL}(n, \mathbb{C})$ and $u_1, \ldots, u_m$ are the eigenvalues of $g_2 \in \text{GL}(m, \mathbb{C})$, then comparing the graded characters in this decomposition for (g_1, g_2) gives the *Cauchy identity*

$$\prod_{i,j}(1 - t_i u_j q)^{-1} = \sum_\lambda s_\lambda(t)\, s_\lambda(u)\, q^{|\lambda|}, \tag{B.2}$$

where the sum is over partitions of length $\leqslant \min\{m, n\}$.

Proof of Theorem B.3. We will first prove the decomposition in the case $n = m$ and $G_1 = G_2$.

Instead of the action on the tensor product, we consider the action of $G = G_1 = G_2$ on the dual vector space $\text{Mat}_n(\mathbb{C})^*$ of $\text{Mat}_n(\mathbb{C})$ by

$$(g_1, g_2)f(X) = f({}^t g_2\, X\, g_1), \qquad f \in \text{Mat}_{n \times m}(\mathbb{C})^*. \tag{B.3}$$

It is not hard to see that this action is equivalent to the action on $\mathbb{C}^n \otimes \mathbb{C}^n$. The symmetric algebra over Mat_n^* is the same as the affine algebra $\mathcal{O}(\text{Mat}_n(\mathbb{C}))$, that is the polynomial ring generated by the coordinate functions g_{ij}. It is this ring that we will analyze.

First let us discuss the related decomposition of the affine algebra $\mathcal{O}(\mathrm{GL}(n,\mathbb{C})) = \mathbb{C}[g_{ij}, \det(g)^{-1}]$ over the affine variety $G = \mathrm{GL}(n,\mathbb{C})$. We consider two slightly different actions of $G \times G$. Under the action of $G \times G$ described in Proposition B.1, that is by $(g_1, g_2)f(x) = f(g_2^{-1}xg_1)$, that proposition says that the decomposition is into the direct sum $\bigoplus \pi \otimes \hat{\pi}$, where π runs over all irreducible representations. It follows from Proposition B.2 that for the action (B.3), the decomposition is as

$$\bigoplus_\lambda \pi_\lambda^{\mathrm{GL}(n)} \otimes \pi_\lambda^{\mathrm{GL}(n)},$$

where the decomposition is over all finite-dimensional irreducible representations, that is, over all dominant weights λ.

Now we are interested in the decomposition of the affine algebra $\mathrm{Mat}_n(\mathbb{C})$. Since G is embedded in $\mathrm{Mat}_n(\mathbb{C})$ as an open subvariety, the complement of the determinant locus, the restriction map $\bigvee \mathrm{Mat}_n(\mathbb{C}) \longrightarrow \mathcal{O}(G)$ is injective, and so we have only to decide for which representations π_λ the regular functions extend to $\mathrm{Mat}_n(\mathbb{C})$. These are the ones that are regular on the determinant locus, namely $\pi_\lambda^{\mathrm{GL}(n)}$ where λ is a partition. The statement is now clear if $n = m$.

To prove the general case, we may assume without loss of generality that $n \geqslant m$. Then we start with the $n = m$ case, and deduce the general case from it. To prove the decomposition it is enough to compare the graded characters on both sides of (B.1). This is (B.2) and if $n = m$ it is aready proved. To deduce the case $n \geqslant m$ from the diagonal case we simply specialize $u_j \mapsto 0$ if $j > m$, making use of (A.11).   □

This theorem gives another example of a correspondence. Thus if Ω is the symmetric algebra over $\mathbb{C}^n \otimes \mathbb{C}^m$, then in the notation (A.18),

$$\pi_\lambda^{\mathrm{GL}(n)} \xleftarrow{\;\;\Omega\;\;} \pi_\lambda^{\mathrm{GL}(m)}.$$

Howe calls this $\mathrm{GL}(n) \times \mathrm{GL}(m)$-duality, but here we call it the *Cauchy correspondence* in reference to its close relationship with the Cauchy identity.

B.2   Three interpretations of Littlewood–Richardson coefficients

We regard $\mathrm{GL}(n)$ as embedded in $\mathrm{GL}(n) \times \mathrm{GL}(n)$ along the diagonal $(g \mapsto (g, g))$, and we regard $\mathrm{GL}(r) \times \mathrm{GL}(s)$ embedded in $\mathrm{GL}(r+s)$ as a *Levi subgroup*

$$(g, h) \mapsto \begin{pmatrix} g & \\ & h \end{pmatrix}, \qquad g \in \mathrm{GL}(r), \; h \in \mathrm{GL}(s).$$

Now let Ω be the symmetric algebra on $\mathrm{Mat}_{(r+s)\times n}(\mathbb{C})$. We consider the following actions of $\mathrm{GL}(r+s, \mathbb{C})$ and $\mathrm{GL}(n, \mathbb{C}) \times \mathrm{GL}(n, \mathbb{C})$ on $\mathrm{Mat}_{(r+s)\times n}(\mathbb{C})$. If

$X = \text{Mat}_{(r+s)\times n}(\mathbb{C})$, we write $X = \begin{pmatrix} X' \\ X'' \end{pmatrix}$. Thus $(X', X'') \mapsto X$ is an isomorphism of $\text{Mat}_{r\times n}(\mathbb{C}) \oplus \text{Mat}_{s\times n}(\mathbb{C})$ with $\text{Mat}_{(r+s)\times n}(\mathbb{C})$. Therefore

$$\Omega = \bigvee \text{Mat}_{(r+s)\times n}(\mathbb{C}) \cong \left(\bigvee \text{Mat}_{r\times n}(\mathbb{C}) \right) \otimes \left(\bigvee \text{Mat}_{s\times n}(\mathbb{C}) \right).$$

We let $\text{GL}(n, \mathbb{C}) \times \text{GL}(n, \mathbb{C})$ act by right matrix multiplication on $\text{Mat}_{r\times n}(\mathbb{C}) \oplus \text{Mat}_{s\times n}(\mathbb{C})$ and hence on Ω. On the other hand, we let $\text{GL}(r+s, \mathbb{C})$ act by left multiplication on $\text{Mat}_{(r+s)\times n}(\mathbb{C})$, and hence on Ω. The action of $\text{GL}(n, \mathbb{C}) \times \text{GL}(n, \mathbb{C})$ does not commute with the action of $\text{GL}(r+s, \mathbb{C})$, but the subgroup $\text{GL}(n, \mathbb{C})$ embedded diagonally in $\text{GL}(n, \mathbb{C}) \times \text{GL}(n, \mathbb{C})$ does commute with $\text{GL}(r+s, \mathbb{C})$, and the Levi subgroup $\text{GL}(r, \mathbb{C}) \times \text{GL}(s, \mathbb{C})$ commutes with $\text{GL}(n, \mathbb{C}) \times \text{GL}(n, \mathbb{C})$. Therefore we have a see-saw:

$$
\begin{array}{ccc}
\text{GL}(n) \times \text{GL}(n) & & \text{GL}(r+s) \\
\uparrow & \times & \uparrow \\
\text{GL}(n) & & \text{GL}(r) \times \text{GL}(s)
\end{array}
\tag{B.4}
$$

The diagonal lines are the Cauchy correspondences for $\text{GL}(n) \times \text{GL}(r+s)$, and the Cauchy correspondence for $\text{GL}(n) \times \text{GL}(r)$ tensored with the correspondence for $\text{GL}(n) \times \text{GL}(s)$.

Theorem B.4. *Let $\lambda \vdash k+\ell$, $\mu \vdash k$, $\nu \vdash \ell$. Let $n \geqslant k+\ell$, $r \geqslant k$ and $s \geqslant \ell$. Then the multiplicity of $\pi_\lambda^{\text{GL}(n)}$ in $\pi_\mu^{\text{GL}(n)} \otimes \pi_\nu^{\text{GL}(n)}$ equals the multiplicity of $\pi_\mu^{\text{GL}(r)} \otimes \pi_\nu^{\text{GL}(s)}$ in the restriction of $\pi_\lambda^{\text{GL}(r+s)}$ to $\text{GL}(r) \times \text{GL}(s)$.*

Proof. This follows by applying Proposition A.13 to the see-saw (B.4). $\qquad\square$

Theorem B.5. *Let μ, ν and λ be partitions. Assume that $\mu \vdash k$, $\nu \vdash \ell$ and $\lambda \vdash k + \ell$. Then the following three quantities are equal:*

(i) *The multiplicity of the character $\pi_\mu^{S_k} \otimes \pi_\nu^{S_\ell}$ in the character $\pi_\lambda^{S_{k+\ell}}$ when it is restricted to the Levi subgroup $S_k \times S_\ell$ of $S_{k+\ell}$.*

(ii) *The multiplicity of the character $\pi_\lambda^{\text{GL}(n)}$ in $\pi_\mu^{\text{GL}(n)} \otimes \pi_\nu^{\text{GL}(n)}$, where $n \geqslant k + \ell$.*

(iii) *The multiplicity of the character $\pi_\mu^{\text{GL}(r)} \otimes \pi_\nu^{\text{GL}(s)}$ in the restriction of $\pi_\lambda^{\text{GL}(r+s)}$ to the Levi subgroup $\text{GL}(r) \times \text{GL}(s)$ of $\text{GL}(r+s)$, where $r \geqslant k$ and $s \geqslant \ell$.*

The coefficient defined to be any one of the three equivalent multiplicities is called the *Littlewood–Richardson coefficient* and is denoted $c_{\mu\nu}^\lambda$.

Proof. The equivalence of (i) and (ii) is explained in the proof of Theorem A.14 as a consequence of the see-saw in (A.20). The equivalence of (ii) and (iii) is Theorem B.4 as a consequence of the see-saw (B.4). $\qquad\square$

B.3 Pieri's formula

The Littlewood–Richardson coefficients may be described by an algorithm known as the *Littlewood–Richardson rule*. An important special case is known as *Pieri's formula*, which describes the coefficients when $\nu = (\ell)$, that is, the partition $(\ell, 0, 0, \ldots)$, or $\nu = (1^\ell)$, that is, the partition $(1, \ldots, 1, 0, \ldots)$ with ℓ parts equal to 1.

We recall that if λ and μ are partitions and $\mathrm{YD}(\mu) \supseteq \mathrm{YD}(\lambda)$, then the pair (μ, λ) is called a *skew shape* and denoted μ/λ. In this case $\mathrm{YD}(\mu/\lambda)$ is defined to be the set-theoretic difference $\mathrm{YD}(\mu) \setminus \mathrm{YD}(\lambda)$.

We recall that the skew shape μ/λ is a *horizontal strip* of length $|\mu| - |\lambda|$ if $\mathrm{YD}(\mu/\lambda)$ has no two boxes in the same column, and a *vertical strip* if $\mathrm{YD}(\mu/\lambda)$ has no two boxes in the same row.

Lemma B.6. *Let λ be a partition and ν be a weight such that all $\nu_i \geqslant 0$. Let $\mu = \lambda + \nu$. A necessary and sufficient condition that μ is a partition and μ/λ is a horizontal strip is that $\lambda_{i+1} + \nu_{i+1} \leqslant \lambda_i$ for all i.*

Proof. Suppose that $\lambda_{i+1} + \nu_{i+1} > \lambda_i$ for some i. If, furthermore, $\mu = \lambda + \nu$ is a partition, then $\lambda_i + \nu_i \geqslant \lambda_{i+1} + \nu_{i+1}$. Now we see that $\mathrm{YD}(\mu)$ has boxes in the positions $(i, \lambda_i + 1)$ and $(i + 1, \lambda_i + 1)$. Since it has two boxes in the same column it is not a horizontal strip. This proves one direction; we leave the other to the reader. $\qquad\square$

Theorem B.7 (Pieri formula). *Let $\lambda \vdash k$ and let $\mu \vdash (k + \ell)$.*

(1) *The multiplicity of $\pi_\mu^{\mathrm{GL}(n)}$ in $\pi_\lambda^{\mathrm{GL}(n)} \otimes \bigwedge^\ell \mathbb{C}^n$ is 1 if μ/λ is a vertical strip, and zero otherwise.*

(2) *The multiplicity of $\pi_\mu^{\mathrm{GL}(n)}$ in $\pi_\lambda^{\mathrm{GL}(n)} \otimes \bigvee^\ell \mathbb{C}^n$ is 1 if μ/λ is a horizontal strip, and zero otherwise.*

For example, let us use this to compute $s_{(2,1)} h_2$ and $s_{(2,1)} e_2$. Applied to the eigenvalues of a matrix $g \in \mathrm{GL}(n, \mathbb{C})$, these symmetric functions describe the characters of $\pi_{(2,1)}^{\mathrm{GL}(n)} \otimes \bigvee^2 \mathbb{C}^n$ and $\pi_{(2,1)}^{\mathrm{GL}(n)} \otimes \bigwedge^2 \mathbb{C}^n$, respectively. So to compute $s_{(2,1)} h_2$ we need to determine which partitions μ of 5 have $\mu/(2,1)$ a horizontal strip. These are $(4,1)$, $(3,2)$, $(3,1,1)$ and $(2,2,1)$. We will represent the Young diagrams of the skew shapes as follows:

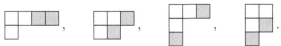

The shaded boxes in each case are the diagrams of the skew shapes. Thus Pieri's formula asserts

$$s_{(2,1)} h_2 = s_{(4,1)} + s_{(3,2)} + s_{(3,1,1)} + s_{(2,2,1)}.$$

Similarly

$$s_{(2,1)} e_2 = s_{(3,2)} + s_{(3,1,1)} + s_{(2,2,1)} + s_{(2,1,1,1)}.$$

Proof of Theorem B.7. We will analyze the characters of $\pi_\lambda^{\mathrm{GL}(n)} \otimes \bigvee^\ell \mathbb{C}^n$ and $\pi_\lambda^{\mathrm{GL}(n)} \otimes \bigwedge^\ell \mathbb{C}^n$.

Let $\pi_\lambda = \pi_\lambda^{\mathrm{GL}(n)}$ and let θ be another irreducible representation which in practice will be either $\bigvee^\ell \mathbb{C}^n$ or $\bigwedge^\ell \mathbb{C}^n$. Let $\mathrm{mult}(\nu)$ be the weight multiplicities for θ. Then by the Weyl character formula for π, the character of $\pi_\lambda \otimes \theta$ is

$$\Delta^{-1} \sum_\nu \mathrm{mult}(\nu) t^\nu \sum_{w \in W} \mathrm{sgn}(w)\, t^{w(\lambda+\rho)}.$$

Here $\Delta = $ is the "Weyl denominator," which is the denominator in (A.10). Interchanging the order of summation, we may replace ν by $w(\nu)$ and since $\mathrm{mult}\big(w(\nu)\big) = \mathrm{mult}(\nu)$, this equals

$$\sum_\nu \mathrm{mult}(\nu) X(\nu) \qquad \text{where} \qquad X(\nu) = \Delta^{-1} \sum_{w \in W} \mathrm{sgn}(w)\, t^{w(\lambda+\nu+\rho)}. \tag{B.5}$$

Let us now specialize to the case where θ is the representation of $\mathrm{GL}(n)$ on $\bigvee^\ell \mathbb{C}^n$. In this case, the weights of θ are the set S of "monomials of weight ℓ," by which we mean weights $(\nu_1, \ldots, \nu_n)$ with $0 \leqslant \nu_n$ and $\sum \nu_i = \ell$. Each weight multiplicity $\mathrm{mult}(\nu)$ equals in θ equals 1 when $\nu \in S$.

Let S_1 be the set of monomials ν of weight ℓ such that $\lambda + \nu$ is a partition and a horizontal strip. Let S_2 be the complement of S_1 in S. By the Weyl character formula, the contribution of S_1 is just the sum of the characters of the $\pi_\mu^{\mathrm{GL}(n)}$ such that μ/λ is a horizontal strip of length ℓ.

Thus (i) will be proved if we show that the contribution of S_2 is zero. We will exhibit a bijection $\nu \mapsto \nu'$ from $S_2 \longrightarrow S_2$ that is an involution (i.e. $\nu'' = \nu$) such that $X(\nu) = -X(\nu')$. This is sufficient, since if $\nu = \nu'$, this means that $X(\nu) = 0$, while if $\nu \neq \nu'$ it means that the contributions of ν and ν' to (B.5) cancel.

To define the bijection $\nu \mapsto \nu'$, observe that by Lemma B.6, if $\nu \in S_2$, there exists an i such that $\lambda_{i+1} + \nu_{i+1} > \lambda_i$. We define $i(\nu)$ to be the largest such i. Then we define ν' by requiring that $\nu'_j = \nu_j$ unless $j = i$ or $i + 1$. We define

$$\nu'_i = \lambda_{i+1} + \nu_{i+1} - \lambda_i - 1, \qquad \nu'_{i+1} = \nu_i - \lambda_{i+1} + \lambda_i + 1.$$

We have $0 \leqslant \nu'_i, \nu'_{i+1}$ and $\sum \nu'_j = \sum \nu_j = \ell$, so $\nu' \in S$. Moreover, $\nu'_{i+1} + \lambda_{i+1} > \lambda_i$ while if $j > i$ then $\nu'_{j+1} = \nu_{j+1}$, so $\nu'_{j+1} + \lambda_{j+1} \leqslant \lambda_j$. Therefore $\nu' \in S_2$ and $i(\nu') = i(\nu)$. From this we may check that $\nu'' = \nu$, so $\nu \mapsto \nu''$ is an involution, and in particular a bijection.

Now if $\nu \in S_2$, then we find that $s_i(\lambda + \nu + \rho) = \lambda + \nu' + \rho$, so making the variable change $w \mapsto w s_i$ in the definition of $X(\nu)$ shows that $X(\nu') = -X(\nu)$, and this proves (i).

Next, turning to (ii), suppose that θ is the representation of $\mathrm{GL}(n)$ on $\bigwedge^\ell \mathbb{C}^n$. In this case, the weights of θ are the set S of "elementary monomials" $\nu = (\nu_1, \ldots, \nu_n)$ such that every ν_i equals 0 or 1, and $\sum_i \nu_i = \ell$. The coefficients $\mathrm{mult}(\nu) = 1$ for all $\nu \in S$.

Let $\nu \in S$. If $\mu = \nu + \lambda$ is a partition, then clearly μ/λ is a vertical strip, and moreover every vertical strip of length $\leqslant n$ may be obtained this way for a unique choice of ν. By the Weyl character formula, $X(\nu)$ equals the character of $\pi_\mu^{\mathrm{GL}(n)}$ in this case. These are the terms we want.

On the other hand, if $\nu + \lambda$ is not a partition, we will show that $X(\nu) = 0$. Since the λ_i are decreasing, and the coefficients of ν_i are either 1 or 0, the only way $\mu = \nu + \lambda$ can fail to be a partition is if $\nu_i = 0$ and $\nu_{i+1} = 1$, while $\lambda_i = \lambda_{i+1}$. If this happens, then $\lambda_i + \nu_i + \rho_i = \lambda_{i+1} + \nu_{i+1} + \rho_{i+1}$ and so the contributions to the sum in braces in (B.5) from w and ws_i cancel, and the term in braces is equal to zero. Thus the character of $\pi_\lambda^{\mathrm{GL}(n)} \otimes \bigwedge^\ell \mathbb{C}^n$ equals the sum of the characters of the $\pi_\mu^{S_{k+\ell}}$, where μ runs through all the partitions of $k + \ell$ such that μ/λ is a vertical strip. $\qquad\square$

We have defined h_k and e_k if k is an integer. If λ is a partition, let us define $h_\lambda = \prod_i h_{\lambda_i}$ and $e_\lambda = \prod_i e_{\lambda_i}$.

Proposition B.8. *Let λ be a partition of k of length $\leqslant n$. Expand h_λ as a product of Schur functions:*

$$h_\lambda = \sum_\mu c(\lambda, \mu)\, s_\mu.$$

Then $c(\lambda, \mu) = 0$ unless $\mu \succcurlyeq \lambda$, and $c(\lambda, \lambda) = 1$. Moreover

$$e_\lambda = \sum_\mu c(\lambda, \mu)\, s_{\mu'}.$$

The matrix $c(\lambda, \mu)$ is invertible.

Proof. We may compute h_λ recursively using Pieri's formula by adding horizontal strips of lengths $\lambda_1, \lambda_2, \ldots$. It is easy to see that λ is minimal with respect to the partial order $\succcurlyeq$ among the partitions that may be obtained this way. It is thus clear that $c(\lambda, \mu) = 0$ unless $\mu \succcurlyeq \lambda$, and $c(\lambda, \lambda) = 1$. We may compute e_λ the same way, but we are adding vertical strips. We obtain the same coefficients, but whereever we get s_μ in computing h_λ, we get $s_{\mu'}$ in computing e_λ.

Refining the partial order $\succcurlyeq$ to a total order, the matrix c is upper triangular and has 1's on the diagonal. Therefore it is invertible. $\qquad\square$

B.4   Symmetric group branching rules

Proposition B.9. *Let $\lambda \vdash k$. Then the restriction of $\pi_\lambda^{S_k}$ to S_{k-1} is the direct sum of the $\pi_\mu^{S_{k-1}}$ where μ runs through the partitions of $k-1$ such that $\mathrm{YD}(\mu) \subset \mathrm{YD}(\lambda)$. Each occurs with multiplicity one.*

Proof. By Frobenius reciprocity, it is sufficient to show that the representation of S_k induced from the representation $\pi_\mu^{S_{k-1}}$ of S_{k-1} equals the direct sum of the $\pi_\lambda^{S_k}$ where λ runs through the partitions of k such that $\mathrm{YD}(\mu) \subset \mathrm{YD}(\lambda)$. We may identify S_{k-1} with the Levi subgroup $S_{k-1} \times S_1$, since S_1 is the trivial subgroup with one element. In the ring $\mathcal{R}$ (see Section A.3), the induced representation is thus the product $[\pi_\mu^{S_{k-1}}] \cdot [\pi_{(1)}^{S_1}]$ and applying the characteristic map, what we need to show is that

$$s_\mu \cdot s_{(1)} = \sum_{\substack{\lambda \vdash k \\ \mathrm{YD}(\lambda) \supset \mathrm{YD}(\mu)}} s_\lambda.$$

Now $s_{(1)} = e_1$ and so this follows from Pieri's formula in Theorem B.7. ☐

Given a partition λ, we have defined the Young diagram $\mathrm{YD}(\lambda)$ as an array of boxes arranged in the lower right quadrant of the plane. The i-th row has λ_i boxes. A *tableau of shape* λ is a filling of these boxes by elements (called *letters*) of some *alphabet*; in this chapter we encounter the alphabet $[n] = \{1, 2, 3, \ldots, n\}$. Tableaux in this alphabet are related to the representations of the general linear groups and the symmetric groups.

We recall that a tableau in the alphabet $[n]$ is called *semistandard* if the columns are strictly increasing, and the rows are weakly increasing. If $\lambda \vdash k$ and the tableau in the alphabet $[k]$ contains each letter exactly once, the tableau is called *standard*.

Theorem B.10. *Let* $\lambda \vdash k$. *Then the dimension of* $\pi_\lambda^{S_k}$ *equals the number of standard tableaux with shape* λ.

Proof. Given a tableau T of shape λ, the location of the unique entry equal to k must be in a box whose removal results in a tableau T' of shape μ, where $\mu \vdash k - 1$ is such that $\mathrm{YD}(\lambda) \supset \mathrm{YD}(\mu)$. The tableau resulting from removing this box is then a standard tableau for μ. The number of such tableaux is, by induction, equal to the dimension of $\pi_\mu^{S_{k-1}}$. The statement now follows from Proposition B.9. ☐

We next observe that the dominance order is reversed by conjugation of partitions.

Lemma B.11. *Let* $\lambda, \mu \vdash k$. *If* $\lambda \succcurlyeq \mu$ *then* $\mu' \succcurlyeq \lambda'$.

Proof. The partial order is generated by "elementary sucessions" which we will now define. Suppose for some j we have $\mu_i = \lambda_i$ for all $i \neq j, j + 1$, while $\mu_i = \lambda_i + 1$ and $\mu_{i+1} = \lambda_{i+1} - 1$. Then $\mu \succcurlyeq \lambda$. This means that $\mathrm{YD}(\mu)$ is obtained from $\mathrm{YD}(\lambda)$ by moving one box up and to the right. In this case we say that μ is an *elementary successor* of λ. To say that the partial order is generated by elementary successions, we mean the following easily established fact. If μ and λ are arbitrary partitions such that $\mu \succcurlyeq \lambda$, then we may interpolate a sequence of other partitions

$$\mu = \mu_N \succcurlyeq \cdots \succcurlyeq \mu_2 \succcurlyeq \mu_1 = \lambda$$

such that μ_{i+1} is an elementary successor of μ_i.

Conjugating this type of operation by partition conjugation, if μ is an elementary successor of λ, then $\mathrm{YD}(\mu')$ is obtained from $\mathrm{YD}(\lambda')$ by moving one box down and to the left. Thus $\lambda' \succcurlyeq \mu'$ for these types of elementary operations. It is now clear that partition conjugation reverses the partial order. ☐

B.5   The involution on symmetric functions

Let us consider the involution ι on the ring of symmetric functions. Since $\mathcal{R} \cong \mathrm{Sym}$ (see Section A.3), ι corresponds to an involution of the ring $\mathcal{R}$, which we also denote ι. We recall that the homogeneous part $\mathcal{R}_k$ of degree k is the Grothendieck group of virtual representations of S_k.

Lemma B.12. *Let S be a finite group and T a subgroup. Let σ be a one-dimensional representation S, and τ a representation of T. Then the representation $\mathrm{Ind}_T^S(\sigma \otimes \tau)$ is isomorphic to $\sigma \otimes \mathrm{Ind}_T^S(\tau)$.*

Proof. We leave this to the reader. This may be proved by comparing their characters, or directly by constructing an isomorphism. The assumption that σ is one-dimensional is unnecessary, but this is all we need. □

Theorem B.13.

(i) *Let θ be an irreducible representation of S_k. Then ι maps $[\theta] \in \mathcal{R}_k$ to $[\mathrm{sgn} \otimes \theta]$.*
(ii) *The involution ι corresponds to conjugation of partitions. In other words if $\lambda \vdash k$, then*

$$\iota[\pi_\lambda^{S_k}] = [\pi_{\lambda'}^{S_k}].$$

Proof. First let us observe that the map $[\theta] \mapsto [\mathrm{sgn} \otimes \theta]$ is a ring homomorphism. Call this map ι'. What has to be checked is that if θ and θ' are representations of S_k and S_ℓ, then

$$\mathrm{sgn} \otimes \mathrm{Ind}_{S_k \times S_\ell}^{S_{k+\ell}}(\theta \otimes \theta') \cong \mathrm{Ind}_{S_k \times S_\ell}^{S_{k+\ell}}(\mathrm{sgn} \otimes (\theta \otimes \theta')),$$

and this follows from Lemma B.12.

Now we make use of the characteristic map to Sym. The maps $\mathrm{ch} \circ \iota'$ and $\iota \circ \mathrm{ch}$ agree on the sign characters, since ι' sends $[\mathrm{sgn}_{S_k}]$ to the class of the trivial character $[1_{S_k}]$, and ch sends this to h_k, while ch sends $[\mathrm{sgn}_{S_k}]$ to e_k, and then ι sends this to h_k. Since the e_k generate Sym, part (i) is now clear.

Since $\iota(e_k) = h_k$ and ι is a ring homomorphism, the involution ι interchanges e_λ and h_λ. It follows from Proposition B.8 that it interchanges s_λ and $s_{\lambda'}$. □

B.6   The GL$(n, \mathbb{C})$ branching rule

Let λ be a partition of length $\leqslant n$ and let μ be a partition of length $\leqslant n - 1$. We say that λ and μ *interleave* if

$$\lambda_1 \geqslant \mu_1 \geqslant \lambda_2 \geqslant \mu_2 \geqslant \cdots \geqslant \mu_{n-1} \geqslant \lambda_n.$$

Theorem B.14. *Let λ be a partition of length $\leqslant n$, and let μ be a partition of length $\leqslant n - 1$. Then the multiplicity of $\pi_\mu^{\mathrm{GL}(n-1)}$ in the restriction of $\pi_\lambda^{\mathrm{GL}(n)}$ from $\mathrm{GL}(n, \mathbb{C})$ to $\mathrm{GL}(n - 1, \mathbb{C})$ equals 1 if λ and μ interleave, and 0 otherwise.*

Proof. We consider the restriction of $\pi_\lambda^{\mathrm{GL}(n)}$ to the Levi subgroup $\mathrm{GL}(n-1) \times \mathrm{GL}(1)$. Since the character of $\pi_\lambda^{\mathrm{GL}(n)}$ is a homogeneous polynomial s_λ its restriction to $\mathrm{GL}(n-1) \times \mathrm{GL}(1)$ must also be homogeneous, and so every summand is of the form $\pi_\mu^{\mathrm{GL}(n-1)} \otimes \pi_{(k)}^{\mathrm{GL}(1)}$, where homogeneity implies that $k = |\lambda| - |\mu|$. The multiplicity is the Littlewood–Richardson coefficient $c^\lambda_{\mu,(k)}$ which by Pieri's formula is 1 if λ/μ is a horizontal strip, and 0 otherwise. In particular we must have $\lambda_i \geqslant \mu_i$. We have proved that a necessary and sufficient condition for $\pi_\mu^{\mathrm{GL}(n-1)}$ to occur in the restriction of $\pi_\lambda^{\mathrm{GL}(n)}$ is that $\mathrm{YD}(\mu) \subseteq \mathrm{YD}(\lambda)$ and that λ/μ is a horizontal strip. It follows from Lemma B.6 that this is equivalent to λ and μ interleaving. $\qquad\square$

Theorem B.15. *Let λ be a partition of length $\leqslant n$. Then the dimension of $\pi_\lambda^{\mathrm{GL}(n)}$ equals the number of semistandard Young tableaux of shape λ in the alphabet $[n]$.*

Proof. Let us consider what happens when we branch $\pi_\lambda^{\mathrm{GL}(n)}$ down through the sequence of subgroups

$$\mathrm{GL}(n) \supset \mathrm{GL}(n-1) \supset \cdots \supset \mathrm{GL}(1). \tag{B.6}$$

Each branching rule is multiplicity free, so we may obtain a basis of the space of $\pi_\lambda^{\mathrm{GL}(n)}$ by specifying a sequence of representations of these groups, each (except the first) contained in the restriction of the previous. The final representation of $\mathrm{GL}(1)$ is one-dimensional, so the cardinality of the basis equals the number of such possible sequences of representations. By Theorem B.14 the highest weights of the sequence of representations of $\mathrm{GL}(n), \mathrm{GL}(n-1), \ldots, \mathrm{GL}(1)$ must interleave, so they form a Gelfand–Tsetlin pattern. Thus the number of basis vectors equals the number of Gelfand–Tsetlin patterns with top row λ, and these are in bijection with the semistandard Young tableaux in the alphabet $[n]$. $\qquad\square$

Remark B.16. We may very easily prove the combinatorial formula for the Schur polynomial, equation (3.3), by the same method. In Chapter 3 that formula was taken as the definition, but in these Appendices we have taken a different definition of the Schur polynomial, as the character of a representation, which is thus given by the Weyl character formula. So we would like to know these two definitions are equivalent. We may prove this by the method of Theorem B.15. Instead of branching down through the subgroups (B.6), we use the sequence

$$\mathrm{GL}(n) \supset \mathrm{GL}(n-1) \times \mathrm{GL}(1) \supset \mathrm{GL}(n-2) \times \mathrm{GL}(1) \times \mathrm{GL}(1) \supset \cdots .$$

The last group in the chain is the diagonal torus $\mathrm{GL}(1) \times \cdots \times \mathrm{GL}(1)$ (n times). Instead of simply counting the number of semistandard Young tableaux of shape λ we keep track of their weights. This requires the $\mathrm{GL}(n) \times \mathrm{GL}(n-1) \times \mathrm{GL}(1)$ branching rule, which we saw how to get from the see-saw as in the proof of Theorem B.14. We can thus compute the character of the representation as easily as we computed its dimension in Theorem B.15, to obtain the combinatorial formula for s_λ. Alternatively, the equivalence of the two definitions of the Schur polynomial is a special case of Corollary 13.9.

B.7   The dual Cauchy identity

Theorem B.17. *Let $G_1 = \mathrm{GL}(n, \mathbb{C})$ and $G_2 = \mathrm{GL}(m, \mathbb{C})$. Let $G_1 \times G_2$ act on $\mathbb{C}^n \otimes \mathbb{C}^m$. Then $G_1 \times G_2$ acts on the exterior algebra $\bigwedge(\mathbb{C}^n \otimes \mathbb{C}^m)$. We have the decomposition*

$$\bigwedge(\mathbb{C}^n \otimes \mathbb{C}^m) = \bigoplus_\lambda \pi_\lambda^{\mathrm{GL}(n)} \otimes \pi_{\lambda'}^{\mathrm{GL}(m)},$$

where λ runs through all partitions such that $\mathrm{YD}(\lambda)$ fits in an $n \times m$ rectangle.

If $t_1, \ldots, t_n$ are the eigenvalues of $g_1 \in \mathrm{GL}(n, \mathbb{C})$, and $u_1, \ldots, u_m$ are the eigenvalues of $g_2 \in \mathrm{GL}(m, \mathbb{C})$, then comparing the graded characters in this decomposition for (g_1, g_2) gives the *dual Cauchy identity*

$$\prod_{i,j}(1 + t_i u_j q) = \sum_\lambda s_\lambda(t)\, s_{\lambda'}(u)\, q^{|\lambda|}, \tag{B.7}$$

where the sum is over partitions such that $\mathrm{YD}(\lambda)$ fits in an $n \times m$ rectangle.

Proof of Theorem B.17. Using the graded characters, it is sufficient to prove the dual Cauchy identity. We will deduce this from the ordinary Cauchy identity by making use of the involution in the ring Sym. We will be working with symmetric polynomials in two sets of variables, that is, in the ring $\mathrm{Sym}^{(n)} \otimes \mathrm{Sym}^{(m)}$. Let us therefore modify the notation (A.5) and (A.6) and write

$$H_t(q) = \prod_i (1 - t_i q)^{-1}, \qquad E_t(q) = \prod_i (1 + t_i q).$$

Thus the Cauchy identity may be written

$$\prod_j H_t(q u_j) = \sum_\lambda q^{|\lambda|} s_\lambda(t)\, s_\lambda(u).$$

In this form, we may pass to the inverse limit and regard this as an identity in $\mathrm{Sym} \otimes \mathrm{Sym}^{(m)}$. Apply the involution in Sym (that is, in the copy corresponding to the t_i) and obtain the identity

$$\prod_j E_t(q u_j) = \sum_\lambda q^{|\lambda|} s_{\lambda'}(t)\, s_\lambda(u).$$

Now specializing back to $\mathrm{Sym}^{(n)} \otimes \mathrm{Sym}^{(m)}$, we obtain the dual Cauchy identity. $\square$

Bibliography

Andersen, H. H. (1985). Schubert varieties and Demazure's character formula, *Invent. Math.* **79**, 3, pp. 611–618.

Anderson, J. (2000). *On Mirkovic and Vilonen's intersection homology cycles for the loop Grassmannian* (ProQuest LLC, Ann Arbor, MI), thesis (Ph.D.)–Princeton University.

Anderson, J. (2003). A polytope calculus for semisimple groups, *Duke Math. J.* **116**, 3, pp. 567–588.

Anderson, J. and Kogan, M. (2004). Mirković-Vilonen cycles and polytopes in Type A, *Int. Math. Res. Not.* , 12, pp. 561–591.

Ariki, S. (1996). On the decomposition numbers of the Hecke algebra of $G(m, 1, n)$, *J. Math. Kyoto Univ.* **36**, 4, pp. 789–808.

Atiyah, M. F. and Pressley, A. N. (1983). Convexity and loop groups, in *Arithmetic and geometry, Vol. II, Progr. Math.*, Vol. 36 (Birkhäuser Boston, Boston, MA), pp. 33–63.

Baker, T. H. (2000). Zero actions and energy functions for perfect crystals, *Publ. Res. Inst. Math. Sci.* **36**, 4, pp. 533–572.

Beineke, J., Brubaker, B., and Frechette, S. (2012). A crystal definition for symplectic multiple Dirichlet series, in *Multiple Dirichlet series, L-functions and automorphic forms, Progr. Math.*, Vol. 300 (Birkhäuser/Springer, New York), pp. 37–63.

Benkart, G., Kang, S.-J., and Kashiwara, M. (2000). Crystal bases for the quantum superalgebra $u_q(\mathfrak{gl}(m, n))$, *J. Amer. Math. Soc.* **13**, 2, pp. 295–331.

Berele, A. and Regev, A. (1987). Hook Young diagrams with applications to combinatorics and to representations of Lie superalgebras, *Adv. in Math.* **64**, 2, pp. 118–175.

Berele, A. and Remmel, J. B. (1985). Hook flag characters and their combinatorics, *J. Pure Appl. Algebra* **35**, 3, pp. 225–245.

Berenstein, A., Fomin, S., and Zelevinsky, A. (1996). Parametrizations of canonical bases and totally positive matrices, *Adv. Math.* **122**, 1, pp. 49–149.

Berenstein, A. and Kazhdan, D. (2000). Geometric and unipotent crystals, *Geom. Funct. Anal.* , Special Volume, Part I, pp. 188–236, gAFA 2000 (Tel Aviv, 1999).

Berenstein, A. and Kazhdan, D. (2007). Geometric and unipotent crystals. II. From unipotent bicrystals to crystal bases, in *Quantum groups, Contemp. Math.*, Vol. 433 (Amer. Math. Soc., Providence, RI), pp. 13–88.

Berenstein, A. and Zelevinsky, A. (1996). Canonical bases for the quantum group of type A_r and piecewise-linear combinatorics, *Duke Math. J.* **82**, 3, pp. 473–502.

Berenstein, A. and Zelevinsky, A. (1997). Total positivity in Schubert varieties, *Comment. Math. Helv.* **72**, 1, pp. 128–166.

Berenstein, A. and Zelevinsky, A. (2001). Tensor product multiplicities, canonical bases and totally positive varieties, *Invent. Math.* **143**, 1, pp. 77–128.

Borel, A. (1970). Properties and linear representations of Chevalley groups, in *Seminar on Algebraic Groups and Related Finite Groups (The Institute for Advanced Study, Princeton, N.J., 1968/69)*, Lecture Notes in Mathematics, Vol. 131 (Springer, Berlin), pp. 1–55.

Borel, A. (1991). *Linear algebraic groups, Graduate Texts in Mathematics*, Vol. 126, 2nd edn. (Springer-Verlag, New York).

Bott, R. (1957). Homogeneous vector bundles, *Ann. of Math. (2)* **66**, pp. 203–248.

Bourbaki, N. (2002). *Lie groups and Lie algebras. Chapters 4–6*, Elements of Mathematics (Berlin) (Springer-Verlag, Berlin), translated from the 1968 French original by Andrew Pressley.

Braverman, A. and Kazhdan, D. (2000). γ-functions of representations and lifting, *Geom. Funct. Anal.* , Special Volume, Part I, pp. 237–278, with an appendix by V. Vologodsky, GAFA 2000 (Tel Aviv, 1999).

Brubaker, B., Buciumas, V., and Bump, D. (2016). A Yang–Baxter equation for metaplectic ice, `arXiv:1604.02206`.

Brubaker, B., Bump, D., Chinta, G., Friedberg, S., and Gunnells, P. E. (2012). Metaplectic ice, in *Multiple Dirichlet series, L-functions and automorphic forms*, *Progr. Math.*, Vol. 300 (Birkhäuser/Springer, New York), pp. 65–92.

Brubaker, B., Bump, D., and Friedberg, S. (2011a). Schur polynomials and the Yang-Baxter equation, *Comm. Math. Phys.* **308**, 2, pp. 281–301.

Brubaker, B., Bump, D., and Friedberg, S. (2011b). *Weyl group multiple Dirichlet series: type A combinatorial theory*, Annals of Mathematics Studies, Vol. 175 (Princeton University Press, Princeton, NJ).

Brubaker, B., Bump, D., Friedberg, S., and Hoffstein, J. (2007). Weyl group multiple Dirichlet series. III. Eisenstein series and twisted unstable A_r, *Ann. of Math. (2)* **166**, 1, pp. 293–316.

Brubaker, B. and Schultz, A. (2015). The six-vertex model and deformations of the Weyl character formula, *J. Algebraic Combin.* **42**, 4, pp. 917–958.

Brundan, J. (2003). Kazhdan-Lusztig polynomials and character formulae for the Lie superalgebra $\mathfrak{gl}(m|n)$, *J. Amer. Math. Soc.* **16**, 1, pp. 185–231.

Bump, D. (2013). *Lie groups, Graduate Texts in Mathematics*, Vol. 225, 2nd edn. (Springer, New York).

Bump, D. and Nakasuji, M. (2010). Integration on p-adic groups and crystal bases, *Proc. Amer. Math. Soc.* **138**, 5, pp. 1595–1605.

Bump, D., Schilling, A., and Salisbury, B. (2015). Lie methods and related combinatorics in Sage, `http://doc.sagemath.org/html/en/thematic_tutorials/lie.html`.

Cain, A. J., Gray, R. D., and Malheiro, A. (2014). Crystal monoids & crystal bases: rewriting systems and biautomatic structures for plactic monoids of types A_n, B_n, C_n, D_n, and G_2, `arXiv:1412.7040`.

Casselman, W. and Shalika, J. (1980). The unramified principal series of p-adic groups. II. The Whittaker function, *Compositio Math.* **41**, 2, pp. 207–231.

Chari, V. and Pressley, A. (1995). Quantum affine algebras and their representations, in *Representations of groups (Banff, AB, 1994)*, CMS Conf. Proc., Vol. 16 (Amer. Math. Soc., Providence, RI), pp. 59–78.

Chari, V. and Pressley, A. (1998). Twisted quantum affine algebras, *Comm. Math. Phys.* **196**, 2, pp. 461–476.

Cheng, S.-J. and Wang, W. (2012). *Dualities and representations of Lie superalgebras, Graduate Studies in Mathematics*, Vol. 144 (American Mathematical Society, Providence, RI).

Chhaibi, R. (2015). Whittaker processes and Landau-Ginzburg potentials for flag manifolds, `arXiv:1504.07321`.

Chinta, G. and Gunnells, P. E. (2012). Littelmann patterns and Weyl group multiple Dirichlet series of type *D*, in *Multiple Dirichlet series, L-functions and automorphic forms, Progr. Math.*, Vol. 300 (Birkhäuser/Springer, New York), pp. 119–130.

Claxton, J. and Tingley, P. (2014). Young tableaux, multisegments, and PBW bases, *Sém. Lothar. Combin.* **73**, pp. Art. B73c, 21.

Cliff, G. (1998). Crystal bases and Young tableaux, *J. Algebra* **202**, 1, pp. 10–35.

Corwin, I., O'Connell, N., Seppäläinen, T., and Zygouras, N. (2014). Tropical combinatorics and Whittaker functions, *Duke Math. J.* **163**, 3, pp. 513–563.

Coxeter, H. S. M. (1934). Discrete groups generated by reflections, *Ann. of Math. (2)* **35**, 3, pp. 588–621.

Danilov, V. I. and Koshevoy, G. A. (2004). Bi-crystals and crystal GL(V), GL(W) duality, *RIMS Preprint* **1458**.

Danilov, V. I. and Koshevoy, G. A. (2005/07). The octahedron recurrence and RSK-correspondence, *Sém. Lothar. Combin.* **54A**.

De Concini, C. (1979). Symplectic standard tableaux, *Adv. in Math.* **34**, 1, pp. 1–27.

Demazure, M. (1974). Désingularisation des variétés de Schubert généralisées, *Ann. Sci. École Norm. Sup. (4)* **7**, pp. 53–88, collection of articles dedicated to Henri Cartan on the occasion of his 70th birthday, I.

Demazure, M. (1976). A very simple proof of Bott's theorem, *Invent. Math.* **33**, 3, pp. 271–272.

Dipper, R. and James, G. (1986). Representations of Hecke algebras of general linear groups, *Proc. London Math. Soc. (3)* **52**, 1, pp. 20–52.

Drinfel'd, V. G. (1985). Hopf algebras and the quantum Yang-Baxter equation, *Dokl. Akad. Nauk SSSR* **283**, 5, pp. 1060–1064.

Drinfeld, V. G. (1989). Quasi-Hopf algebras, *Algebra i Analiz* **1**, 6, pp. 114–148.

Edelman, P. and Greene, C. (1987). Balanced tableaux, *Adv. in Math.* **63**, 1, pp. 42–99.

Fomin, S. (1995). Schensted algorithms for dual graded graphs, *J. Algebraic Combin.* **4**, 1, pp. 5–45.

Fomin, S. and Greene, C. (1998). Noncommutative Schur functions and their applications, *Discrete Math.* **193**, 1-3, pp. 179–200, selected papers in honor of Adriano Garsia (Taormina, 1994).

Fourier, G., Okado, M., and Schilling, A. (2009). Kirillov-Reshetikhin crystals for nonexceptional types, *Adv. Math.* **222**, 3, pp. 1080–1116.

Fourier, G., Okado, M., and Schilling, A. (2010). Perfectness of Kirillov-Reshetikhin crystals for nonexceptional types, in *Quantum affine algebras, extended affine Lie algebras, and their applications, Contemp. Math.*, Vol. 506 (Amer. Math. Soc., Providence, RI), pp. 127–143.

Fourier, G., Schilling, A., and Shimozono, M. (2007). Demazure structure inside Kirillov-Reshetikhin crystals, *J. Algebra* **309**, 1, pp. 386–404.

Frenkel, E. and Reshetikhin, N. (1999). The *q*-characters of representations of quantum affine algebras and deformations of $\mathcal{W}$-algebras, in *Recent developments in quantum affine algebras and related topics (Raleigh, NC, 1998), Contemp. Math.*, Vol. 248 (Amer. Math. Soc., Providence, RI), pp. 163–205.

Friedberg, S. and Zhang, L. (2014). Tokuyama-type formulas for type *B*, `arXiv:1409.0464`.

Frieden, G. (2016). Affine type A geometric crystal structure on the Grassmannian, *DMTCS proc.*, pp. 503–514.

Friedlander, H., Gaudet, L., and Gunnells, P. E. (2015). Crystal graphs, Tokuyama's theorem, and the Gindikin-Karpelevič formula for G_2, *J. Algebraic Combin.* **41**, 4, pp. 1089–1102.

Fulton, W. (1997). *Young tableaux, London Mathematical Society Student Texts*, Vol. 35 (Cambridge University Press, Cambridge), with applications to representation theory and geometry.

Fulton, W. and Harris, J. (1991). *Representation theory, Graduate Texts in Mathematics*, Vol. 129 (Springer-Verlag, New York), a first course, Readings in Mathematics.

Gaussent, S. and Littelmann, P. (2005). LS galleries, the path model, and MV cycles, *Duke Math. J.* **127**, 1, pp. 35–88.

Ginzburg, V. (1995). Perverse sheaves on a loop group and Langlands' duality, `arXiv:alg-geom/9511007`.

Grantcharov, D., Jung, J. H., Kang, S.-J., Kashiwara, M., and Kim, M. (2014). Crystal bases for the quantum queer superalgebra and semistandard decomposition tableaux, *Trans. Amer. Math. Soc.* **366**, 1, pp. 457–489.

Grantcharov, D., Jung, J. H., Kang, S.-J., and Kim, M. (2010). Highest weight modules over quantum queer superalgebra $u_q(\mathfrak{q}(n))$, *Comm. Math. Phys.* **296**, 3, pp. 827–860.

Grojnowski, I. (1999). Affine sl_p controls the representation theory of the symmetric group and related Hecke algebras, `arXiv:math/9907129`.

Haiman, M. D. (1992). Dual equivalence with applications, including a conjecture of Proctor, *Discrete Math.* **99**, 1-3, pp. 79–113.

Hamaker, Z. and Young, B. (2014). Relating Edelman-Greene insertion to the Little map, *J. Algebraic Combin.* **40**, 3, pp. 693–710.

Hamel, A. M. and King, R. C. (2002). Symplectic shifted tableaux and deformations of Weyl's denominator formula for sp($2n$), *J. Algebraic Combin.* **16**, 3, pp. 269–300 (2003).

Hamel, A. M. and King, R. C. (2005). U-turn alternating sign matrices, symplectic shifted tableaux and their weighted enumeration, *J. Algebraic Combin.* **21**, 4, pp. 395–421.

Hatayama, G., Kuniba, A., Okado, M., Takagi, T., and Tsuboi, Z. (2002). Paths, crystals and fermionic formulae, in *MathPhys odyssey, 2001, Prog. Math. Phys.*, Vol. 23 (Birkhäuser Boston, Boston, MA), pp. 205–272.

Hatayama, G., Kuniba, A., Okado, M., Takagi, T., and Yamada, Y. (1999). Remarks on fermionic formula, in *Recent developments in quantum affine algebras and related topics (Raleigh, NC, 1998), Contemp. Math.*, Vol. 248 (Amer. Math. Soc., Providence, RI), pp. 243–291.

Henriques, A. and Kamnitzer, J. (2006). Crystals and coboundary categories, *Duke Math. J.* **132**, 2, pp. 191–216.

Hernandez, D. (2006). The Kirillov-Reshetikhin conjecture and solutions of T-systems, *J. Reine Angew. Math.* **596**, pp. 63–87.

Hernandez, D. (2010). Kirillov-Reshetikhin conjecture: the general case, *Int. Math. Res. Not. IMRN* , 1, pp. 149–193.

Hernandez, D. and Nakajima, H. (2006). Level 0 monomial crystals, *Nagoya Math. J.* **184**, pp. 85–153.

Hodge, W. V. D. (1943). Some enumerative results in the theory of forms, *Proc. Cambridge Philos. Soc.* **39**, pp. 22–30.

Hong, J. and Kang, S.-J. (2002). *Introduction to quantum groups and crystal bases, Graduate Studies in Mathematics*, Vol. 42 (American Mathematical Society, Providence, RI).

Hong, J. and Lee, H. (2008). Young tableaux and crystal $B(\infty)$ for finite simple Lie algebras, *J. Algebra* **320**, 10, pp. 3680–3693.

Hong, J. and Lee, H. (2012). Young tableaux and crystal $B(\infty)$ for the exceptional Lie algebra types, *J. Combin. Theory Ser. A* **119**, 2, pp. 397–419.

Howe, R. (1989). Remarks on classical invariant theory, *Trans. Amer. Math. Soc.* **313**, 2, pp. 539–570.

Howe, R. (1995). Perspectives on invariant theory: Schur duality, multiplicity-free actions and beyond, in *The Schur lectures (1992) (Tel Aviv), Israel Math. Conf. Proc.*, Vol. 8 (Bar-Ilan Univ., Ramat Gan), pp. 1–182.

Howe, R., Tan, E.-C., and Willenbring, J. F. (2005). Stable branching rules for classical symmetric pairs, *Trans. Amer. Math. Soc.* **357**, 4, pp. 1601–1626.

Humphreys, J. E. (1990). *Reflection groups and Coxeter groups, Cambridge Studies in Advanced Mathematics*, Vol. 29 (Cambridge University Press, Cambridge).

Jeong, K. (2001). Crystal bases for Kac-Moody superalgebras, *J. Algebra* **237**, 2, pp. 562–590.

Jiang, Y. and Sheng, J. (2016). An insight into the description of the crystal structure for Mirković–Vilonen polytopes, `arXiv:1501.07628`.

Jimbo, M. (1985). A q-difference analogue of $U(\mathfrak{g})$ and the Yang-Baxter equation, *Lett. Math. Phys.* **10**, 1, pp. 63–69.

Jimbo, M. (1986). A q-analogue of $U(\mathfrak{gl}(N+1))$, Hecke algebra, and the Yang-Baxter equation, *Lett. Math. Phys.* **11**, 3, pp. 247–252.

Jones, B. and Schilling, A. (2010). Affine structures and a tableau model for E_6 crystals, *J. Algebra* **324**, 9, pp. 2512–2542.

Joseph, A. (1985). On the Demazure character formula, *Ann. Sci. École Norm. Sup. (4)* **18**, 3, pp. 389–419.

Joseph, A. (1995). *Quantum groups and their primitive ideals, Ergebnisse der Mathematik und ihrer Grenzgebiete (3) [Results in Mathematics and Related Areas (3)]*, Vol. 29 (Springer-Verlag, Berlin), ISBN 3-540-57057-8.

Joseph, A. (2012). Consequences of the Littelmann path theory for the structure of the Kashiwara $B(\infty)$ crystal, in *Highlights in Lie algebraic methods, Progr. Math.*, Vol. 295 (Birkhäuser/Springer, New York), pp. 25–64.

Jung, J. H. and Kang, S.-J. (2012). Quantum queer superalgebras, in *Algebraic groups and quantum groups, Contemp. Math.*, Vol. 565 (Amer. Math. Soc., Providence, RI), pp. 81–104.

Kac, V. G. (1977). Lie superalgebras, *Advances in Math.* **26**, 1, pp. 8–96.

Kac, V. G. (1990). *Infinite-dimensional Lie algebras*, 3rd edn. (Cambridge University Press, Cambridge), ISBN 0-521-37215-1; 0-521-46693-8.

Kamnitzer, J. (2007). The crystal structure on the set of Mirković-Vilonen polytopes, *Adv. Math.* **215**, 1, pp. 66–93.

Kamnitzer, J. (2010). Mirković-Vilonen cycles and polytopes, *Ann. of Math. (2)* **171**, 1, pp. 245–294.

Kamnitzer, J. and Tingley, P. (2009a). The crystal commutor and Drinfeld's unitarized R-matrix, *J. Algebraic Combin.* **29**, 3, pp. 315–335.

Kamnitzer, J. and Tingley, P. (2009b). A definition of the crystal commutor using Kashiwara's involution, *J. Algebraic Combin.* **29**, 2, pp. 261–268.

Kang, S.-J., Kashiwara, M., Misra, K. C., Miwa, T., Nakashima, T., and Nakayashiki, A. (1992a). Affine crystals and vertex models, *Int. J. Mod. Phys.* **A7**, S1A, pp. 449–484.

Kang, S.-J., Kashiwara, M., Misra, K. C., Miwa, T., Nakashima, T., and Nakayashiki, A. (1992b). Perfect crystals of quantum affine Lie algebras, *Duke Math. J.* **68**, 3, pp. 499–607.

Kang, S.-J., Kim, J.-A., and Shin, D.-U. (2007). Modified Nakajima monomials and the crystal $B(\infty)$, *J. Algebra* **308**, 2, pp. 524–535.

Kang, S.-J. and Misra, K. C. (1994). Crystal bases and tensor product decompositions of $U_q(G_2)$-modules, *J. Algebra* **163**, 3, pp. 675–691.

Kashiwara, M. (1990). Crystalizing the q-analogue of universal enveloping algebras, *Comm. Math. Phys.* **133**, 2, pp. 249–260.

Kashiwara, M. (1991). On crystal bases of the Q-analogue of universal enveloping algebras, *Duke Math. J.* **63**, 2, pp. 465–516.

Kashiwara, M. (1993). The crystal base and Littelmann's refined Demazure character formula, *Duke Math. J.* **71**, 3, pp. 839–858.

Kashiwara, M. (1994). Crystal bases of modified quantized enveloping algebra, *Duke Math. J.* **73**, 2, pp. 383–413.

Kashiwara, M. (1995). On crystal bases, in *Representations of groups (Banff, AB, 1994)*, *CMS Conf. Proc.*, Vol. 16 (Amer. Math. Soc., Providence, RI), pp. 155–197.

Kashiwara, M. (1996). Similarity of crystal bases, in *Lie algebras and their representations (Seoul, 1995)*, *Contemp. Math.*, Vol. 194 (Amer. Math. Soc., Providence, RI), pp. 177–186.

Kashiwara, M. (2002). *Bases cristallines des groupes quantiques*, *Cours Spécialisés [Specialized Courses]*, Vol. 9 (Société Mathématique de France, Paris), edited by Charles Cochet.

Kashiwara, M. (2003). Realizations of crystals, in *Combinatorial and geometric representation theory (Seoul, 2001)*, *Contemp. Math.*, Vol. 325 (Amer. Math. Soc., Providence, RI), pp. 133–139.

Kashiwara, M., Misra, K. C., Okado, M., and Yamada, D. (2007). Perfect crystals for $U_q(D_4^{(3)})$, *J. Algebra* **317**, 1, pp. 392–423.

Kashiwara, M. and Nakashima, T. (1994). Crystal graphs for representations of the q-analogue of classical Lie algebras, *J. Algebra* **165**, 2, pp. 295–345.

Kashiwara, M., Nakashima, T., and Okado, M. (2008). Affine geometric crystals and limit of perfect crystals, *Trans. Amer. Math. Soc.* **360**, 7, pp. 3645–3686.

Kerov, S. V., Kirillov, A. N., and Reshetikhin, N. Y. (1986). Combinatorics, the Bethe ansatz and representations of the symmetric group, *Zap. Nauchn. Sem. Leningrad. Otdel. Mat. Inst. Steklov. (LOMI)* **155**, Differentsialnaya Geometriya, Gruppy Li i Mekh. VIII, pp. 50–64, 193.

Kim, J.-A. and Shin, D.-U. (2004). Insertion scheme for the classical Lie algebras, *Comm. Algebra* **32**, 8, pp. 3139–3167.

King, R. C. (1976). Weight multiplicities for the classical groups, in *Group theoretical methods in physics (Fourth Internat. Colloq., Nijmegen, 1975)* (Springer, Berlin), pp. 490–499. Lecture Notes in Phys., Vol. 50.

Kirillov, A. N. (2001). Introduction to tropical combinatorics, in *Physics and combinatorics, 2000 (Nagoya)* (World Sci. Publ., River Edge, NJ), pp. 82–150.

Kirillov, A. N. and Berenstein, A. D. (1995). Groups generated by involutions, Gelfand-Tsetlin patterns, and combinatorics of Young tableaux, *Algebra i Analiz* **7**, 1, pp. 92–152.

Kirillov, A. N. and Reshetikhin, N. Y. (1986). The Bethe ansatz and the combinatorics of Young tableaux, *Zap. Nauchn. Sem. Leningrad. Otdel. Mat. Inst. Steklov. (LOMI)* **155**, Differentsialnaya Geometriya, Gruppy Li i Mekh. VIII, pp. 65–115, 194.

Kirillov, A. N. and Reshetikhin, N. Y. (1987). Representations of Yangians and multiplicities of the inclusion of the irreducible components of the tensor product of representations of simple Lie algebras, *Zap. Nauchn. Sem. Leningrad. Otdel. Mat. Inst. Steklov. (LOMI)* **160**, Anal. Teor. Chisel i Teor. Funktsii. 8, pp. 211–221, 301.

Kirillov, A. N., Schilling, A., and Shimozono, M. (2002). A bijection between Littlewood-Richardson tableaux and rigged configurations, *Selecta Math. (N.S.)* **8**, 1, pp. 67–135.

Kleshchev, A. (2005). *Linear and projective representations of symmetric groups, Cambridge Tracts in Mathematics*, Vol. 163 (Cambridge University Press, Cambridge).

Kleshchev, A. (2010). Representation theory of symmetric groups and related Hecke algebras, *Bull. Amer. Math. Soc. (N.S.)* **47**, 3, pp. 419–481.

Kleshchev, A. S. (1995). Branching rules for modular representations of symmetric groups. I, *J. Algebra* **178**, 2, pp. 493–511.

Knuth, D. E. (1970). Permutations, matrices, and generalized Young tableaux, *Pacific J. Math.* **34**, pp. 709–727.

Knuth, D. E. (1998). *The art of computer programming. Vol. 3* (Addison-Wesley, Reading, MA), sorting and searching, Second edition.

Kudla, S. S. (1984). Seesaw dual reductive pairs, in *Automorphic forms of several variables (Katata, 1983), Progr. Math.*, Vol. 46 (Birkhäuser Boston, Boston, MA), pp. 244–268.

Kumar, S. (1987). Demazure character formula in arbitrary Kac-Moody setting, *Invent. Math.* **89**, 2, pp. 395–423.

Kuniba, A., Nakanishi, T., and Tsuboi, Z. (2002). The canonical solutions of the Q-systems and the Kirillov-Reshetikhin conjecture, *Comm. Math. Phys.* **227**, 1, pp. 155–190.

Kus, D. (2013). Realization of affine type A Kirillov-Reshetikhin crystals via polytopes, *J. Combin. Theory Ser. A* **120**, 8, pp. 2093–2117.

Kvinge, H. and Vazirani, M. (2015). Categorifying the tensor product of the Kirillov-Reshetikhin crystal $B^{1,1}$ and a fundamental crystal, arXiv:1508.04182.

Kwon, J.-H. (2007). Crystal graphs for Lie superalgebras and Cauchy decomposition, *J. Algebraic Combin.* **25**, 1, pp. 57–100.

Kwon, J.-H. (2013). RSK correspondence and classically irreducible Kirillov-Reshetikhin crystals, *J. Combin. Theory Ser. A* **120**, 2, pp. 433–452.

Kwon, J.-H. (2015). Combinatorial extension of stable branching rules for classical groups, to appear in *Trans. Amer. Math. Soc.*, arXiv:1512.01877.

Lakshmibai, V., Musili, C., and Seshadri, C. S. (1979). Geometry of G/P. IV. Standard monomial theory for classical types, *Proc. Indian Acad. Sci. Sect. A Math. Sci.* **88**, 4, pp. 279–362.

Lakshmibai, V. and Seshadri, C. S. (1991). Standard monomial theory, in *Proceedings of the Hyderabad Conference on Algebraic Groups (Hyderabad, 1989)* (Manoj Prakashan, Madras), pp. 279–322.

Lam, T. (2006). Affine Stanley symmetric functions, *Amer. J. Math.* **128**, 6, pp. 1553–1586.

Lang, S. (2002). *Algebra, Graduate Texts in Mathematics*, Vol. 211, 3rd edn. (Springer-Verlag, New York).

Lascoux, A., Leclerc, B., and Thibon, J.-Y. (1996). Hecke algebras at roots of unity and crystal bases of quantum affine algebras, *Comm. Math. Phys.* **181**, 1, pp. 205–263.

Lascoux, A. and Schützenberger, M.-P. (1981). Le monoïde plaxique, in *Noncommutative structures in algebra and geometric combinatorics (Naples, 1978), Quad. "Ricerca Sci."*, Vol. 109 (CNR, Rome), pp. 129–156.

Lascoux, A. and Schützenberger, M.-P. (1985). Schubert polynomials and the Littlewood-Richardson rule, *Lett. Math. Phys.* **10**, 2-3, pp. 111–124.

Lascoux, A. and Schützenberger, M.-P. (1990). Keys & standard bases, in *Invariant theory and tableaux (Minneapolis, MN, 1988), IMA Vol. Math. Appl.*, Vol. 19 (Springer, New York), pp. 125–144.

Lecouvey, C. (2002). Schensted-type correspondence, plactic monoid, and jeu de taquin for type C_n, *J. Algebra* **247**, 2, pp. 295–331.

Lecouvey, C. (2003). Schensted-type correspondences and plactic monoids for types B_n and D_n, *J. Algebraic Combin.* **18**, 2, pp. 99–133.

Lenart, C. (2007). On the combinatorics of crystal graphs. I. Lusztig's involution, *Adv. Math.* **211**, 1, pp. 204–243.

Lenart, C., Naito, S., Sagaki, D., Schilling, A., and Shimozono, M. (2015a). A uniform model for Kirillov-Reshetikhin crystals I: Lifting the parabolic quantum Bruhat graph, *Int. Math. Res. Not. IMRN* , 7, pp. 1848–1901.

Lenart, C., Naito, S., Sagaki, D., Schilling, A., and Shimozono, M. (2015b). A uniform model for Kirillov-Reshetikhin crystals III: Nonsymmetric Macdonald polynomials at $t = 0$ and Demazure characters, `arXiv:1511.00465`.

Lenart, C., Naito, S., Sagaki, D., Schilling, A., and Shimozono, M. (2016). A uniform model for Kirillov-Reshetikhin crystals II: Alcove model, path model, and $p = x$, *Int. Math. Res. Not. IMRN* .

Lenart, C. and Postnikov, A. (2008). A combinatorial model for crystals of Kac-Moody algebras, *Trans. Amer. Math. Soc.* **360**, 8, pp. 4349–4381.

Littelmann, P. (1994). A Littlewood-Richardson rule for symmetrizable Kac-Moody algebras, *Invent. Math.* **116**, 1-3, pp. 329–346.

Littelmann, P. (1995a). Crystal graphs and Young tableaux, *J. Algebra* **175**, 1, pp. 65–87.

Littelmann, P. (1995b). Paths and root operators in representation theory, *Ann. of Math. (2)* **142**, 3, pp. 499–525.

Littelmann, P. (1997). Characters of representations and paths in $\mathfrak{H}_{\mathbf{R}}^*$, in *Representation theory and automorphic forms (Edinburgh, 1996)*, Proc. Sympos. Pure Math., Vol. 61 (Amer. Math. Soc., Providence, RI), pp. 29–49.

Littelmann, P. (1998). Cones, crystals, and patterns, *Transform. Groups* **3**, 2, pp. 145–179.

Little, D. P. (2005). Factorization of the Robinson-Schensted-Knuth correspondence, *J. Combin. Theory Ser. A* **110**, 1, pp. 147–168.

Littlewood, D. E. (1938). The Construction of Invariant Matrices, *Proc. London Math. Soc.* **S2-43**, 3, p. 226.

Littlewood, D. E. (1940). *The Theory of Group Characters and Matrix Representations of Groups* (Oxford University Press, New York).

Lothaire, M. (2002). *Algebraic combinatorics on words, Encyclopedia of Mathematics and its Applications*, Vol. 90 (Cambridge University Press, Cambridge), a collective work by Jean Berstel, Dominique Perrin, Patrice Seebold, Julien Cassaigne, Aldo De Luca, Steffano Varricchio, Alain Lascoux, Bernard Leclerc, Jean-Yves Thibon, Veronique Bruyere, Christiane Frougny, Filippo Mignosi, Antonio Restivo, Christophe Reutenauer, Dominique Foata, Guo-Niu Han, Jacques Desarmenien, Volker Diekert, Tero Harju, Juhani Karhumaki and Wojciech Plandowski, With a preface by Berstel and Perrin.

Lusztig, G. (1983). Singularities, character formulas, and a q-analog of weight multiplicities, in *Analysis and topology on singular spaces, II, III (Luminy, 1981)*, Astérisque, Vol. 101 (Soc. Math. France, Paris), pp. 208–229.

Lusztig, G. (1990a). Canonical bases arising from quantized enveloping algebras, *J. Amer. Math. Soc.* **3**, 2, pp. 447–498.

Lusztig, G. (1990b). Canonical bases arising from quantized enveloping algebras. II, *Progr. Theoret. Phys. Suppl.* , 102, pp. 175–201 (1991), common trends in mathematics and quantum field theories (Kyoto, 1990).

Lusztig, G. (1992). Introduction to quantized enveloping algebras, in *New developments in Lie theory and their applications (Córdoba, 1989)*, Progr. Math., Vol. 105 (Birkhäuser Boston, Boston, MA), pp. 49–65.

Lusztig, G. (1993). *Introduction to quantum groups, Progress in Mathematics*, Vol. 110 (Birkhäuser Boston, Inc., Boston, MA).

Lusztig, G. (1994). Total positivity in reductive groups, in *Lie theory and geometry, Progr. Math.*, Vol. 123 (Birkhäuser Boston, Boston, MA), pp. 531–568.

Lusztig, G. (1997). Total positivity and canonical bases, in *Algebraic groups and Lie groups, Austral. Math. Soc. Lect. Ser.*, Vol. 9 (Cambridge Univ. Press, Cambridge), pp. 281–295.

Lusztig, G. (2011). Piecewise linear parametrization of canonical bases, *Pure Appl. Math. Q.* **7**, 3, Special Issue: In honor of Jacques Tits, pp. 783–796.

Mac Lane, S. (1971). *Categories for the working mathematician* (Springer-Verlag, New York-Berlin), graduate Texts in Mathematics, Vol. 5.

Macdonald, I. G. (1972). Affine root systems and Dedekind's η-function, *Invent. Math.* **15**, pp. 91–143.

Macdonald, I. G. (1992). Schur functions: theme and variations, in *Séminaire Lotharingien de Combinatoire (Saint-Nabor, 1992), Publ. Inst. Rech. Math. Av.*, Vol. 498 (Univ. Louis Pasteur, Strasbourg), pp. 5–39.

Macdonald, I. G. (1995). *Symmetric functions and Hall polynomials*, 2nd edn., Oxford Mathematical Monographs (The Clarendon Press, Oxford University Press, New York), with contributions by A. Zelevinsky, Oxford Science Publications.

Maclagan, D. and Sturmfels, B. (2015). *Introduction to tropical geometry, Graduate Studies in Mathematics*, Vol. 161 (American Mathematical Society, Providence, RI), ISBN 978-0-8218-5198-2.

Mathieu, O. (1988). Formules de caractères pour les algèbres de Kac-Moody générales, *Astérisque*, 159-160, p. 267.

Matsumoto, H. (1964). Générateurs et relations des groupes de Weyl généralisés, *C. R. Acad. Sci. Paris* **258**, pp. 3419–3422.

McNamara, P. J. (2011). Metaplectic Whittaker functions and crystal bases, *Duke Math. J.* **156**, 1, pp. 1–31.

Mirković, I. and Vilonen, K. (2000). Perverse sheaves on affine Grassmannians and Langlands duality, *Math. Res. Lett.* **7**, 1, pp. 13–24.

Mirković, I. and Vilonen, K. (2007). Geometric Langlands duality and representations of algebraic groups over commutative rings, *Ann. of Math. (2)* **166**, 1, pp. 95–143.

Misra, K. and Miwa, T. (1990). Crystal base for the basic representation of $U_q(\mathfrak{sl}(n))$, *Comm. Math. Phys.* **134**, 1, pp. 79–88.

Morier-Genoud, S. (2008). Geometric lifting of the canonical basis and semitoric degenerations of Richardson varieties, *Trans. Amer. Math. Soc.* **360**, 1, pp. 215–235 (electronic).

Morse, J. and Schilling, A. (2016). Crystal approach to affine Schubert calculus, *Int. Math. Res. Not.* **2016**, pp. 2239–2294.

Naito, S. and Sagaki, D. (2005). An approach to the branching rule from $\mathfrak{sl}_{2n}(\mathbb{C})$ to $\mathfrak{sp}_{2n}(\mathbb{C})$ via Littelmann's path model, *J. Algebra* **286**, 1, pp. 187–212.

Naito, S. and Sagaki, D. (2008). A modification of the Anderson-Mirković conjecture for Mirković-Vilonen polytopes in types B and C, *J. Algebra* **320**, 1, pp. 387–416.

Nakajima, H. (2003a). t-analogs of q-characters of Kirillov-Reshetikhin modules of quantum affine algebras, *Represent. Theory* **7**, pp. 259–274 (electronic).

Nakajima, H. (2003b). t-analogs of q-characters of quantum affine algebras of type A_n, D_n, in *Combinatorial and geometric representation theory (Seoul, 2001), Contemp. Math.*, Vol. 325 (Amer. Math. Soc., Providence, RI), pp. 141–160.

Noumi, M. and Yamada, Y. (2004). Tropical Robinson-Schensted-Knuth correspondence and birational Weyl group actions, in *Representation theory of algebraic groups and quantum groups, Adv. Stud. Pure Math.*, Vol. 40 (Math. Soc. Japan, Tokyo), pp. 371–442.

Okado, M. (2007). Existence of crystal bases for Kirillov-Reshetikhin modules of type D, *Publ. Res. Inst. Math. Sci.* **43**, 4, pp. 977–1004.

Okado, M. and Schilling, A. (2008). Existence of Kirillov-Reshetikhin crystals for nonexceptional types, *Represent. Theory* **12**, pp. 186–207.

Okado, M., Schilling, A., and Shimozono, M. (2003a). Virtual crystals and fermionic formulas of type $D_{n+1}^{(2)}, A_{2n}^{(2)}$, and $C_n^{(1)}$, *Represent. Theory* **7**, pp. 101–163 (electronic).

Okado, M., Schilling, A., and Shimozono, M. (2003b). Virtual crystals and Kleber's algorithm, *Comm. Math. Phys.* **238**, 1-2, pp. 187–209.

Olshanski, G. I. (1992). Quantized universal enveloping superalgebra of type Q and a super-extension of the Hecke algebra, *Lett. Math. Phys.* **24**, 2, pp. 93–102.

Patnaik, M. M. and Puskás, A. (2015). On the Casselman–Shalika formula for metaplectic groups, `arXiv:1509.01594`.

Puskás, A. (2016). Whittaker functions on metaplectic covers of $GL(r)$, `arXiv:1605.05400`.

Reiner, V. and Shimozono, M. (1998). Percentage-avoiding, northwest shapes and peelable tableaux, *J. Combin. Theory Ser. A* **82**, 1, pp. 1–73.

Sagan, B. E. (1987). Shifted tableaux, Schur Q-functions, and a conjecture of R. Stanley, *J. Combin. Theory Ser. A* **45**, 1, pp. 62–103.

Sagan, B. E. (2001). *The symmetric group, Graduate Texts in Mathematics*, Vol. 203, 2nd edn. (Springer-Verlag, New York), representations, combinatorial algorithms, and symmetric functions.

Sage (2016). *Sage Mathematics Software (Version 7.2)*, The Sage Developers, `http://www.sagemath.org`.

Sage-Combinat community, T. (2008). Sage-Combinat: enhancing Sage as a toolbox for computer exploration in algebraic combinatorics, `http://combinat.sagemath.org`.

Salisbury, B., Schultze, A., and Tingley, P. (2016a). Combinatorial descriptions of the crystal structure on certain PBW bases, `arXiv:1606.01978`.

Salisbury, B., Schultze, A., and Tingley, P. (2016b). PBW bases and marginally large tableaux in type D, `arXiv:1606.02517`.

Salisbury, B. and Scrimshaw, T. (2015). A rigged configuration model for $B(\infty)$, *J. Combin. Theory Ser. A* **133**, pp. 29–57.

Salisbury, B. and Scrimshaw, T. (2016). Rigged configurations and the $*$-involution, `arXiv:1601.06137`.

Savage, A. (2009). Braided and coboundary monoidal categories, in *Algebras, representations and applications, Contemp. Math.*, Vol. 483 (Amer. Math. Soc., Providence, RI), pp. 229–251.

Schilling, A. (2006). Crystal structure on rigged configurations, *Int. Math. Res. Not.* , pp. Art. ID 97376, 27.

Schilling, A. and Scrimshaw, T. (2015). Crystal structure on rigged configurations and the filling map, *Electron. J. Combin.* **22**, 1, pp. Paper 1.73, 56.

Schilling, A. and Shimozono, M. (2006). $X = M$ for symmetric powers, *J. Algebra* **295**, 2, pp. 562–610.

Schilling, A. and Tingley, P. (2012). Demazure crystals, Kirillov-Reshetikhin crystals, and the energy function, *Electron. J. Combin.* **19**, 2, pp. Paper 4, 42.

Schumann, B. and Torres, J. (2016). A non-Levi branching rule in terms of Littelmann paths, `arXiv:1607.08225`.

Schützenberger, M. P. (1972). Promotion des morphismes d'ensembles ordonnés, *Discrete Math.* **2**, pp. 73–94.

Serganova, V. (1996). Kazhdan-Lusztig polynomials and character formula for the Lie superalgebra $\mathfrak{gl}(m|n)$, *Selecta Math. (N.S.)* **2**, 4, pp. 607–651.

Sergeev, A. N. (1984). Tensor algebra of the identity representation as a module over the Lie superalgebras Gl(n, m) and $Q(n)$, *Mat. Sb. (N.S.)* **123(165)**, 3, pp. 422–430.

Sheats, J. T. (1999). A symplectic jeu de taquin bijection between the tableaux of King and of De Concini, *Trans. Amer. Math. Soc.* **351**, 9, pp. 3569–3607.

Shimozono, M. (2002). Affine type A crystal structure on tensor products of rectangles, Demazure characters, and nilpotent varieties, *J. Algebraic Combin.* **15**, 2, pp. 151–187.

Stanley, R. P. (1984). On the number of reduced decompositions of elements of Coxeter groups, *European J. Combin.* **5**, 4, pp. 359–372.

Stanley, R. P. (1999). *Enumerative combinatorics. Vol. 2, Cambridge Studies in Advanced Mathematics*, Vol. 62 (Cambridge University Press, Cambridge), with a foreword by Gian-Carlo Rota and appendix 1 by Sergey Fomin.

Steinberg, R. (1968). *Lectures on Chevalley groups* (Yale University, New Haven, Conn.), notes prepared by John Faulkner and Robert Wilson.

Stembridge, J. R. (1989). Shifted tableaux and the projective representations of symmetric groups, *Adv. Math.* **74**, 1, pp. 87–134.

Stembridge, J. R. (2003). A local characterization of simply-laced crystals, *Trans. Amer. Math. Soc.* **355**, 12, pp. 4807–4823.

Tingley, P. (2016). Elementary construction of Lusztig's canonical basis, `arXiv:1602.04895`.

Tingley, P. and Webster, B. (2012). Mirkovic-Vilonen polytopes and Khovanov-Lauda-Rouquier algebras, `arXiv:1210.6921`.

Tits, J. (1966). Normalisateurs de tores. I. Groupes de Coxeter étendus, *J. Algebra* **4**, pp. 96–116.

Tokuyama, T. (1988). A generating function of strict Gel'fand patterns and some formulas on characters of general linear groups, *J. Math. Soc. Japan* **40**, 4, pp. 671–685.

van Leeuwen, M. A. A. (2006). Double crystals of binary and integral matrices, *Electron. J. Combin.* **13**, 1, pp. Research Paper 86, 93 pp. (electronic).

Varagnolo, M. and Vasserot, E. (1999). On the decomposition matrices of the quantized Schur algebra, *Duke Math. J.* **100**, 2, pp. 267–297.

Vazirani, M. (2002). Parameterizing Hecke algebra modules: Bernstein-Zelevinsky multi-segments, Kleshchev multipartitions, and crystal graphs, *Transform. Groups* **7**, 3, pp. 267–303.

Viennot, G. (1977). Une forme géométrique de la correspondance de Robinson-Schensted, in *Combinatoire et représentation du groupe symétrique (Actes Table Ronde CNRS, Univ. Louis-Pasteur Strasbourg, Strasbourg, 1976)* (Springer, Berlin), pp. 29–58. Lecture Notes in Math., Vol. 579.

Worley, D. R. (1984). *A theory of shifted Young tableaux* (ProQuest LLC, Ann Arbor, MI), thesis (Ph.D.)–Massachusetts Institute of Technology.

Yamane, H. (1994). Quantized enveloping algebras associated with simple Lie superalgebras and their universal R-matrices, *Publ. Res. Inst. Math. Sci.* **30**, 1, pp. 15–87.

Yamane, S. (1998). Perfect crystals of $U_q(G_2^{(1)})$, *J. Algebra* **210**, 2, pp. 440–486.

Index

Printed in the United States
By Bookmasters